COMPACT NUMERICAL METHODS FOR COMPUTERS

linear algebra and function minimisation

Second Edition

COMPACT NUMERICAL METHODS FOR COMPUTERS

linear algebra and function minimisation

Second Edition

J C NASH

Adam Hilger, Bristol and New York

British Library Cataloguing in Publication Data

Nash, J. C.
Compact numerical methods for computers: linear algebra and function minimisation – 2nd ed.
1. Numerical analysis. Applications of microcomputer & minicomputer systems. Algorithms
I. Title
519.4

ISBN 0-85274-318-1
ISBN 0-85274-319-X (pbk)
ISBN 0-7503-0036-1 ($5\frac{1}{4}''$ IBM disc)
ISBN 0-7503-0043-4 ($3\frac{1}{2}''$ IBM disc)

Library of Congress Cataloging-in-Publication Data are available

First published, 1979
Reprinted, 1980
Second edition, 1990

Published under the Adam Hilger imprint by IOP Publishing Ltd
Techno House, Redcliffe Way, Bristol BS1 6NX, England
335 East 45th Street, New York, NY 10017-3483, USA

Filmset by Bath Typesetting Ltd, Bath, Avon
Printed in Great Britain by Page Bros (Norwich) Ltd

CONTENTS

PREFACE TO THE SECOND EDITION

The first edition of this book was written between 1975 and 1977. It may come as a surprise that the material is still remarkably useful and applicable in the solution of numerical problems on computers. This is perhaps due to the interest of researchers in the development of quite complicated computational methods which require considerable computing power for their execution. More modest techniques have received less time and effort of investigators. However, it has also been the case that the algorithms presented in the first edition have proven to be reliable yet simple.

The need for simple, compact numerical methods continues, even as software packages appear which relieve the user of the task of programming. Indeed, such methods are needed to implement these packages. They are also important when users want to perform a numerical task within their own programs.

The most obvious difference between this edition and its predecessor is that the algorithms are presented in Turbo Pascal, to be precise, in a form which will operate under Turbo Pascal 3.01a. I decided to use this form of presentation for the following reasons:

(i) Pascal is quite similar to the Step-and-Description presentation of algorithms used previously;

(ii) the codes can be typeset directly from the executable Pascal code, and the very difficult job of proof-reading and correction avoided;

(iii) the Turbo Pascal environment is very widely available on microcomputer systems, and a number of similar systems exist.

Section 1.6 and appendix 4 give some details about the codes and especially the driver and support routines which provide examples of use.

The realization of this edition was not totally an individual effort. My research work, of which this book represents a product, is supported in part by grants from the Natural Sciences and Engineering Research Council of Canada. The Mathematics Department of the University of Queensland and the Applied Mathematics Division of the New Zealand Department of Scientific and Industrial Research provided generous hospitality during my 1987–88 sabbatical year, during which a great part of the code revision was accomplished. Thanks are due to Mary Walker-Smith for reading early versions of the codes, to Maureen Clarke of IOP Publishing Ltd for reminders and encouragement, and to the Faculty of Administration of the University of Ottawa for use of a laser printer to prepare the program codes. Mary Nash has been a colleague and partner for two decades, and her contribution to this project in many readings, edits, and innumerable other tasks has been a large one.

In any work on computation, there are bound to be errors, or at least program structures which operate in unusual ways in certain computing environments. I encourage users to report to me any such observations so that the methods may be improved.

J. C. Nash

Ottawa, 12 June 1989

PREFACE TO THE FIRST EDITION

This book is designed to help people solve numerical problems. In particular, it is directed to those who wish to solve numerical problems on 'small' computers, that is, machines which have limited storage in their main memory for program and data. This may be a programmable calculator—even a pocket model—or it may be a subsystem of a monster computer. The algorithms that are presented in the following pages have been used on machines such as a Hewlett–Packard 9825 programmable calculator and an IBM 370/168 with Floating Point Systems Array Processor. That is to say, they are designed to be used anywhere that a problem exists for them to attempt to solve. In some instances, the algorithms will not be as efficient as others available for the job because they have been chosen and developed to be 'small'. However, I believe users will find them surprisingly economical to employ because their size and/or simplicity reduces errors and human costs compared with equivalent 'larger' programs.

Can this book be used as a text to teach numerical methods? I believe it can. The subject areas covered are, principally, numerical linear algebra, function minimisation and root-finding. Interpolation, quadrature and differential equations are largely ignored as they have not formed a significant part of my own work experience. The instructor in numerical methods will find perhaps too few examples and no exercises. However, I feel the examples which are presented provide fertile ground for the development of many exercises. As much as possible, I have tried to present examples from the real world. Thus the origins of the mathematical problems are visible in order that readers may appreciate that these are not merely interesting diversions for those with time and computers available.

Errors in a book of this sort, especially in the algorithms, can depreciate its value severely. I would very much appreciate hearing from anyone who discovers faults and will do my best to respond to such queries by maintaining an errata sheet. In addition to the inevitable typographical errors, my own included, I anticipate that some practitioners will take exception to some of the choices I have made with respect to algorithms, convergence criteria and organisation of calculations. Out of such differences, I have usually managed to learn something of value in improving my subsequent work, either by accepting new ideas or by being reassured that what I was doing had been through some criticism and had survived.

There are a number of people who deserve thanks for their contribution to this book and who may not be mentioned explicitly in the text:

(i) in the United Kingdom, the many members of the Numerical Algorithms Group, of the Numerical Optimization Centre and of various university departments with whom I discussed the ideas from which the algorithms have condensed;

(ii) in the United States, the members of the Applied Mathematics Division of the Argonne National Laboratory who have taken such an interest in the algorithms, and Stephen Nash who has pointed out a number of errors and faults; and
(iii) in Canada, the members of the Economics Branch of Agriculture Canada for presenting me with such interesting problems to solve, Kevin Price for careful and detailed criticism, Bob Henderson for trying out most of the algorithms, Richard Wang for pointing out several errors in chapter 8, John Johns for trying (and finding errors in) eigenvalue algorithms, and not least Mary Nash for a host of corrections and improvements to the book as a whole.

It is a pleasure to acknowledge the very important roles of Neville Goodman and Geoff Amor of Adam Hilger Ltd in the realisation of this book.

J. C. Nash

Ottawa, 22 December 1977

Chapter 1

A STARTING POINT

1.1. PURPOSE AND SCOPE

This monograph is written for the person who has to solve problems with (small) computers. It is a handbook to help him or her obtain reliable answers to specific questions, posed in a mathematical way, using limited computational resources. To this end the solution methods proposed are presented not only as formulae but also as algorithms, those recipes for solving problems which are more than merely a list of the mathematical ingredients.

There has been an attempt throughout to give examples of each type of calculation and in particular to give examples of cases which are prone to upset the execution of algorithms. No doubt there are many gaps in the treatment where the experience which is condensed into these pages has not been adequate to guard against all the pitfalls that confront the problem solver. The process of learning is continuous, as much for the teacher as the taught. Therefore, the user of this work is advised to think for him/herself and to use his/her own knowledge and familiarity of particular problems as much as possible. There is, after all, barely a working career of experience with automatic computation and it should not seem surprising that satisfactory methods do not exist as yet for many problems. Throughout the sections which follow, this underlying novelty of the art of solving numerical problems by automatic algorithms finds expression in a conservative design policy. Reliability is given priority over speed and, from the title of the work, space requirements for both the programs and the data are kept low.

Despite this policy, it must be mentioned immediately and with some emphasis that the algorithms may prove to be surprisingly efficient from a cost-of-running point of view. In two separate cases where explicit comparisons were made, programs using the algorithms presented in this book cost less to run than their large-machine counterparts. Other tests of execution times for algebraic eigenvalue problems, roots of a function of one variable and function minimisation showed that the eigenvalue algorithms were by and large 'slower' than those recommended for use on large machines, while the other test problems were solved with notable efficiency by the compact algorithms. That 'small' programs may be more frugal than larger, supposedly more efficient, ones based on different algorithms to do the same job has at least some foundation in the way today's computers work.

Since the first edition of this work appeared, a large number and variety of inexpensive computing machines have appeared. Often termed the 'microcomputer revolution', the widespread availability of computing power in forms as diverse as programmable calculators to desktop workstations has increased the need for suitable software of all types, including numerical methods. The present work is directed at the user who needs, for whatever reason, to program a numerical method to solve a problem. While software packages and libraries exist to provide for the solution of numerical problems, financial, administrative or other obstacles may render their use impossible or inconvenient. For example, the programming tools available on the chosen computer may not permit the packaged software to be used.

Firstly, most machines are controlled by operating systems which control (and sometimes charge for) the usage of memory, storage, and other machine resources. In both compilation (translation of the program into machine code) and execution, a smaller program usually will make smaller demands on resources than a larger one. On top of this, the time of compilation is usually related to the size of the source code.

Secondly, once the program begins to execute, there are housekeeping operations which must be taken care of:

(i) to keep programs and data belonging to one task or user separate from those belonging to others in a time-sharing environment, and

(ii) to access the various parts of the program and data within the set of resources allocated to a single user.

Studies conducted some years ago by Dr Maurice Cox of the UK National Physical Laboratory showed that (ii) requires about 90% of the time a computer spends with a typical scientific computation. Only about 10% of the effort goes to actual arithmetic. This mix of activity will vary greatly with the machine and problem under consideration. However, it is not unreasonable that a small program can use simpler structures, such as address maps and decision tables, than a larger routine. It is tempting to suggest that the computer may be able to perform useful work with a small program while deciding what to do with a larger one. Gathering specific evidence to support such conjectures requires the fairly tedious work of benchmarking. Moreover, the results of the exercise are only valid as long as the machine, operating system, programming language translators and programs remain unchanged. Where performance is critical, as in the case of real-time computations, for example in air traffic control, then benchmarking will be worthwhile. In other situations, it will suffice that programs operate correctly and sufficiently quickly that the user is not inconvenienced.

This book is one of the very few to consider algorithms which have very low storage requirements. The first edition appeared just as programmable calculators and the first microcomputers were beginning to make their presence felt. These brought to the user's desk a quantum improvement in computational power. Comparisons with the earliest digital computers showed that even a modest microcomputer was more powerful. It should be noted, however, that the programmer did not have to handle all the details of arithmetic and data storage, as on the early computers, thanks to the quick release of programming language translators. There is unfortunately still a need to be vigilant for errors in the floating-point arithmetic and the special function routines. Some aids to such checking are mentioned later in §1.2.

Besides the motivation of cost savings or the desire to use an available and possibly under-utilised small computer, this work is directed to those who share my philosophy that human beings are better able to comprehend and deal with small programs and systems than large ones. That is to say, it is anticipated that the costs involved in implementing, modifying and correcting a small program will be lower for small algorithms than for large ones, though this comparison will depend greatly on the structure of the algorithms. By way of illustration, I implemented and tested the eigenvalue/vector algorithm (algorithm 13) in under half an hour from a 10 character/second terminal in Aberystwyth using a Xerox Sigma 9 computer in Birmingham. The elapsed time includes my instruction in the use of the system which was of a type I had not previously encountered. I am grateful to Mrs Lucy Tedd for showing me this system. Dr John Johns of the Herzberg Institute of Astrophysics was able to obtain useful eigensolutions from the same algorithm within two hours of delivery of a Hewlett–Packard 9825 programmable calculator. He later discovered a small error in the prototype of the algorithm.

The topics covered in this work are numerical linear algebra and function minimisation. Why not differential equations? Quite simply because I have had very little experience with the numerical solution of differential equations except by techniques using linear algebra or function minimisation. Within the two broad areas, several subjects are given prominence. Linear equations are treated in considerable detail with separate methods given for a number of special situations. The algorithms given here are quite similar to those used on larger machines. The algebraic eigenvalue problem is also treated quite extensively, and in this edition, a method for complex matrices is included. Computing the eigensolutions of a general square matrix is a problem with many inherent difficulties, but we shall not dwell on these at length.

Constrained optimisation is still a topic where I would be happier to offer more material, but general methods of sufficient simplicity to include in a handbook of this sort have eluded my research efforts. In particular, the mathematical programming problem is not treated here.

Since the aim has been to give a problem-solving person some tools with which to work, the mathematical detail in the pages that follow has been mostly confined to that required for explanatory purposes. No claim is made to rigour in any 'proof', though a sincere effort has been made to ensure that the statement of theorems is correct and precise.

1.2. MACHINE CHARACTERISTICS

In the first edition, a 'small computer' was taken to have about 6000 characters of main memory to hold both programs and data. This logical machine, which might be a part of a larger physical computer, reflected the reality facing a quite large group of users in the mid- to late-1970s.

A more up-to-date definition of 'small computer' could be stated, but it is not really necessary. Users of this book are likely to be those scientists, engineers, and statisticians who must, for reasons of financial or administrative necessity or convenience, carry out their computations in environments where the programs cannot be acquired simply and must, therefore, be written in-house. There are also a number of portable computers now available. This text is being entered on a Tandy Radio Shack TRS-80 Model 100, which is only the size of a large book and is powered by four penlight batteries.

Users of the various types of machines mentioned above often do not have much choice as to the programming tools available. On 'borrowed' computers, one has to put up with the compilers and interpreters provided by the user who has paid for the resource. On portables, the choices may be limited by the decisions of the manufacturer. In practice, I have, until recently, mostly programmed in BASIC, despite its limitations, since it has at least been workable on most of the machines available to me.

Another group of users of the material in this book is likely to be software developers. Some scenarios which have occurred are:

—software is being developed in a particular computing environment (e.g. LISP for artificial intelligence) and a calculation is required for which suitable off-the-shelf routines are not available;

—standard routines exist but when linked into the package cause the executable code to be too large for the intended disk storage or memory;

—standard routines exist, but the coding is incompatible with the compiler or interpreter at hand.

It is worth mentioning that programming language standards have undergone considerable development in the last decade. Nevertheless, the goal of portable source codes of numerical methods is still only won by careful and conservative programming practices.

Because of the emphasis on the needs of the user to program the methods, there is considerable concern in this book to keep the length and complexity of the algorithms to a minimum. Ideally, I like program codes for the algorithms to be no longer than a page of typed material, and at worse, less than three pages. This makes it feasible to implement the algorithms in a single work session. However, even this level of effort is likely to be considered tedious and it is unnecessary if the code can be provided in a suitable form. Here we provide source code in Turbo Pascal for the algorithms in this book and for the driver and support routines to run them (under Turbo Pascal version 3.01a).

The philosophy behind this book remains one of working with available tools rather than complaining that better ones exist, albeit not easily accessible. This should not lead to complacency in dealing with the machine but rather to an active wariness of any and every feature of the system. A number of these can and should be checked by using programming devices which force the system to reveal itself in spite of the declarations in the manual(s). Others will have to be determined by exploring every error possibility when a program fails to produce expected results. In most cases programmer error is to blame, but I have encountered at least one system error in each of the systems I have used seriously. For instance, trigonometric functions are usually computed by power series approximation. However, these approximations have validity over specified domains, usually $[0, \pi/4]$ or $[0, \pi/2]$ (see Abramowitz and Stegun 1965, p 76). Thus the argument of the function must first be transformed to bring it into the appropriate range. For example

$$\sin(\pi-\phi)=\sin\phi \tag{1.1}$$

or

$$\sin(\pi/2-\phi)=\cos\phi. \tag{1.2}$$

Unless this range reduction is done very carefully the results may be quite unexpected. On one system, hosted by a Data General NOVA, I have observed that the sine of an angle near π and the cosine of an angle near $\pi/2$ were both computed as unity instead of a small value, due to this type of error. Similarly, on some early models of Hewlett–Packard pocket calculators, the rectangular-to-polar coordinate transformation may give a vector 180° from the correct direction. (This appears to have been corrected now.)

Testing the quality of the floating-point arithmetic and special functions is technically difficult and tedious. However, some developments which aid the user have been made by public-spirited scientists. Of these, I consider the most worthy example to be PARANOIA, a program to examine the floating-point arithmetic provided by a programming language translator. Devised originally by Professor W Kahan of the University of California, Berkeley, it has been developed and distributed in a number of programming languages (Karpinski 1985). Its output is didactic, so that one does not have to be a numerical analyst to interpret the results. I have used the BASIC, FORTRAN, Pascal and C versions of PARANOIA, and have seen reports of Modula-2 and ADA®† versions.

In the area of special functions, Cody and Waite (1980) have devised software to both calculate and test a number of the commonly used transcendental functions (sin, cos, tan, log, exp, sqrt, x^y). The ELEFUNT testing software is available in their book, written in FORTRAN. A considerable effort would be needed to translate it into other programming languages.

An example from our own work is the program DUBLTEST, which is designed to determine the precision to which the standard special functions in BASIC are computed (Nash and Walker-Smith 1987). On the IBM PC, many versions of Microsoft BASIC (GWBASIC, BASICA) would only compute such functions in single precision, even if the variables involved as arguments or results were double precision. For some nonlinear parameter estimation problems, the resulting low precision results caused unsatisfactory operation of our software.

Since most algorithms are in some sense iterative, it is necessary that one has some criterion for deciding when sufficient progress has been made that the execution of a program can be halted. While, in general, I avoid tests which require knowledge of the machine, preferring to use the criterion that no progress has been made in an iteration, it is sometimes convenient or even necessary to employ tests involving tolerances related to the structure of the computing device at hand.

The most useful property of a system which can be determined systematically is the machine precision. This is the smallest number, eps, such that

$$1+\text{eps}>1 \tag{1.3}$$

† ADA is a registered name of the US Department of Defense.

within the arithmetic of the system. Two programs in FORTRAN for determining the machine precision, the radix or base of the arithmetic, and machine rounding or truncating properties have been given by Malcolm (1972). The reader is cautioned that, since these programs make use of tests of conditions like (1.3), they may be frustrated by optimising compilers which are able to note that (1.3) in exact arithmetic is equivalent to

$$\text{eps} > 0. \tag{1.4}$$

Condition (1.4) is not meaningful in the present context. The Univac compilers have acquired some notoriety in this regard, but they are by no means the only offenders.

To find the machine precision and radix by using arithmetic of the computer itself, it is first necessary to find a number q such that $(1+q)$ and q are represented identically, that is, the representation of 1 having the same exponent as q has a digit in the $(t+1)$th radix position where t is the number of radix digits in the floating-point mantissa. As an example, consider a four decimal digit machine. If $q = 10\,000$ or greater, then q is represented as (say)

$$0{\cdot}1 * 1\text{E}5$$

while 1 is represented as

$$0{\cdot}00001 * 1\text{E}5.$$

The action of storing the five-digit sum

$$0{\cdot}10001 * 1\text{E}5$$

in a four-digit word causes the last digit to be dropped. In the example, $q = 10\,000$ is the smallest number which causes $(1+q)$ and q to be represented identically, but any number

$$q > 9999$$

will have the same property. If the machine under consideration has radix R, then any

$$q \geqslant R^t \tag{1.5}$$

will have the desired property. If, moreover, q and R^{t+1} are represented so that

$$q < R^{t+1} \tag{1.6}$$

then

$$q + R > q. \tag{1.7}$$

In our example, $R = 10$ and $t = 4$ so the largest q consistent with (1.6) is

$$q = 10^5 - 10 = 99\,990 = 0{\cdot}9999 * 1\text{E}5$$

and

$$99\,990 + 10 = 100\,000 = 0{\cdot}1000 * 1\text{E}6 > q.$$

Starting with a trial value, say $q = 1$, successive doubling will give some number

$$q = 2^k$$

such that $(q+1)$ and q are represented identically. By then setting r to successive integers 2, 3, 4, . . . , a value such that

$$q+r>q \tag{1.8}$$

will be found. On a machine which truncates, r is then the radix R. However, if the machine rounds in some fashion, the condition (1.8) may be satisfied for $r<R$. Nevertheless, the representations of q and $(q+r)$ will differ by R. In the example, doubling will produce $q=16\,384$ which will be represented as

$$0{\cdot}1638 * 1\text{E}5$$

so $q+r$ is represented as

$$0{\cdot}1639 * 1\text{E}5$$

for some $r \leqslant 10$. Then subtraction of these gives

$$0{\cdot}0001 * 1\text{E}5 = 10.$$

Unfortunately, it is possible to foresee situations where this will not work. Suppose that $q=99\,990$, then we have

$$0{\cdot}9999 * 1\text{E}5 + 10 = 0{\cdot}1000 * 1\text{E}6$$

and

$$0{\cdot}1000 * 1\text{E}6 - 0{\cdot}9999 * 1\text{E}5 = R'.$$

But if the second number in this subtraction is first transformed to

$$0{\cdot}0999 * 1\text{E}6$$

then R' is assigned the value 100. Successive doubling should not, unless the machine arithmetic is extremely unusual, give q this close to the upper bound of (1.6).

Suppose that R has been found and that it is greater than two. Then if the representation of $q+(R-1)$ is greater than that of q, the machine we are using *rounds*, otherwise it *chops* or *truncates* the results of arithmetic operations.

The number of radix digits t is now easily found as the smallest integer such that

$$R^t+1$$

is represented identically to R^t. Thus the machine precision is given as

$$\text{eps} = R^{1-t} = R^{-(t-1)}. \tag{1.9}$$

In the example, $R=10$, $t=4$, so

$$R^{-3}=0{\cdot}001.$$

Thus

$$1+0{\cdot}001 = 1{\cdot}001 > 1$$

but $1+0{\cdot}0009$ is, on a machine which truncates, represented as 1.

In all of the previous discussion concerning the computation of the machine precision it is important that the representation of numbers be that in the memory, not in the working registers where extra digits may be carried. On a Hewlett–Packard 9830, for instance, it was necessary when determining the so-called 'split precision' to store numbers specifically in array elements to force the appropriate truncation.

The above discussion has assumed a model of floating-point arithmetic which may be termed an additive form in that powers of the radix are added together and the entire sum multiplied by some power of the radix (the exponent) to provide the final quantity representing the desired real number. This representation may or may not be exact. For example, the fraction $\frac{1}{5}$ cannot be exactly represented in additive binary (radix 2) floating-point arithmetic. While there are other models of floating-point arithmetic, the additive form is the most common, and is used in the IEEE binary and radix-free floating-point arithmetic standards. (The March, 1981, issue of *IEEE Computer* magazine, volume 3, number 4, pages 51–86 contains a lucid description of the binary standard and its motivations.)

If we are concerned with having absolute upper and lower bounds on computed quantities, interval arithmetic is possible, but not commonly supported by programming languages (e.g. Pascal SC (Kulisch 1987)). Despite the obvious importance of assured bounds on results, the perceived costs of using interval arithmetic have largely prevented its widespread use.

The development of standards for floating-point arithmetic has the great benefit that results of similar calculations on different machinery should be the same. Furthermore, manufacturers have been prompted to develop hardware implementations of these standards, notably the Intel 80 × 87 family and the Motorola 68881 of circuit devices. Hewlett–Packard implemented a decimal version of the IEEE 858 standard in their HP 71B calculator.

Despite such developments, there continues to be much confusion and misinformation concerning floating-point arithmetic. Because an additive decimal form of arithmetic can represent fractions such as $\frac{1}{5}$ exactly, and in general avoid input–output conversion errors, developers of software products using such arithmetic (usually in binary coded decimal or BCD form) have been known to claim that it has 'no round-off error', which is patently false. I personally prefer decimal arithmetic, in that data entered into a calculation can generally be represented exactly, so that a display of the stored raw data reproduces the input familiar to the user. Nevertheless, the differences between good implementations of floating-point arithmetic, whether binary or decimal, are rarely substantive.

While the subject of machine arithmetic is still warm, note that the mean of two numbers may be calculated to be smaller or greater than either! An example in four-figure decimal arithmetic will serve as an illustration of this.

	Exact	Rounded	Truncated
a	5008	5008	5008
b	5007	5007	5007
$a+b$	10015	1002 * 10	1001 * 10
$(a+b)/2$	5007·5	501·0 * 10	500·5 * 10
		= 5010	= 5005

That this can and does occur should be kept in mind whenever averages are computed. For instance, the calculations are quite stable if performed as

$$(a+b)/2=5000+[(a-5000)+(b-5000)]/2.$$

Taking account of every eventuality of this sort is nevertheless extremely tedious.

Another annoying characteristic of small machines is the frequent absence of extended precision, also referred to as double precision, in which extra radix digits are made available for calculations. This permits the user to carry out arithmetic operations such as accumulation, especially of inner products of vectors, and averaging with less likelihood of catastrophic errors. On equipment which functions with number representations similar to the IBM/360 systems, that is, six hexadecimal ($R=16$) digits in the mantissa of each number, many programmers use the so-called 'double precision' routinely. Actually $t=14$, which is not double six. In most of the calculations that I have been asked to perform, I have not found such a sledgehammer policy necessary, though the use of this feature in appropriate situations is extremely valuable. The fact that it does not exist on most small computers has therefore coloured much of the development which follows.

Finally, since the manufacturers' basic software has been put in question above, the user may also wonder about their various application programs and packages. While there are undoubtedly some 'good' programs, my own experience is that the quality of such software is very mixed. Badly written and poorly documented programs may take longer to learn and understand than a satisfactory homegrown product takes to code and debug. A major fault with many software products is that they lack references to the literature and documentation of their pedigree and authorship. Discussion of performance and reliability tests may be restricted to very simple examples. Without such information, it may be next to impossible to determine the methods used or the level of understanding of the programmer of the task to be carried out, so that the user is unable to judge the quality of the product. Some developers of mathematical and statistical software are beginning to recognise the importance of background documentation, and their efforts will hopefully be rewarded with sales.

1.3. SOURCES OF PROGRAMS

When the first edition of this book was prepared, there were relatively few sources of mathematical software in general, and in essence none (apart from a few manufacturers' offerings) for users of small computers. This situation has changed remarkably, with some thousands of suppliers. Source codes of numerical methods, however, are less widely available, yet many readers of this book may wish to conduct a search for a suitable program already coded in the programming language to be used before undertaking to use one of the algorithms given later.

How should such a search be conducted? I would proceed as follows.

First, if FORTRAN is the programming language, I would look to the major collections of mathematical software in the *Collected Algorithms of the Association for Computing Machinery (ACM)*. This collection, abbreviated as CALGO, is comprised of all the programs published in the *Communications of the ACM* (up to 1975) and the *ACM Transactions on Mathematical Software* (since 1975). Other important collections are EISPACK, LINPACK, FUNPACK and MINPACK, which concern algebraic eigenproblems, linear equations, special functions and nonlinear least squares minimisation problems. These and other packages are, at time of writing, available from the Mathematics and Computer Sciences Division of the Argonne National Laboratory of the US Department of Energy. For those users fortunate enough to have access to academic and governmental electronic mail networks, an index of software available can be obtained by sending the message

SEND INDEX

to the pseudo-user NETLIB at node ANL-MCS on the ARPA network (Dongarra and Grosse 1987). The software itself may be obtained by a similar mechanism.

Suppliers such as the Numerical Algorithms Group (NAG), International Mathematical and Statistical Libraries (IMSL), C Abaci, and others, have packages designed for users of various computers and compilers, but provide linkable object code rather than the FORTRAN source. C Abaci, in particular, allows users of the Scientific Desk to also operate the software within what is termed a 'problem solving environment' which avoids the need for programming.

For languages other than FORTRAN, less software is available. Several collections of programs and procedures have been published as books, some with accompanying diskettes, but once again, the background and true authorship may be lacking. The number of truly awful examples of badly chosen, badly coded algorithms is alarming, and my own list of these too long to include here.

Several sources I consider worth looking at are the following.

Maindonald (1984)
: —A fairly comprehensive collection of programs in BASIC (for a Digital Equipment Corporation VAX computer) are presented covering linear estimation, statistical distributions and pseudo-random numbers.

Nash and Walker-Smith (1987)
: —Source codes in BASIC are given for six nonlinear minimisation methods and a large selection of examples. The algorithms correspond, by and large, to those presented later in this book.

LEQB05 (Nash 1984b, 1985)
: —This single 'program' module (actually there are three starting points for execution) was conceived as a joke to show how small a linear algebra package could be made. In just over 300 lines of BASIC is the capability to solve linear equations, linear least squares, matrix inverse and generalised inverse, symmetric matrix eigenproblem and nonlinear least squares problems. The joke back-fired in that the capability of this program, which ran on the Sinclair ZX81 computer among other machines, is quite respectable.

Kahaner, Moler and Nash (1989)
: —This numerical analysis textbook includes FORTRAN codes which illustrate the material presented. The authors have taken pains to choose this software for

quality. The user must, however, learn how to invoke the programs, as there is no user interface to assist in problem specification and input.

Press *et al* (1986) Numerical Recipes
—This is an ambitious collection of methods with wide capability. Codes are offered in FORTRAN, Pascal, and C. However, it appears to have been only superficially tested and the examples presented are quite simple. It has been heavily advertised.

Many other products exist and more are appearing every month. Finding out about them requires effort, the waste of which can sometimes be avoided by using modern online search tools. Sadly, more effort is required to determine the quality of the software, often after money has been spent.

Finally on sources of software, readers should be aware of the Association for Computing Machinery (ACM) *Transactions on Mathematical Software* which publishes research papers and reports algorithms. The algorithms themselves are available after a delay of approximately 1 year on NETLIB and are published in full in the *Collected Algorithms of the ACM*. Unfortunately, many are now quite large programs, and the *Transactions on Mathematical Software (TOMS)* usually only publishes a summary of the codes, which is insufficient to produce a working program. Moreover, the programs are generally in FORTRAN.

Other journals which publish algorithms in some form or other are *Applied Statistics (Journal of the Royal Statistical Society, Part C)*, the Society for Industrial and Applied Mathematics (SIAM) journals on *Numerical Analysis* and on *Scientific and Statistical Computing*, the *Computer Journal* (of the British Computer Society), as well as some of the specialist journals in computational statistics, physics, chemistry and engineering. Occasionally magazines, such as *Byte* or *PC Magazine*, include articles with interesting programs for scientific or mathematical problems. These may be of very variable quality depending on the authorship, but some exceptionally good material has appeared in magazines, which sometimes offer the codes in machine-readable form, such as the Byte Information Exchange (BIX) and disk ordering service. The reader has, however, to be assiduous in verifying the quality of the programs.

1.4. PROGRAMMING LANGUAGES USED AND STRUCTURED PROGRAMMING

The algorithms presented in this book are designed to be coded quickly and easily for operation on a diverse collection of possible target machines in a variety of programming languages. Originally, in preparing the first edition of the book, I considered presenting programs in BASIC, but found at the time that the various dialects of this language were not uniform in syntax. Since then, International Standard Minimal BASIC (ISO 6373/1984) has been published, and most commonly available BASICs will run Minimal BASIC without difficulty. The obstacle for the user is that Minimal BASIC is too limited for most serious computing tasks, in that it lacks string and file handling capabilities. Nevertheless, it is capable of demonstrating all the algorithms in this book.

As this revision is being developed, efforts are ongoing to agree an international standard for Full BASIC. Sadly, in my opinion, these efforts do not reflect the majority of existing commercial and scientific applications, which are coded in a dialect of BASIC compatible with language processors from Microsoft Corporation or Borland International (Turbo BASIC).

Many programmers and users do not wish to use BASIC, however, for reasons quite apart from capability. They may prefer FORTRAN, APL, C, Pascal, or some other programming language. On certain machinery, users may be forced to use the programming facilities provided. In the 1970s, most Hewlett–Packard desktop computers used exotic programming languages of their own, and this has continued to the present on programmable calculators such as the HP 15C. Computers offering parallel computation usually employ programming languages with special extensions to allow the extra functionality to be exploited.

As an author trying to serve this fragmented market, I have therefore wanted to keep to my policy of presenting the algorithms in step-and-description form. However, implementations of the algorithms allow their testing, and direct publication of a working code limits the possibilities for typographical errors. Therefore, in this edition, the step-and-description codes have been replaced by Turbo Pascal implementations. A coupon for the diskette of the codes is included. Turbo Pascal has a few disadvantages, notably some differences from International Standard Pascal, but one of its merits (others are discussed in §1.6) is that it allows the algorithms to be presented in a manner which is readable and structured.

In recent years, the concepts of structured and modular programming have become very popular, to the extent that one programming language (Modula-2) is founded on such principles. The interested reader is referred to Kernighan and Plauger (1974) or Yourdon (1975) for background, and to Riley (1988) for a more modern exposition of these ideas. In my own work, I have found such concepts extremely useful, and I recommend them to any practitioner who wishes to keep his debugging and re-programming efforts to a minimum. Nevertheless, while modularity is relatively easy to impose at the level of individual tasks such as the decomposition of a matrix or the finding of the minimum of a function along a line, it is not always reasonable to insist that the program avoid GOTO instructions. After all, in aimimg to keep memory requirements as low as possible, any program code which can do double duty is desirable. If possible, this should be collected into a subprogram. In a number of cases this will not be feasible, since the code may have to be entered at several points. Here the programmer has to make a judgement between compactness and readability of his program. I have opted for the former goal when such a decision has been necessary and have depended on comments and the essential shortness of the code to prevent it from becoming incomprehensible.

The coding of the algorithms in the book is *not* as compact as it might be in a specific application. In order to maintain a certain generality, I have chosen to allow variables and parameters to be passed to procedures and functions from fairly general driver programs. If algorithms are to be placed in-line in applications, it is possible to remove some of this program 'glue'. Furthermore, some features may not always be necessary, for example, computation of eigenvectors in the Jacobi method for eigensolutions of a real symmetric matrix (algorithm 14).

It should also be noted that I have taken pains to make it easy to save a 'copy' of the screen output to a file by duplicating all the output statements, that is the 'write' and 'writeln' commands, so that output is copied to a file which the user may name. (These statements are on the disk files, but deleted from the listings to reduce space and improve readability.) Input is allowed from an input file to allow examples to be presented without the user needing to type the appropriate response other than the name of the relevant 'example' file.

Furthermore, I have taken advantage of features within the MS-DOS operating system, and supported by compiler directives in Turbo Pascal, which allow for pipelining of input and output. This has allowed me to use batch files to automate the running of tests.

In the driver programs I have tried to include tests of the results of calculations, for example, the residuals in eigenvalue computations. In practice, I believe it is worthwhile to perform these calculations. When memory is at a premium, they can be performed 'off-line' in most cases. That is, the results can be saved to disk (backing storage) and the tests computed as a separate task, with data brought in from the disk only as needed.

These extra features use many extra bytes of code, but are, of course, easily deleted. Indeed, for specific problems, 75% or more of the code can be removed.

1.5. CHOICE OF ALGORITHMS

The algorithms in this book have been chosen for their utility in solving a variety of important problems in computational mathematics and their ease of implementation to short programs using relatively little working storage. Many topics are left out, despite their importance, because I feel that there has been insufficient development in directions relevant to compact numerical methods to allow for a suitable algorithm to be included. For example, over the last 15 years I have been interested in methods for the mathematical programming problem which do not require a tableau to be developed either explicitly or implicitly, since these techniques are generally quite memory and code intensive. The appearance of the interior point methods associated with the name of Karmarkar (1984) hold some hope for the future, but currently the programs are quite complicated.

In the solution of linear equations, my exposition has been confined to Gauss elimination and the Choleski decomposition. The literature on this subject is, however, vast and other algorithms exist. These can and should be used if special circumstances arise to make them more appropriate. For instance, Zambardino (1974) presents a form of Gauss elimination which uses less space than the one presented here. This procedure, in ALGOL, is called QUARTERSOLVE because only $n^2/4$ elements are stored, though an integer vector is needed to store pivot information and the program as given by Zambardino is quite complicated.

Many special methods can also be devised for matrices having special structures such as diagonal bands. Wilkinson and Reinsch (1971) give several such algorithms for both linear equations and the eigenvalue problem. The programmer with many problems of a special structure should consider these. However, I have found that most users want a reliable general-purpose method for linear equations because their day-to-day problems vary a great deal. I have deliberately avoided including a great many algorithms in this volume because most users will likely be their own implementors and not have a great deal of time to spend choosing, coding, testing and, most of all, maintaining programs.

Another choice which has been made is that of only including algorithms which are relatively 'small' in the sense that they fit into the machine all at once. For instance, in the solution of the algebraic eigenvalue problem on large computers, conventionally one reduces the matrix to a special form such as a tridiagonal or a Hessenberg matrix, solves the eigenproblem of the simpler system then back-transforms the solutions. Each of the three phases of this procedure could be fitted into a small machine. Again, for the practitioner with a lot of matrices to solve or a special requirement for only partial solution, such methods should be employed. For the one-at-a-time users, however, there is three times as much program code to worry about.

The lack of double-precision arithmetic on the machines I used to develop the algorithms which are included has no doubt modified my opinions concerning algorithms. Certainly, any algorithm requiring inner products of vectors, that is

$$\boldsymbol{u} \cdot \boldsymbol{v} = \sum_{i=1}^{n} u_i v_i \tag{1.10}$$

cannot be executed as accurately without extended-precision arithmetic (Wilkinson 1963). This has led to my abandonment of algorithms which seek to find the minimum of a function along a line by use of gradient information. Such algorithms require the derivative along the line and employ an inner product to compute this derivative. While these methods are distinctly unreliable on a machine having only a single, low-precision arithmetic, they can no doubt be used very effectively on other machines.

From the above discussion it will be evident that the principles guiding algorithm selection have been:

(i) shortness of program which results from implementation and low storage requirement, and
(ii) general utility of the method and importance of the problem which it solves.
To these points should be added:
(iii) proven reliability in a number of tests
(iv) the ease and speed with which a user can obtain useful results from the algorithms.

The third point is very important. No program should be considered acceptable until it has been tested fairly extensively. If possible, any method which gives solutions that can be checked by computing diagnostics should compute such information routinely. For instance, I have had users of my eigenvalue/eigenvector programs call me to say, 'Your program doesn't work!' In all cases to date they have been premature in their judgement, since the residuals computed as a routine adjunct to the eigensolution formation have shown the output to be reasonable even though it might be very different from what the user expected. Furthermore, I have saved myself the effort of having to duplicate their calculation to prove the correctness of the results. Therefore, if at all possible, such checks are always built into my programs.

The fourth point is important if users are to be able to try out the ideas presented in this book. As a user myself, I do not wish to spend many hours mastering the details of a code. The programs are to be treated as tools, not an end in themselves.

These principles lead to the choice of the Givens' reduction in algorithm 4 as a method for the solution of least-squares problems where the amount of data is too great to allow all of it to be stored in the memory at once. Similarly, algorithms 24 and 25 require the user to provide a rule for the calculation of the product of a matrix and a vector as a step in the solution of linear equations or the algebraic eigenproblem. However, the matrix itself need not be stored explicitly. This avoids the problem of designing a special method to take advantage of one type of matrix, though it requires rather more initiative from the user as it preserves this measure of generality.

In designing the particular forms of the algorithms which appear, a conscious effort has been made to avoid the requirement for many tolerances. Some machine-dependent quantities are unfortunately needed (they can in some cases be calculated by the program but this does lengthen the code), but as far as possible, and especially in determining when iterative algorithms have converged, devices are used which attempt to ensure that as many digits are determined as the machine is able to store. This may lead to programs continuing to execute long after acceptable answers have been obtained. However, I prefer to sin on the side of excess rather than leave results wanting in digits. Typically, the convergence test requires that the last and present iterates be identical to the precision of the machine by means of a test such as

```
if x + delta + offset = x + offset then halt;
```

where offset is some modest number such as 10. On machines which have an accumulator with extra digits, this type of test may never be satisfied, and must be replaced by

```
y: = x + delta + offset;
z: = x + offset;
if y = z then halt;
```

The 'tolerance' in this form of test is provided by the offset: within the computer the representations of y and z must be equal to halt execution. The simplicity of this type of test usually saves code though, as mentioned, at the possible expense of execution time.

1.6. A METHOD FOR EXPRESSING ALGORITHMS

In the first edition of this work, algorithms were expressed in step-and-description form. This allowed users to program them in a variety of programming languages. Indeed, one of the strengths of the first edition was the variety of implementations. At the time it appeared, a number of computers had their own languages or dialects, and specialisation to one programming language would have inhibited users of these special machines. Now, however, computer users are unlikely to be willing to type in code if a machine-readable form of an algorithm exists. Even if the programming language must be translated, having a workable form is useful as a starting point.

The original codes for the first edition were in BASIC for a Data General NOVA. Later these codes were made to run on a North Star Horizon. Some were made to work on a Hewlett–Packard 9830A. Present BASIC versions run on various common microcomputers under the Microsoft BASIC dialect; however, since I have used very conservative coding practices, apart from occasional syntactic deviations, they conform to ISO Minimal BASIC (ISO 6373-1984).

Rather than proof-reading the algorithms for the first edition, I re-coded them in FORTRAN. These codes exist as NASHLIB, and were and are commercially available from the author. I have not, however, been particularly satisfied that the FORTRAN implementation shows the methods to advantage, since the structure of the algorithms seems to get lost in the detail of FORTRAN code. Also, the working parts of the codes are overshadowed by the variable definitions and subroutine calls. Compact methods are often best placed in-line rather than called as standard subprograms as I have already indicated.

In the current edition, I want to stay close to the original step-and-description form of the algorithms, but nevertheless wish to offer working codes which could be distributed in machine-readable form. I have chosen to present the algorithms in Borland Turbo Pascal. This has the following justification.

(i) Pascal allows comments to be placed anywhere in the code, so that the original style for the algorithms, except for the typographic conventions, could be kept.

(ii) Turbo Pascal is available for many machines and is relatively inexpensive. It is used as a teaching tool at many universities and colleges, including the University of Ottawa. Version 3.01a of the Turbo Pascal system was used in developing the codes which appear here. I intend to prepare versions of the codes to run under later versions of this programming environment.

(iii) The differences between Turbo and Standard Pascal are unlikely to be important for the methods, so that users of other Pascal compilers can also use these codes.

(iv) Pascal is 'close enough' to many other programming languages to allow for straightforward translation of the codes.

A particular disadvantage of Turbo Pascal for my development work is that I have yet to find a convenient mechanism allowing automatic compilation and execution of codes, which would permit me to check a complete set of code via batch execution. From the perspective of the physical length of the listings, the present algorithms are also longer than I would like because Pascal requires program headings and declarations. In the procedural parts of the codes, 'begin' and 'end' statements also add to the line count to some extent.

From the user perspective, the requirement that matrix sizes be explicitly specified can be a nuisance. For problems with varying matrix sizes it may be necessary to compile separate versions of programs.

Section 1.8 notes some other details of algorithm expression which relate to the ease of use of the codes.

1.7. GENERAL NOTATION

I have not attempted to avoid re-use of symbols within this work since this would have required an over-large set of symbols. In fact, I have used greek letters as little as possible to save my typists' and typesetters' effort. However, within chapters and within a subject area the symbols should be consistent. There follow some brief general points on notation.

(i) Absolute value is denoted by vertical bars about a quantity, $| \ |$.
(ii) The norm of a quantity is denoted by double vertical bars, $\| \ \|$. The form of this must be found, where necessary, from the context.
(iii) A closed interval $[u, v]$ comprises all points x such that $u \leqslant x \leqslant v$. An open interval (u, v) comprises all points x such that $u < x < v$.
(iv) The exponents of decimal numbers will be expressed using the symbol E as in

$$1{\cdot}234 * 10^{-5} = 1{\cdot}234\text{E}-5$$

and

$$6{\cdot}78 * 10^{2} = 678 = 6{\cdot}78\text{E}2.$$

This notation has already appeared in §1.2.

1.8. SOFTWARE ENGINEERING ISSUES

The development of microcomputer software for users who are not trained in computer science or related subjects has given rise to user interfaces which are much more sophisticated and easy to use than were common when the first edition appeared. Mathematical software has not escaped such developments, but source code collections have generally required the user to consolidate program units, add driver and user routines, and compile and run the programs. In my experience, the lack of convenience implied by this requirement is a major obstacle to users learning about software published in source code form. In our nonlinear estimation software (Nash and Walker-Smith 1987), we were careful to include batch command files to allow for easy consolidation and execution of programs. This philosophy is continued here, albeit adapted to Turbo Pascal.

1. All driver programs include code (from the fragment in file startup.pas) to allow the user to specify a file from which the program control input and the problem data are to be input. We refer to this as a 'control input file'. It has a name stored in the global string variable **infname**, and is referred to by the global text variable **infile**. Algorithms which need input get it by **read** or **readln** statements from **infile**. The input file can be 'con', the console keyboard.

WARNING: errors in input control files may cause source code files to be destroyed. I believe this is a 'bug' in Turbo Pascal 3.01a, the version used to develop the codes.

The use of an include file which is not a complete procedure or function is not permitted by Turbo Pascal 5.0.

2. The same program code (startup.pas) allows an output file to be specified so that all output which appears on the console screen is copied to a file. The name for this file is stored in the global variable **confname**, and the file is referred to in programs by the global text variable **confile**. Output is saved by the crude but effective means of duplicating every **write**(. . .) and **writeln**(. . .) statement with equivalent **write(confile**,. . .) and **writeln(confile**,. . .) statements.
3. To make the algorithms less cluttered, these **write** and **writeln** statements to **confile** do not appear in the listings. They are present in the files on diskette.
4. To discourage unauthorised copying of the diskette files, all commentary and documentation of the algorithm codes has been deleted.
5. To allow for execution of the programs from operating system commands (at least in MS-DOS), compiler directives have been included at the start of all driver programs. Thus, if a compiled form of code dr0102.pas exists as dr0102.com, and a file dr0102x contains text equivalent to the keyboard input needed to *correctly* run this program, the command

dr0102 <dr0102x

will execute the program for the given data.

Chapter 2

FORMAL PROBLEMS IN LINEAR ALGEBRA

2.1. INTRODUCTION

A great many practical problems in the scientific and engineering world give rise to models or descriptions of reality which involve matrices. In consequence, a very large proportion of the literature of numerical mathematics is devoted to the solution of various matrix equations. In the following sections, the major formal problems in numerical linear algebra will be introduced. Some examples are included to show how these problems may arise directly in practice. However, the formal problems will in most cases occur as steps in larger, more difficult computations. In fact, the algorithms of numerical linear algebra are the keystones of numerical methods for solving real problems.

Matrix computations have become a large area for mathematical and computational research. Textbooks on this subject, such as Stewart (1973) and Strang (1976), offer a foundation useful for understanding the uses and manipulations of matrices and vectors. More advanced works detail the theorems and algorithms for particular situations. An important collection of well-referenced material is Golub and Van Loan (1983). Kahaner, Moler and Nash (1989) contains a very readable treatment of numerical linear algebra.

2.2. SIMULTANEOUS LINEAR EQUATIONS

If there are n known relationships

$$A_{i1}x_1 + A_{i2}x_2 + \ldots + A_{in}x_n = b_i \qquad i = 1, 2, \ldots, n \tag{2.1}$$

between the n quantities x_j with the coefficients A_{ij} and right-hand side elements b_i, $i = 1, 2, \ldots, n$, then (2.1) is a set of n simultaneous linear equations in n unknowns x_j, $j = 1, 2, \ldots, n$. It is simpler to write this problem in matrix form

$$\mathbf{A}\boldsymbol{x} = \boldsymbol{b} \tag{2.2}$$

where the coefficients have been collected into the matrix $\mathbf{A}$, the right-hand side is now the vector $\boldsymbol{b}$ and the unknowns have been collected as the vector $\boldsymbol{x}$. A further way to write the problem is to collect each column of $\mathbf{A}$ (say the jth) into a column vector (i.e. $\boldsymbol{a}_j$). Then we obtain

$$\sum_{j=1}^{n} \boldsymbol{a}_j x_j = \boldsymbol{b}. \tag{2.3}$$

Numerous textbooks on linear algebra, for instance Mostow and Sampson (1969) or Finkbeiner (1966), will provide suitable reading for anyone wishing to learn theorems and proofs concerning the existence of solutions to this problem. For the purposes of this monograph, it will suffice to outline a few basic properties of matrices as and when required.

Consider a set of n vectors of length n, that is

$$\boldsymbol{a}_1, \boldsymbol{a}_2, \ldots, \boldsymbol{a}_n. \tag{2.4}$$

These vectors are linearly independent if there exists no set of parameters x_j, $j = 1, 2, \ldots, n$ (not all zero), such that

$$\sum_{j=1}^{n} \boldsymbol{a}_j x_j = \mathbf{0} \tag{2.5}$$

where $\mathbf{0}$ is the null vector having all components zero. If the vectors $\boldsymbol{a}_j$ are now assembled to make the matrix $\mathbf{A}$ and are linearly independent, then it is always possible to find an $\boldsymbol{x}$ such that (2.2) is satisfied. Other ways of stating the condition that the columns of $\mathbf{A}$ are linearly independent are that $\mathbf{A}$ has full rank or

$$\operatorname{rank}(\mathbf{A}) = n \tag{2.6}$$

or that $\mathbf{A}$ is non-singular.

If only $k < n$ of the vectors are linearly independent, then

$$\operatorname{rank}(\mathbf{A}) = k \tag{2.7}$$

and $\mathbf{A}$ is singular. In general (2.2) cannot be solved if $\mathbf{A}$ is singular, though *consistent* systems of equations exist where $\boldsymbol{b}$ belongs to the space spanned by $\{\boldsymbol{a}_j : j = 1, 2, \ldots, n\}$.

In practice, it is useful to separate linear-equation problems into two categories. (The same classification will, in fact, apply to all problems involving matrices.)

(i) The matrix $\mathbf{A}$ is of modest order with probably few zero elements (*dense*).
(ii) The matrix $\mathbf{A}$ is of high order and has most of its elements zero (*sparse*).

The methods presented in this monograph for large matrices do not specifically require sparsity. The question which must be answered when computing on a small machine is, 'Does the matrix fit in the memory available?'

Example 2.1. Mass-spectrograph calibration

To illustrate a use for the solution of a system of linear equations, consider the determination of the composition of a mixture of four hydrocarbons using a mass spectrograph. Four lines will be needed in the spectrum. At these lines the intensity for the sample will be b_i, $i = 1, 2, 3, 4$. To calibrate the instrument, intensities A_{ij} for the ith line using a pure sample of the jth hydrocarbon are measured. Assuming additive line intensities, the composition of the mixture is then given by the solution $\boldsymbol{x}$ of

$$\mathbf{A}\boldsymbol{x} = \boldsymbol{b}.$$

Example 2.2. Ordinary differential equations: a two-point boundary-value problem

Large sparse sets of linear equations arise in the numerical solution of differential equations. Fröberg (1965, p 256) considers the differential equation

$$y'' + y/(1+x^2) = 7x \tag{2.8}$$

with the boundary conditions

$$y = \begin{cases} 0 & \text{for } x = 0 \quad (2.9) \\ 2 & \text{for } x = 1. \quad (2.10) \end{cases}$$

To solve this problem numerically, Fröberg replaces the continuum in x on the interval $[0, 1]$ with a set of $(n+1)$ points, that is, the step size on the grid is $h = 1/n$. The second derivative is therefore replaced by the second difference at point j

$$(y_{j+1} - 2y_j + y_{j-1})/h^2. \tag{2.11}$$

The differential equation (2.8) is therefore approximated by a set of linear equations of which the jth is

$$\frac{y_{j+1} - 2y_j + y_{j-1}}{h^2} + \frac{y_j}{1 + j^2h^2} = 7jh \tag{2.12}$$

or

$$y_{j+1} - \frac{2 - h^2 + 2j^2h^2}{1 + j^2h^2} y_j + y_{j-1} = 7jh^3. \tag{2.13}$$

Because $y_0 = 0$ and $y_n = 2$, this set of simultaneous linear equations is of order $(n-1)$. However, each row involves at most three of the values y_j. Thus, if the order of the set of equations is large, the matrix of coefficients is sparse.

2.3. THE LINEAR LEAST-SQUARES PROBLEM

As described above, n linear equations give relationships which permit n parameters to be determined if the equations give rise to linearly independent coefficient vectors. If there are more than n conditions, say m, then all of them may not necessarily be satisfied at once by any set of parameters $\boldsymbol{x}$. By asking a somewhat different question, however, it is possible to determine solutions $\boldsymbol{x}$ which in some way approximately satisfy the conditions. That is, we wish to write

$$\mathbf{A}\boldsymbol{x} \simeq \boldsymbol{b} \tag{2.14}$$

where the sense of the sign $\simeq$ is yet to be defined.

By defining the *residual* vector

$$\boldsymbol{r} = \boldsymbol{b} - \mathbf{A}\boldsymbol{x} \tag{2.15}$$

we can express the lack of approximation for a given $\boldsymbol{x}$ by the *norm* of $\boldsymbol{r}$

$$\|\boldsymbol{r}\|. \tag{2.16}$$

This must fulfil the following conditions:

$$\|\boldsymbol{r}\| > 0 \tag{2.17}$$

for $\boldsymbol{r} \neq \mathbf{0}$, and $\|\mathbf{0}\| = 0$,

$$\|c\boldsymbol{r}\| = |c| \cdot \|\boldsymbol{r}\| \tag{2.18}$$

for an arbitrary complex number c, and

$$\|\boldsymbol{r}+\boldsymbol{s}\| \leqslant \|\boldsymbol{r}\|+\|\boldsymbol{s}\| \tag{2.19}$$

where $\boldsymbol{s}$ is a vector of the same order as $\boldsymbol{r}$ (that is, m).

Condition (2.19) is called the triangle inequality since the lengths of the sides of a triangle satisfy this relationship. While there exist many norms, only a few are of widespread utility, and by and large in this work only the *Euclidean norm*

$$\|\boldsymbol{r}\|_{\mathrm{E}} = (\boldsymbol{r}^{\mathrm{T}}\boldsymbol{r})^{1/2} \tag{2.20}$$

will be used. The superscript T denotes transposition, so the norm is a scalar. The square of the Euclidean norm of $\boldsymbol{r}$

$$\boldsymbol{r}^{\mathrm{T}}\boldsymbol{r} = \sum_{i=1}^{m} r_i^2 \tag{2.21}$$

is appropriately called the *sum of squares*. The *least-squares solution* $\boldsymbol{x}$ of (2.14) is that set of parameters which minimises this sum of squares. In cases where $\operatorname{rank}(\mathbf{A})<n$ this solution is not unique. However, further conditions may be imposed upon the solution to ensure uniqueness. For instance, it may be required that in the non-unique case, $\boldsymbol{x}$ shall be that member of the set of vectors which minimises $\boldsymbol{r}^{\mathrm{T}}\boldsymbol{r}$ which has $\boldsymbol{x}^{\mathrm{T}}\boldsymbol{x}$ a minimum also. In this case $\boldsymbol{x}$ is the unique minimum-length least-squares solution.

If the minimisation of $\boldsymbol{r}^{\mathrm{T}}\boldsymbol{r}$ with respect to $\boldsymbol{x}$ is attempted directly, then using (2.15) and elementary calculus gives

$$\mathbf{A}^{\mathrm{T}}\mathbf{A}\boldsymbol{x} = \mathbf{A}^{\mathrm{T}}\boldsymbol{b} \tag{2.22}$$

as the set of conditions which $\boldsymbol{x}$ must satisfy. These are simply n simultaneous linear equations in n unknowns $\boldsymbol{x}$ and are called the *normal equations*. Solution of the least-squares problem via the normal equations is the most common method by which such problems are solved. Unfortunately, there are several objections to such an approach if it is not carefully executed, since the special structure of $\mathbf{A}^{\mathrm{T}}\mathbf{A}$ and the numerical instabilities which attend its formation are ignored at the peril of meaningless computed values for the parameters $\boldsymbol{x}$.

Firstly, any matrix $\mathbf{B}$ such that

$$\boldsymbol{x}^{\mathrm{T}}\mathbf{B}\boldsymbol{x} > 0 \tag{2.23}$$

for all $\boldsymbol{x} \neq \mathbf{0}$ is called *positive definite*. If

$$\boldsymbol{x}^{\mathrm{T}}\mathbf{B}\boldsymbol{x} \geqslant 0 \tag{2.24}$$

for all $\boldsymbol{x}$, $\mathbf{B}$ is *non-negative definite* or *positive semidefinite*. The last two terms are synonymous. The matrix $\mathbf{A}^{\mathrm{T}}\mathbf{A}$ gives the quadratic form

$$Q = \boldsymbol{x}^{\mathrm{T}}\mathbf{A}^{\mathrm{T}}\mathbf{A}\boldsymbol{x} \tag{2.25}$$

for any vector $\boldsymbol{x}$ of order n. By setting

$$\mathbf{y} = \mathbf{A}\boldsymbol{x} \tag{2.26}$$

$$Q = \mathbf{y}^{\mathrm{T}}\mathbf{y} \geqslant 0 \tag{2.27}$$

so that $\mathbf{A}^T\mathbf{A}$ is non-negative definite. In fact, if the columns of $\mathbf{A}$ are linearly independent, it is not possible for $\mathbf{y}$ to equal the order-m null vector $\mathbf{0}$, so that in this case $\mathbf{A}^T\mathbf{A}$ is positive definite. This is also called the full-rank case.

Secondly, many computer programs for solving the linear least-squares problem ignore the existence of special algorithms for the solution of linear equations having a symmetric, positive definite coefficient matrix. Above it has already been established that $\mathbf{A}^T\mathbf{A}$ is positive definite and symmetry is proved trivially. The special algorithms have advantages in efficiency and reliability over the methods for the general linear-equation problem.

Thirdly, in chapter 5 it will be shown that the formation of $\mathbf{A}^T\mathbf{A}$ can lead to loss of information. Techniques exist for the solution of the least-squares problem without recourse to the normal equations. When there is any question as to the true linear independence of the columns of $\mathbf{A}$, these have the advantage that they permit the minimum-length least-squares solution to be computed.

It is worth noting that the linear-equation problem of equation (2.2) can be solved by treating it as a least-squares problem. Then for singular matrices $\mathbf{A}$ there is still a least-squares solution $\boldsymbol{x}$ which, if the system of equations is consistent, has a zero sum of squares $\boldsymbol{r}^T\boldsymbol{r}$. For small-computer users who do not regularly need solutions to linear equations or whose equations have coefficient matrices which are near-singular (*ill conditioned* is another way to say this), it is my opinion that a least-squares solution method which avoids the formation of $\mathbf{A}^T\mathbf{A}$ is useful as a general approach to the problems in both equations (2.2) and (2.14).

As for linear equations, linear least-squares problems are categorised by whether or not they can be stored in the main memory of the computing device at hand. Once again, the traditional terms dense and sparse will be used, though some problems having m large and n reasonably small will have very few zero entries in the matrix $\mathbf{A}$.

Example 2.3. Least squares

It is believed that in the United States there exists a linear relationship between farm money income and the agricultural use of nitrogen, phosphate, potash and petroleum. A model is therefore formulated using, for simplicity, a linear form

$$\text{(money income)} = x_1 + x_2\,\text{(nitrogen)} + x_3\,\text{(phosphate)} + x_4\,\text{(potash)} + x_5\,\text{(petroleum)}. \tag{2.28}$$

For this problem the data are supplied as index numbers (1940 = 100) to avoid difficulties associated with the units in which the variables are measured. By collecting the values for the dependent variable (money income) as a vector $\boldsymbol{b}$ and the values for the other variables as the columns of a matrix $\mathbf{A}$ including the constant unity which multiplies x_1, a problem

$$\mathbf{A}\boldsymbol{x} \simeq \boldsymbol{b} \tag{2.14}$$

is obtained. The data and solutions for this problem are given as table 3.1 and example 3.2.

Example 2.4. Surveying-data fitting

Consider the measurement of height differences by levelling (reading heights off a vertical pole using a levelled telescope). This enables the difference between the heights of two points i and j to be measured as

$$b_{ij} = h_i - h_j + r_{ij} \tag{2.29}$$

where r_{ij} is the error made in taking the measurement. If m height differences are measured, the best set of heights $\boldsymbol{h}$ is obtained as the solution to the least-squares problem

$$\text{minimise} \quad \boldsymbol{r}^{\mathrm{T}}\boldsymbol{r} \tag{2.30}$$

where

$$\boldsymbol{r} = \boldsymbol{b} - \mathbf{A}\boldsymbol{h}$$

and each row of $\mathbf{A}$ has only two non-zero elements, 1 and -1, corresponding to the indices of the two points involved in the height-difference measurement. Sometimes the error is defined as the *weighted* residual

$$r_{ij} = [b_{ij} - (h_i - h_j)]d_{ij}$$

where d_{ij} is the horizontal distance between the two points (that is, the measurement error increases with distance).

A special feature of this particular problem is that the solution is only determined to within a constant, h_0, because no origin for the height scale has been specified. In many instances, only relative heights are important, as in a study of subsidence of land. Nevertheless, the matrix $\mathbf{A}$ is rank-deficient, so any method chosen to solve the problem as it has been presented should be capable of finding a least-squares solution despite the singularity (see example 19.2).

2.4. THE INVERSE AND GENERALISED INVERSE OF A MATRIX

An important concept is that of the *inverse* of a *square* matrix $\mathbf{A}$. It is defined as that square matrix, labelled $\mathbf{A}^{-1}$, such that

$$\mathbf{A}^{-1}\mathbf{A} = \mathbf{A}\mathbf{A}^{-1} = \mathbf{1}_n \tag{2.31}$$

where $\mathbf{1}_n$ is the unit matrix of order n. The inverse exists only if $\mathbf{A}$ has full rank. Algorithms exist which compute the inverse of a matrix explicitly, but these are of value only if the matrix inverse itself is useful. These algorithms should not be used, for instance, to solve equation (2.2) by means of the formal expression

$$\boldsymbol{x} = \mathbf{A}^{-1}\boldsymbol{b} \tag{2.32}$$

since this is inefficient. Furthermore, the inverse of a matrix $\mathbf{A}$ can be computed by setting the right-hand side $\boldsymbol{b}$ in equation (2.2) to the n successive columns of the unit matrix $\mathbf{1}_n$. Nonetheless, for positive definite symmetric matrices, this monograph presents a very compact algorithm for the inverse in §8.2.

When $\mathbf{A}$ is *rectangular*, the concept of an inverse must be generalised. Corresponding to (2.32) consider solving equation (2.14) by means of a matrix $\mathbf{A}^{+}$, yet to be defined, such that

$$\boldsymbol{x} = \mathbf{A}^{+}\boldsymbol{b}. \tag{2.33}$$

In other words, we have

$$\mathbf{A}^+\mathbf{A} = \mathbf{1}_n. \tag{2.34}$$

When $\mathbf{A}$ has only k linearly independent columns, it will be satisfactory if

$$\mathbf{A}^+\mathbf{A} = \begin{bmatrix} \mathbf{1}_k & & & & & \\ & 0 & & & & \\ & & 0 & & & \\ & & & \cdot & & \\ & & & & \cdot & \\ & & & & & 0 \end{bmatrix} \tag{2.35}$$

with the bottom-right block of zeros spanning $(n-k)$,

but in this case x is not defined uniquely since it can contain arbitrary components from the orthogonal complement of the space spanned by the columns of $\mathbf{A}$. That is, we have

$$x = \mathbf{A}^+ b + (\mathbf{1}_n - \mathbf{A}^+\mathbf{A})\mathbf{g} \tag{2.36}$$

where $\mathbf{g}$ is any vector of order n.

The normal equations (2.22) must still be satisfied. Thus in the full-rank case, it is straightforward to identify

$$\mathbf{A}^+ = (\mathbf{A}^{\mathrm{T}}\mathbf{A})^{-1}\mathbf{A}^{\mathrm{T}}. \tag{2.37}$$

In the rank-deficient case, the normal equations (2.22) imply by substitution of (2.36) that

$$\begin{aligned} \mathbf{A}^{\mathrm{T}}\mathbf{A}x &= \mathbf{A}^{\mathrm{T}}\mathbf{A}\mathbf{A}^+ b + (\mathbf{A}^{\mathrm{T}}\mathbf{A} - \mathbf{A}^{\mathrm{T}}\mathbf{A}\mathbf{A}^+\mathbf{A})\mathbf{g} \\ &= \mathbf{A}^{\mathrm{T}} b. \end{aligned} \tag{2.38}$$

If

$$\mathbf{A}^{\mathrm{T}}\mathbf{A}\mathbf{A}^+ = \mathbf{A}^{\mathrm{T}} \tag{2.39}$$

then equation (2.38) is obviously true. By requiring $\mathbf{A}^+$ to satisfy

$$\mathbf{A}\mathbf{A}^+\mathbf{A} = \mathbf{A} \tag{2.40}$$

and

$$(\mathbf{A}\mathbf{A}^+)^{\mathrm{T}} = \mathbf{A}\mathbf{A}^+ \tag{2.41}$$

this can indeed be made to happen. The proposed solution (2.36) is therefore a least-squares solution under the conditions (2.40) and (2.41) on $\mathbf{A}^+$. In order that (2.36) gives the minimum-length least-squares solution, it is necessary that $x^{\mathrm{T}}x$ be minimal also. But from equation (2.36) we find

$$x^{\mathrm{T}}x = b^{\mathrm{T}}(\mathbf{A}^+)^{\mathrm{T}}\mathbf{A}^+ b + \mathbf{g}^{\mathrm{T}}(\mathbf{1} - \mathbf{A}^+\mathbf{A})^{\mathrm{T}}(\mathbf{1} - \mathbf{A}^+\mathbf{A})\mathbf{g} + 2\mathbf{g}^{\mathrm{T}}(\mathbf{1} - \mathbf{A}^+\mathbf{A})^{\mathrm{T}}\mathbf{A}^+ b \tag{2.42}$$

which can be seen to have a minimum at

$$\mathbf{g} = \mathbf{0} \tag{2.43}$$

if

$$(\mathbf{1} - \mathbf{A}^+\mathbf{A})^{\mathrm{T}}$$

is the annihilator of $\mathbf{A}^+\boldsymbol{b}$, thus ensuring that the two contributions (that is, from $\boldsymbol{b}$ and $\boldsymbol{g}$) to $\boldsymbol{x}^T\boldsymbol{x}$ are orthogonal. This requirement imposes on $\mathbf{A}^+$ the further conditions

$$\mathbf{A}^+\mathbf{A}\mathbf{A}^+ = \mathbf{A}^+ \tag{2.44}$$

$$(\mathbf{A}^+\mathbf{A})^T = \mathbf{A}^+\mathbf{A}. \tag{2.45}$$

The four conditions (2.40), (2.41), (2.44) and (2.45) were proposed by Penrose (1955). The conditions are not, however, the route by which $\mathbf{A}^+$ is computed.

Here attention has been focused on one generalised inverse, called the Moore–Penrose inverse. It is possible to relax some of the four conditions and arrive at other types of generalised inverse. However, these will require other conditions to be applied if they are to be specified uniquely. For instance, it is possible to consider any matrix which satisfies (2.40) and (2.41) as a generalised inverse of $\mathbf{A}$ since it provides, via (2.33), a least-squares solution to equation (2.14). However, in the rank-deficient case, (2.36) allows arbitrary components from the null space of $\mathbf{A}$ to be added to this least-squares solution, so that the two-condition generalised inverse is specified incompletely.

Over the years a number of methods have been suggested to calculate 'generalised inverses'. Having encountered some examples of dubious design, coding or applications of such methods, I strongly recommend testing computed generalised inverse matrices to ascertain the extent to which conditions (2.40), (2.41), (2.44) and (2.45) are satisfied (Nash and Wang 1986).

2.5. DECOMPOSITIONS OF A MATRIX

In order to carry out computations with matrices, it is common to decompose them in some way to simplify and speed up the calculations. For a real m by n matrix $\mathbf{A}$, the $\mathbf{QR}$ decomposition is particularly useful. This equates the matrix $\mathbf{A}$ with the product of an orthogonal matrix $\mathbf{Q}$ and a right- or upper-triangular matrix $\mathbf{R}$, that is

$$\mathbf{A} = \mathbf{QR} \tag{2.46}$$

where $\mathbf{Q}$ is m by m and

$$\mathbf{Q}^T\mathbf{Q} = \mathbf{Q}\mathbf{Q}^T = \mathbf{1}_m \tag{2.47}$$

and $\mathbf{R}$ is m by n with all elements

$$R_{ij} = 0 \qquad \text{for } i > j. \tag{2.48}$$

The $\mathbf{QR}$ decomposition leads to the *singular-value decomposition* of the matrix $\mathbf{A}$ if the matrix $\mathbf{R}$ is identified with the product of a diagonal matrix $\mathbf{S}$ and an orthogonal matrix $\mathbf{V}^T$, that is

$$\mathbf{R} = \mathbf{S}\mathbf{V}^T \tag{2.49}$$

where the m by n matrix $\mathbf{S}$ is such that

$$S_{ij} = 0 \qquad \text{for } i \neq j \tag{2.50}$$

and $\mathbf{V}$, n by n, is such that

$$\mathbf{V}^T\mathbf{V} = \mathbf{V}\mathbf{V}^T = \mathbf{1}_n. \tag{2.51}$$

Note that the zeros below the diagonal in both **R** and **S** imply that, apart from orthogonality conditions imposed by (2.47), the elements of columns $(n+1)$, $(n+2), \ldots, m$ of **Q** are arbitrary. In fact, they are not needed in most calculations, so will be dropped, leaving the m by n matrix **U**, where

$$\mathbf{U}^{\mathrm{T}}\mathbf{U} = \mathbf{1}_n. \tag{2.52}$$

Note that it is no longer possible to make any statement regarding $\mathbf{U}\mathbf{U}^{\mathrm{T}}$. Omitting rows $(n+1)$ to m of both **R** and **S** allows the decompositions to be written as

$$\mathbf{A} = \mathbf{U}\mathbf{R} = \mathbf{U}\mathbf{S}\mathbf{V}^{\mathrm{T}} \tag{2.53}$$

where **A** is m by n, **U** is m by n and $\mathbf{U}^{\mathrm{T}}\mathbf{U} = \mathbf{1}_n$, **R** is n by n and upper-triangular, **S** is n by n and diagonal, and **V** is n by n and orthogonal. In the singular-value decomposition the diagonal elements of **S** are chosen to be non-negative.

Both the **QR** and singular-value decompositions can also be applied to square matrices. In addition, an n by n matrix **A** can be decomposed into a product of a lower- and an upper-triangular matrix, thus

$$\mathbf{A} = \mathbf{L}\mathbf{R}. \tag{2.54}$$

In the literature this is also known as the **LU** decomposition from the use of 'U' for 'upper triangular'. Here another mnemonic, 'U' for 'unitary' has been employed. If the matrix **A** is symmetric and positive definite, the decomposition

$$\mathbf{A} = \mathbf{L}\mathbf{L}^{\mathrm{T}} \tag{2.55}$$

is possible and is referred to as the Choleski decomposition.

A scaled form of this decomposition with unit diagonal elements for **L** can be written

$$\mathbf{A} = \mathbf{L}\ \mathbf{D}\ \mathbf{L}^{\mathrm{T}}$$

where **D** is a diagonal matrix.

To underline the importance of decompositions, it can be shown by direct substitution that if

$$\mathbf{A} = \mathbf{U}\mathbf{S}\mathbf{V}^{\mathrm{T}} \tag{2.53}$$

then the matrix

$$\mathbf{A}^{+} = \mathbf{V}\mathbf{S}^{+}\mathbf{U}^{\mathrm{T}} \tag{2.56}$$

where

$$S_{ii}^{+} = \begin{cases} 1/S_{ii} & \text{for } S_{ii} \neq 0 \\ 0 & \text{for } S_{ii} = 0 \end{cases} \tag{2.57}$$

satisfies the four conditions (2.40), (2.41), (2.44) and (2.45), that is

$$\begin{aligned} \mathbf{A}\mathbf{A}^{+}\mathbf{A} &= \mathbf{U}\mathbf{S}\mathbf{V}^{\mathrm{T}}\mathbf{V}\mathbf{S}^{+}\mathbf{U}^{\mathrm{T}}\mathbf{U}\mathbf{S}\mathbf{V}^{\mathrm{T}} \\ &= \mathbf{U}\mathbf{S}\mathbf{S}^{+}\mathbf{S}\mathbf{V}^{\mathrm{T}} \\ &= \mathbf{U}\mathbf{S}\mathbf{V}^{\mathrm{T}} = \mathbf{A} \end{aligned} \tag{2.58}$$

$$\begin{aligned} (\mathbf{A}\mathbf{A}^{+})^{\mathrm{T}} &= (\mathbf{U}\mathbf{S}\mathbf{V}^{\mathrm{T}}\mathbf{V}\mathbf{S}^{+}\mathbf{U}^{\mathrm{T}})^{\mathrm{T}} \\ &= (\mathbf{U}\mathbf{S}\mathbf{S}^{+}\mathbf{U}^{\mathrm{T}})^{\mathrm{T}} = \mathbf{U}\mathbf{S}^{+}\mathbf{S}\mathbf{U}^{\mathrm{T}} \\ &= \mathbf{U}\mathbf{S}\mathbf{S}^{+}\mathbf{U}^{\mathrm{T}} = \mathbf{A}\mathbf{A}^{+} \end{aligned} \tag{2.59}$$

$$\mathbf{A}^+\mathbf{A}\mathbf{A}^+ = \mathbf{V}\mathbf{S}^+\mathbf{U}^T\mathbf{U}\mathbf{S}\mathbf{V}^T\mathbf{V}\mathbf{S}^+\mathbf{U}^T$$
$$= \mathbf{V}\mathbf{S}^+\mathbf{S}\mathbf{S}^+\mathbf{U}^T = \mathbf{V}\mathbf{S}^+\mathbf{U}^T = \mathbf{A}^+ \tag{2.60}$$

and

$$(\mathbf{A}^+\mathbf{A})^T = (\mathbf{V}\mathbf{S}^+\mathbf{U}^T\mathbf{U}\mathbf{S}\mathbf{V}^T)^T = (\mathbf{V}\mathbf{S}^+\mathbf{S}\mathbf{V}^T)^T$$
$$= \mathbf{V}\mathbf{S}^+\mathbf{S}\mathbf{V}^T = \mathbf{A}^+\mathbf{A}. \tag{2.61}$$

Several of the above relationships depend on the diagonal nature of $\mathbf{S}$ and $\mathbf{S}^+$ and on the fact that diagonal matrices commute under multiplication.

2.6. THE MATRIX EIGENVALUE PROBLEM

An eigenvalue e and eigenvector $\boldsymbol{x}$ of an n by n matrix $\mathbf{A}$, real or complex, are respectively a scalar and vector which together satisfy the equation

$$\mathbf{A}\boldsymbol{x} = e\boldsymbol{x}. \tag{2.62}$$

There will be up to n *eigensolutions* $(e, \boldsymbol{x})$ for any matrix (Wilkinson 1965) and finding them for various types of matrices has given rise to a rich literature. In many cases, solutions to the generalised eigenproblem

$$\mathbf{A}\boldsymbol{x} = e\mathbf{B}\boldsymbol{x} \tag{2.63}$$

are wanted, where $\mathbf{B}$ is another n by n matrix. For matrices which are of a size that the computer can accommodate, it is usual to transform (2.63) into type (2.62) if this is possible. For large matrices, an attempt is usually made to solve (2.63) itself for one or more eigensolutions. In all the cases where the author has encountered equation (2.63) with large matrices, $\mathbf{A}$ and $\mathbf{B}$ have fortunately been symmetric, which provides several convenient simplifications, both theoretical and computational.

Example 2.5. Illustration of the matrix eigenvalue problem

In quantum mechanics, the use of the variation method to determine approximate energy states of physical systems gives rise to matrix eigenvalue problems if the trial functions used are linear combinations of some basis functions (see, for instance, Pauling and Wilson 1935, p 180ff).

If the trial function is F, and the energy of the physical system in question is described by the Hamiltonian operator H, then the variation principle seeks stationary values of the energy functional

$$C = \frac{(F, HF)}{(F, F)} \tag{2.64}$$

subject to the normalisation condition

$$(F, F) = 1 \tag{2.65}$$

where the symbol (,) represents an inner product between the elements separated by the comma within the parentheses. This is usually an integral over all the dimensions of the system. If a linear combination of some functions f_j, $j = 1, 2, \ldots, n$, is used for F, that is

$$F = \sum_{j=1}^{n} x_j f_j \tag{2.66}$$

then the variation method gives rise to the eigenvalue problem

$$\mathbf{A}x = e\mathbf{B}x \tag{2.63}$$

with

$$A_{ij} = (f_j, Hf_j) \tag{2.67}$$

and

$$B_{ij} = (f_i, f_j). \tag{2.68}$$

It is obvious that if **B** is a unit matrix, that is, if

$$(f_i, f_j) = \delta_{ij} = \begin{cases} 1 & \text{for } i = j \\ 0 & \text{for } i \neq j \end{cases} \tag{2.69}$$

a problem of type (2.56) arises. A specific example of such a problem is equation (11.1).

Chapter 3

THE SINGULAR-VALUE DECOMPOSITION AND ITS USE TO SOLVE LEAST-SQUARES PROBLEMS

3.1. INTRODUCTION

This chapter presents an algorithm for accomplishing the powerful and versatile singular-value decomposition. This allows the solution of a number of problems to be realised in a way which permits instabilities to be identified at the same time. This is a general strategy I like to incorporate into my programs as much as possible since I find succinct diagnostic information invaluable when users raise questions about computed answers—users do not in general raise too many idle questions! They may, however, expect the computer and my programs to produce reliable results from very suspect data, and the information these programs generate together with a solution can often give an idea of how trustworthy are the results. This is why the singular values are useful. In particular, the appearance of singular values differing greatly in magnitude implies that our data are nearly collinear. Collinearity introduces numerical problems simply because small changes in the data give large changes in the results. For example, consider the following two-dimensional vectors:

$$\mathbf{A} = (1, 0)^{\mathrm{T}}$$

$$\boldsymbol{B} = (1, 0{\cdot}1)^{\mathrm{T}}$$

$$\boldsymbol{C} = (0{\cdot}95, 0{\cdot}1)^{\mathrm{T}}.$$

Vector $\boldsymbol{C}$ is very close to vector $\boldsymbol{B}$, and both form an angle of approximately 6° with vector $\mathbf{A}$. However, while the angle between the vector sums $(\mathbf{A}+\boldsymbol{B})$ and $(\mathbf{A}+\boldsymbol{C})$ is only about 0·07°, the angle between $(\boldsymbol{B}-\mathbf{A})$ and $(\boldsymbol{C}-\mathbf{A})$ is greater than 26°. On the other hand, the set of vectors

$$\mathbf{A} = (1, 0)^{\mathrm{T}}$$

$$\boldsymbol{D} = (0, 1)^{\mathrm{T}}$$

$$\boldsymbol{E} = (0, 0{\cdot}95)^{\mathrm{T}}$$

gives angles between $(\mathbf{A}+\boldsymbol{D})$ and $(\mathbf{A}+\boldsymbol{E})$ and between $(\boldsymbol{D}-\mathbf{A})$ and $(\boldsymbol{E}-\mathbf{A})$ of approximately 1·5°. In summary, the sum of collinear vectors is well determined, the difference is not. Both the sum and difference of vectors which are not collinear are well determined.

3.2. A SINGULAR-VALUE DECOMPOSITION ALGORITHM

It may seem odd that the first algorithm to be described in this work is designed to compute the singular-value decomposition (svd) of a matrix. Such computations are topics well to the back of most books on numerical linear algebra. However, it was the algorithm below which first interested the author in the capabilities of small computers. Moreover, while the svd is somewhat of a sledgehammer method for many nutshell problems, its versatility in finding the eigensolutions of a real symmetric matrix, in solving sets of simultaneous linear equations or in computing minimum-length solutions to least-squares problems makes it a valuable building block in programs used to tackle a variety of real problems.

This versatility has been exploited in a single small program suite of approximately 300 lines of BASIC code to carry out the above problems as well as to find inverses and generalised inverses of matrices and to solve nonlinear least-squares problems (Nash 1984b, 1985).

The mathematical problem of the svd has already been stated in §2.5. However, for computational purposes, an alternative viewpoint is more useful. This considers the possibility of finding an orthogonal matrix $\mathbf{V}$, n by n, which transforms the real m by n matrix $\mathbf{A}$ into another real m by n matrix $\mathbf{B}$ whose columns are orthogonal. That is, it is desired to find $\mathbf{V}$ such that

$$\mathbf{B} = \mathbf{AV} = (\boldsymbol{b}_1, \boldsymbol{b}_2, \ldots, \boldsymbol{b}_n) \tag{3.1}$$

where

$$\boldsymbol{b}_i^{\mathrm{T}} \boldsymbol{b}_j = S_i^2 \delta_{ij} \tag{3.2}$$

and

$$\mathbf{VV}^{\mathrm{T}} = \mathbf{V}^{\mathrm{T}}\mathbf{V} = \mathbf{1}_n. \tag{3.3}$$

The Kronecker delta takes values

$$\delta_{ij} = \begin{cases} 0 & \text{for } i \neq j \\ 1 & \text{for } i = j. \end{cases} \tag{3.4}$$

The quantities S_i may, as yet, be either positive or negative, since only their square is defined by equation (3.2). They will henceforth be taken arbitrarily as positive and will be called *singular values* of the matrix $\mathbf{A}$. The vectors

$$\boldsymbol{u}_j = \boldsymbol{b}_j / S_j \tag{3.5}$$

which can be computed when none of the S_j is zero, are unit orthogonal vectors. Collecting these vectors into a real m by n matrix, and the singular values into a diagonal n by n matrix, it is possible to write

$$\mathbf{B} = \mathbf{US} \tag{3.6}$$

where

$$\mathbf{U}^{\mathrm{T}}\mathbf{U} = \mathbf{1}_n \tag{3.7}$$

is a unit matrix of order n.

In the case that some of the S_j are zero, equations (3.1) and (3.2) are still valid, but the columns of $\mathbf{U}$ corresponding to zero singular values must now be constructed such that they are orthogonal to the columns of **U** computed via equation (3.5) and to each other. Thus equations (3.6) and (3.7) are also satisfied. An alternative approach is to set the columns of **U** corresponding to zero singular values to null vectors. By choosing the first k of the singular values to be the non-zero ones, which is always possible by simple permutations within the matrix **V**, this causes the matrix $\mathbf{U}^{\mathrm{T}}\mathbf{U}$ to be a unit matrix of order k augmented to order n with zeros. This will be written

$$\mathbf{U}^{\mathrm{T}}\mathbf{U} = \begin{pmatrix} \mathbf{1}_k & \\ & \mathbf{0}_{n-k} \end{pmatrix}. \tag{3.8}$$

While not part of the commonly used definition of the svd, it is useful to require the singular values to be sorted, so that

$$S_{11} \geqslant S_{22} \geqslant S_{33} \geqslant \ldots \geqslant S_{kk} \geqslant \ldots \geqslant S_{nn}.$$

This allows (2.53) to be recast as a summation

$$\mathbf{A} = \sum_{j=1}^{n} u_j S_{jj} v_j^{\mathrm{T}}. \tag{2.53a}$$

Partial sums of this series give a sequence of approximations

$$\tilde{\mathbf{A}}_1, \tilde{\mathbf{A}}_2, \ldots, \tilde{\mathbf{A}}_n$$

where, obviously, the last member of the sequence

$$\tilde{\mathbf{A}}_n = \mathbf{A}$$

since it corresponds to a complete reconstruction of the svd. The rank-one matrices

$$u_j S_{jj} v_j^{\mathrm{T}}$$

can be referred to as singular planes, and the partial sums (in order of decreasing singular values) are partial svds (Nash and Shlien 1987).

A combination of (3.1) and (3.6) gives

$$\mathbf{AV} = \mathbf{US} \tag{3.9}$$

or, using (3.3), the orthogonality of **V**,

$$\mathbf{A} = \mathbf{USV}^{\mathrm{T}} \tag{2.53}$$

which expresses the svd of **A**.

The preceding discussion is conditional on the existence and computability of a suitable matrix **V**. The next section shows how this task may be accomplished.

3.3. ORTHOGONALISATION BY PLANE ROTATIONS

The matrix **V** sought to accomplish the orthogonalisation (3.1) will be built up as a product of simpler matrices

$$\mathbf{V} = \prod_{k=1}^{z} \mathbf{V}^{(k)} \tag{3.10}$$

where z is some index not necessarily related to the dimensions m and n of $\mathbf{A}$, the matrix being decomposed. The matrices used in this product will be plane rotations. If $\mathbf{V}^{(k)}$ is a rotation of angle ϕ in the ij plane, then all elements of $\mathbf{V}^{(k)}$ will be the same as those in a unit matrix of order n except for

$$\begin{aligned} V_{ii}^{(k)} &= \cos\phi = V_{jj}^{(k)} \\ -V_{ij}^{(k)} &= \sin\phi = V_{ji}^{(k)}. \end{aligned} \tag{3.11}$$

Thus $\mathbf{V}^{(k)}$ affects only two columns of any matrix it multiplies from the right. These columns will be labelled $\boldsymbol{x}$ and $\mathbf{y}$. Consider the effect of a single rotation involving these two columns

$$(\boldsymbol{x}, \mathbf{y})\begin{pmatrix} \cos\phi & -\sin\phi \\ \sin\phi & \cos\phi \end{pmatrix} = (\boldsymbol{X}, \boldsymbol{Y}). \tag{3.12}$$

Thus we have

$$\begin{aligned} \boldsymbol{X} &= \boldsymbol{x}\cos\phi + \mathbf{y}\sin\phi \\ \boldsymbol{Y} &= -\boldsymbol{x}\sin\phi + \mathbf{y}\cos\phi. \end{aligned} \tag{3.13}$$

If the resulting vectors $\boldsymbol{X}$ and $\boldsymbol{Y}$ are to be orthogonal, then

$$\boldsymbol{X}^{\mathrm{T}}\boldsymbol{Y} = 0 = -(\boldsymbol{x}^{\mathrm{T}}\boldsymbol{x} - \mathbf{y}^{\mathrm{T}}\mathbf{y})\sin\phi\cos\phi + \boldsymbol{x}^{\mathrm{T}}\mathbf{y}(\cos^2\phi - \sin^2\phi). \tag{3.14}$$

There is a variety of choices for the angle ϕ, or more correctly for the sine and cosine of this angle, which satisfy (3.14). Some of these are mentioned by Hestenes (1958), Chartres (1962) and Nash (1975). However, it is convenient if the rotation can order the columns of the orthogonalised matrix $\mathbf{B}$ by length, so that the singular values are in decreasing order of size and those which are zero (or infinitesimal) are found in the lower right-hand corner of the matrix $\mathbf{S}$ as in equation (3.8). Therefore, a further condition on the rotation is that

$$\boldsymbol{X}^{\mathrm{T}}\boldsymbol{X} - \boldsymbol{x}^{\mathrm{T}}\boldsymbol{x} \geqslant 0. \tag{3.15}$$

For convenience, the columns of the product matrix

$$\mathbf{A}\prod_{j=1}^{k-1} \mathbf{V}^{(j)} \tag{3.16}$$

will be denoted $\boldsymbol{a}_i$, $i = 1, 2, \ldots, n$. The progress of the orthogonalisation is then observable if a measure Z of the non-orthogonality is defined

$$Z = \sum_{i=1}^{n-1} \sum_{j=i+1}^{n} (\boldsymbol{a}_i^{\mathrm{T}}\boldsymbol{a}_j)^2. \tag{3.17}$$

Since two columns orthogonalised in one rotation may be made non-orthogonal in subsequent rotations, it is essential that this measure be reduced at each rotation.

Because only two columns are involved in the kth rotation, we have

$$Z^{(k)} = Z^{(k-1)} + (\mathbf{X}^{\mathrm{T}}\mathbf{Y})^2 - (\mathbf{x}^{\mathrm{T}}\mathbf{y})^2. \tag{3.18}$$

But condition (3.14) implies

$$Z^{(k)} = Z^{(k-1)} - (\mathbf{x}^{\mathrm{T}}\mathbf{y})^2 \tag{3.19}$$

so that the non-orthogonality is reduced at each rotation.

The specific formulae for the sine and cosine of the angle of rotation are (see e.g. Nash 1975) given in terms of the quantities

$$p = \mathbf{x}^{\mathrm{T}}\mathbf{y} \tag{3.20}$$

$$q = \mathbf{x}^{\mathrm{T}}\mathbf{x} - \mathbf{y}^{\mathrm{T}}\mathbf{y} \tag{3.21}$$

and

$$v = (4p^2 + q^2)^{1/2}. \tag{3.22}$$

They are

$$\cos\phi = [(v+q)/(2v)]^{1/2} \quad \text{for } q \geqslant 0 \tag{3.23}$$

$$\sin\phi = p/(v\cos\phi) \tag{3.24}$$

$$\sin\phi = \operatorname{sgn}(p)[(v-q)/(2v)]^{1/2} \quad \text{for } q < 0 \tag{3.25}$$

$$\cos\phi = p/(v\sin\phi) \tag{3.26}$$

where

$$\operatorname{sgn}(p) = \begin{cases} 1 & \text{for } p \geqslant 0 \\ -1 & \text{for } p < 0. \end{cases} \tag{3.27}$$

Note that having two forms for the calculation of the functions of the angle of rotation permits the subtraction of nearly equal numbers to be avoided. As the matrix nears orthogonality p will become small, so that q and v are bound to have nearly equal magnitudes.

In the first edition of this book, I chose to perform the computed rotation only when $q \geqslant r$, and to use

$$\sin(\phi) = 1 \qquad \cos(\phi) = 0 \tag{3.28}$$

when $q < 0$. This effects an interchange of the columns of the current matrix **A**. However, I now believe that it is more efficient to perform the rotations as defined in the code presented. The rotations (3.28) were used to force nearly null columns of the final working matrix to the right-hand side of the storage array. This will occur when the original matrix **A** suffers from linear dependencies between the columns (that is, is rank deficient). In such cases, the rightmost columns of the working matrix eventually reflect the lack of information in the data in directions corresponding to the null space of the matrix **A**. The current methods cannot do much about this lack of information, and it is not sensible to continue computations on these columns. In the current implementation of the method (Nash and Shlien 1987), we prefer to ignore columns at the right of the working matrix which become smaller than a specified tolerance. This has a side effect of speeding the calculations significantly when rank deficient matrices are encountered.

3.4. A FINE POINT

Equations (3.15) and (3.19) cause the algorithm just described obviously to proceed *towards* both an orthogonalisation and an ordering of the columns of the resulting matrix $\mathbf{A}^{(z)}$. However, the rotations must be arranged in some sequence to carry this task to completion. Furthermore, it remains to be shown that some sequences of rotations will not place the columns in disorder again. For suppose $\boldsymbol{a}_1$ is orthogonal to all other columns and larger than any of them individually. A sequential arrangement of the rotations to operate first on columns (1, 2), then (1, 3), (1,4), . . . , (1, n), followed by (2, 3), . . . , (2, n), (3, 4), . . . , (($n-1$), n) will be called a *cycle* or *sweep*. Such a sweep applied to the matrix described can easily yield a new $\boldsymbol{a}_2$ for which

$$\boldsymbol{a}_2^{\mathrm{T}}\boldsymbol{a}_2 > \boldsymbol{a}_1^{\mathrm{T}}\boldsymbol{a}_1 \tag{3.29}$$

if, for instance, the original matrix has $\boldsymbol{a}_2 = \boldsymbol{a}_3$ and the norm of these vectors is greater than $2^{-1/2}$ times the norm of $\boldsymbol{a}_1$. Another sweep of rotations will put things right in this case by exchanging $\boldsymbol{a}_1$ and $\boldsymbol{a}_2$. However, once two columns have achieved a separation related in a certain way to the non-orthogonality measure (3.17), it can be shown that no subsequent rotation can exchange them.

Suppose that the algorithm has proceeded so far that the non-orthogonality measure Z satisfies the inequality

$$Z < t^2 \tag{3.30}$$

where t is some positive tolerance. Then, for any subsequent rotation the parameter p, equation (3.21), must obey

$$p^2 < t^2. \tag{3.31}$$

Suppose that all adjacent columns are separated in size so that

$$\boldsymbol{a}_{k-1}^{\mathrm{T}}\boldsymbol{a}_{k-1} - \boldsymbol{a}_k^{\mathrm{T}}\boldsymbol{a}_k > t. \tag{3.32}$$

Then a rotation which changes $\boldsymbol{a}_k$ (but not $\boldsymbol{a}_{k-1}$) cannot change the ordering of the two columns. If $\boldsymbol{x} = \boldsymbol{a}_k$, then straightforward use of equations (3.23) and (3.24) or (3.25) and (3.26) gives

$$\mathbf{X}^{\mathrm{T}}\mathbf{X} - \boldsymbol{x}^{\mathrm{T}}\boldsymbol{x} = (v - q)/2 \geqslant 0. \tag{3.33}$$

Using (3.31) and (3.22) in (3.33) gives

$$\mathbf{X}^{\mathrm{T}}\mathbf{X} - \boldsymbol{x}^{\mathrm{T}}\boldsymbol{x} \leqslant [(4t^2 + q^2)^{1/2} - q]/2 \leqslant [(2t + q) - q]/2 \leqslant t. \tag{3.34}$$

Thus, once columns become sufficiently separated by size and the non-orthogonality sufficiently diminished, the column ordering is stable. When some columns are equal in norm but orthogonal, the above theorem can be applied to columns separated by size.

The general question of convergence in the case of equal singular values has been investigated by T Hoy Booker (Booker 1985). The proof in exact arithmetic is incomplete. However, for a method such as the algorithm presented here, which uses tolerances for zero, Booker has shown that the cyclic sweeps must eventually terminate.

Algorithm 1. Singular-value decomposition

```
procedure NashSVD(nRow, nCol: integer; {size of problem}
                  var W: wmatrix; {working matrix}
                  var Z: rvector); {squares of singular values}
{alg01.pas ==
      form a singular value decomposition of matrix A which is stored in the
      first nRow rows of working array W and the nCol columns of this array.
      The first nRow rows of W will become the product U * S of a
      conventional svd, where S is the diagonal matrix of singular values.
      The last nCol rows of W will be the matrix V of a conventional svd.
      On return, Z will contain the squares of the singular values. An
      extended form of this commentary can be displayed on the screen by
      removing the comment braces on the writeln statements below.

                  Copyright 1988 J. C. Nash
}
var
   i, j, k, EstColRank, RotCount, SweepCount, slimit : integer;
   eps, e2, tol, vt, p, h2, x0, y0, q, r, c0, s0, c2, d1, d2 : real;
procedure rotate; {STEP 10 as a procedure}
{This rotation acts on both U and V, by storing V at the bottom of U}
begin {<< rotation }
   for i := 1 to nRow+nCol do
   begin
      D1 := W[i,j]; D2 := W[i,k];
      W[i,j] := D1*c0+D2*s0; W[i,k] := -D1*s0+D2*c0
   end; { rotation >>}
end; { rotate }
begin { procedure SVD }
{ -- remove the comment braces to allow message to be displayed --
   writeln(' Nash Singular Value Decomposition (NashSVD).');
   writeln;
   writeln('The program takes as input a real matrix A.');
   writeln;
   writeln('Let U and V be orthogonal matrices, & S');
   writeln('a diagonal matrix, such that U'' A V = S .');
   writeln('Then A = U S V'' is the decomposition.');
   writeln('A is assumed to have more rows than columns. If it');
   writeln('does not, the svd of the transpose A'' gives the svd ');
   writeln('of A, since A'' = V S U''.');
   writeln;
   writeln('If A has nRow rows and nCol columns, then the matrix');
   writeln('is supplied to the program in a working array W large');
   writeln('enough to hold nRow+nCol rows and nCol columns.');
   writeln('Output comprises the elements of Z, which are the ');
   writeln('squares of the elements of the vector S, together');
   writeln('with columns of W that correspond to non-zero elements');
```

Algorithm 1. Singular-value decomposition (cont.)

```
    writeln('of Z. The final array W contains the decomposition in a');
    writeln('special form, namely,');
    writeln;
    writeln(' ( U S ) ');
    writeln(' W = ( ) ');
    writeln(' ( V ) ');
    writeln;
    writeln('The matrices U and V are extracted from W, and S is');
    writeln('found from Z. However, the (U S) matrix and V matrix may');
    writeln('also be used directly in calculations, which we prefer');
    writeln('since fewer arithmetic operations are then needed.');
    writeln;
}
{STEP 0 Enter nRow, nCol, the dimensions of the matrix to be decomposed.}
    writeln('alg01.pas -- NashSVD');
    eps := Calceps; {Set eps, the machine precision.}
    slimit := nCol div 4; if slimit<6 then slimit := 6;
    {Set slimit, a limit on the number of sweeps allowed. A suggested
    limit is max([nCol/4], 6).}
    SweepCount := 0; {to count the number of sweeps carried out}
    e2 := 10.0*nRow*eps*eps;
    tol := eps*0.1;
    { Set the tolerances used to decide if the algorithm has converged.
        For further discussion of this, see the commentary under STEP 7.}
    EstColRank := nCol; {current estimate of rank};
    {Set V matrix to the unit matrix of order nCol.
    V is stored in rows (nRow+1) to (nRow+nCol) of array W.}
    for i := 1 to nCol do
    begin
        for j := 1 to nCol do
        W[nRow+i,j] := 0.0; W[nRow+i,i] := 1.0;
    end; {loop on i, and initialization of V matrix}
    { Main SVD calculations }
    repeat {until convergence is achieved or too many sweeps are carried out}
        RotCount := EstColRank*(EstColRank-1) div 2; {STEP 1 -- rotation counter}
        SweepCount := SweepCount+1;
        for j := 1 to EstColRank-1 do {STEP 2 -- main cyclic Jacobi sweep}
            begin {STEP 3}
            for k := j+1 to EstColRank do
            begin {STEP 4}
                p := 0.0; q := 0.0; r := 0.0;
                for i := 1 to nRow do {STEP 5}
                begin
                    x0 := W[i,j]; y0 := W[i,k];
                    p := p+x0*y0; q := q+x0*x0; r := r+y0*y0;
                end;
                Z[j] := q; Z[k] := r;
                {Now come important convergence test considerations. First we
                will decide if rotation will exchange order of columns.}
                if q >= r then {STEP 6 -- check if the columns are ordered.}
                begin {STEP 7 Columns are ordered, so try convergence test.}
                    if (q<=e2*Z[1]) or (abs(p)<= tol*q) then RotCount := RotCount-1
                    {There is no more work on this particular pair of columns in the
```

Algorithm 1. Singular-value decomposition (cont.)

```
                        current sweep. That is, we now go to STEP 11. The first
                        condition checks for very small column norms in BOTH columns, for
                        which no rotation makes sense. The second condition determines
                        if the inner product is small with respect to the larger of the
                        columns, which implies a very small rotation angle.}
                    else {columns are in order, but their inner product is not small}
                    begin {STEP 8}
                        p := p/q; r := 1-r/q; vt := sqrt(4*p*p + r*r);
                        c0 := sqrt(0.5*(1+r/vt)); s0 := p/(vt*c0);
                        rotate;
                    end
                end {columns in order with q>=r}
                else { columns out of order -- must rotate}
                begin {STEP 9}
                    {note: r > q, and cannot be zero since both are sums of squares for
                    the svd. In the case of a real symmetric matrix, this assumption
                    must be questioned.}
                    p := p/r; q := q/r-1; vt := sqrt(4*p*p + q*q);
                    s0 := sqrt(0.5*(1-q/vt));
                    if p<0 then s0 := -s0;
                    c0 := p/(vt*s0);
                    rotate; {The rotation is STEP 10.}
                end;
                {Both angle calculations have been set up so that large numbers do
                not occur in intermediate quantities. This is easy in the svd case,
                since quantities x2,y2 cannot be negative. An obvious scaling for
                the eigenvalue problem does not immediately suggest itself.}
            end; {loop on K -- end-loop is STEP 11}
        end; {loop on j -- end-loop is STEP 12}
        writeln('End of Sweep #', SweepCount,
            '- no. of rotations performed =', RotCount);
        {STEP 13 -- Set EstColRank to largest column index for which
            Z[column index] > (Z[1]*tol + tol*tol)
            Note how Pascal expresses this more precisely.}
        while (EstColRank >= 3) and (Z[EstColRank] <= Z[1]*tol + tol*tol)
            do EstColRank := EstColRank-1;
            {STEP 14 -- Goto STEP 1 to repeat sweep if rotations have been
                performed and the sweep limit has not been reached.}
    until (RotCount=0) or (SweepCount>slimit);
    {STEP 15 -- end SVD calculations }
    if (SweepCount > slimit) then writeln('**** SWEEP LIMIT EXCEEDED');
    if (SweepCount > slimit) then
    {Note: the decomposition may still be useful, even if the sweep
    limit has been reached.}
end; {alg01.pas == NashSVD}
```

3.5. AN ALTERNATIVE IMPLEMENTATION OF THE SINGULAR-VALUE DECOMPOSITION

One of the most time-consuming steps in algorithm 1 is the loop which comprises

STEP 5. While the inner product used to compute $p = x^T y$ must still be performed, it is possible to use equation (3.33) and the corresponding result for $\mathbf{Y}$, that is

$$\mathbf{Y}^T\mathbf{Y} - \mathbf{y}^T\mathbf{y} = -(v-q)/2 \tag{3.35}$$

to compute the updated column norms after each rotation. There is a danger that nearly equal magnitudes may be subtracted, with the resultant column norm having a large relative error. However, if the application requires information from the largest singular values and vectors, this approach offers some saving of effort. The changes needed are:

(1) an initial loop to compute the Z[i], that is, the sum of squares of the elements of each column of the original matrix **A**;
(2) the addition of two statements to the end of the main svd loop on k, which, if a rotation has been performed, update the column norms Z[j] and Z[k] via formulae (3.34) and (3.35). Note that in the present algorithm the quantities needed for these calculations have *not* been preserved. Alternatively, add at the end of STEP 8 (after the rotation) the statements

```
Z[j] := Z[j] + 0·5*q*(vt-r);
Z[k] := Z[k] - 0·5*q*(vt-r);
if Z[k] < 0·0 then Z[k] := 0·0;
```

and at the end of STEP 9 the statements

```
Z[j] := Z[j] + 0·5*r*(vt-q);
Z[k] := Z[k] - 0·5*r*(vt-q);
if Z[k] < 0·0 then Z[k] := 0·0;
```

(3) the deletion of the assignments

```
Z[j] := q;        Z[k] := r;
```

at the end of STEP 5.

As an illustration of the consequences of such changes, the singular-value decompositions of a 6 by 4 matrix derived from the Frank symmetric test matrix and an 8 by 5 portion of the Hilbert matrix were calculated. In the latter test the Turbo-87 Pascal compiler was used rather than the regular Turbo Pascal compiler (versions 3.01a of both systems). The results below present the modified algorithm result(s) above the corresponding figure(s) for the regular method.

```
Frank matrix:

Column orthogonality of U
Largest inner product is 4,4 =  -4.7760764232E-09
Largest inner product is 4,4 =   3.4106051316E-12
```

```
Singular values
  3.3658407311E+00   1.0812763036E+00   6.7431328720E-01   5.3627598567E-01
  3.3658407311E+00   1.0812763036E+00   6.7431328701E-01   5.3627598503E-01


Hilbert segment:

Column orthogonality of U
Largest inner product is 5,5 =  -1.44016460160157E-006
Largest inner product is 3,3 =   5.27355936696949E-016

Singular values
1.27515004411E+000 4.97081651063E-001 1.30419686491E-001 2.55816892287E-002
1.27515004411E+000 4.97081651063E-001 1.30419686491E-001 2.55816892259E-002

3.60194233367E-003
3.60194103682E-003
```

3.6. USING THE SINGULAR-VALUE DECOMPOSITION TO SOLVE LEAST-SQUARES PROBLEMS

By combining equations (2.33) and (2.56), the singular-value decomposition can be used to solve least-squares problems (2.14) via

$$\boldsymbol{x} = \mathbf{V}\mathbf{S}^{+}\mathbf{U}^{\mathrm{T}}\boldsymbol{b}. \tag{3.36}$$

However, the definition (2.57) of $\mathbf{S}^+$ is too strict for practical computation, since a real-world calculation will seldom give singular values which are identically zero. Therefore, for the purposes of an algorithm it is appropriate to define

$$S_{ii}^{+} = \begin{cases} 1/S_{ii} & \text{for } S_{ii} > q \\ 0 & \text{for } S_{ii} \leqslant q \end{cases} \tag{3.37}$$

where q is some tolerance set by the user. The use of the symbol for the tolerance is not coincidental. The previous employment of this symbol in computing the rotation parameters and the norm of the orthogonalised columns of the resulting matrix is finished, and it can be re-used.

Permitting $\mathbf{S}^+$ to depend on a user-defined tolerance places upon him/her the responsibility for deciding the degree of linear dependence in his/her data. In an economic modelling situation, for instance, columns of $\mathbf{U}$ corresponding to small singular values are almost certain to be largely determined by errors or noise in the original data. On the other hand, the same columns when derived from the tracking of a satellite may contain very significant information about orbit perturbations. Therefore, it is not only difficult to provide an automatic definition for $\mathbf{S}^+$, it is inappropriate. Furthermore, the matrix $\mathbf{B} = \mathbf{US}$ contains the *principal components* (Kendall and Stewart 1958–66, vol 3, p 286). By appropriate choices of q in equation (3.37), the solutions $\boldsymbol{x}$ corresponding to only a few of the dominant principal components can be computed. Furthermore, at this stage in the calculation $\mathbf{U}^T\boldsymbol{b}$ should already have been computed and saved, so that only a simple matrix–vector multiplication is involved in finding each of the solutions.

Another way to look at this is to consider the least-squares problem

$$\mathbf{B}\boldsymbol{w} \simeq \boldsymbol{b} \tag{3.38}$$

where $\mathbf{B}$ is the matrix having orthogonal columns and is given in equations (3.1) and (3.6). Thus the normal equations corresponding to (3.38) are

$$\mathbf{B}^T\mathbf{B}\boldsymbol{w} = \mathbf{S}^2\boldsymbol{w} = \mathbf{B}^T\boldsymbol{b}. \tag{3.39}$$

But $\mathbf{S}^2$ is diagonal so that the solutions are easily obtained as

$$\boldsymbol{w} = \mathbf{S}^{-2}\mathbf{B}^T\boldsymbol{b} \tag{3.40}$$

and substitution of (3.6) gives

$$\boldsymbol{w} = \mathbf{S}^{-1}\mathbf{U}^T\boldsymbol{b}. \tag{3.41}$$

Should the problem be singular, then

$$\boldsymbol{w} = \mathbf{S}^{+}\mathbf{U}^T\boldsymbol{b}. \tag{3.42}$$

can be used. Now note that because

$$\mathbf{B}\mathbf{V}^T = \mathbf{A} \tag{3.43}$$

from (3.1), the solution $\boldsymbol{w}$ allows $\boldsymbol{x}$ to be computed via

$$\boldsymbol{x} = \mathbf{V}\boldsymbol{w}. \tag{3.44}$$

The coefficients $\boldsymbol{w}$ are important as the solution of the least-squares problem in terms of the orthogonal combinations of the original variables called the principal components. The normalised components are contained in $\mathbf{U}$. It is easy to rearrange the residual sum of squares so that

$$\boldsymbol{r}^T\boldsymbol{r} = (\boldsymbol{b} - \mathbf{A}\boldsymbol{x})^T(\boldsymbol{b} - \mathbf{A}\boldsymbol{x}) = (\boldsymbol{b} - \mathbf{B}\boldsymbol{w})^T(\boldsymbol{b} - \mathbf{B}\boldsymbol{w}) = \boldsymbol{b}^T\boldsymbol{b} - \boldsymbol{b}^T\mathbf{B}\boldsymbol{w} \tag{3.45}$$

by virtue of the normal equations (3.39). However, substituting (3.37) in (3.42) and noting the ordering of $\mathbf{S}$, it is obvious that if

$$S_{k+1,k+1} \leqslant q \tag{3.46}$$

is the first singular value less than or equal to the tolerance, then

$$w_i = 0 \qquad \text{for } i > k. \tag{3.47}$$

The components corresponding to small singular values are thus dropped from the solution. But it is these components which are the least accurately determined since they arise as differences. Furthermore, from (3.6) and (3.45)

$$\begin{aligned} \boldsymbol{r}^T\boldsymbol{r} &= \boldsymbol{b}^T\boldsymbol{b} - \boldsymbol{b}^T\mathbf{U}\mathbf{S}\mathbf{S}^{+}\mathbf{U}^T\boldsymbol{b} \\ &= \boldsymbol{b}^T\boldsymbol{b} - \sum_{j=1}^{k} (\mathbf{U}^T\boldsymbol{b})_j^2 \end{aligned} \tag{3.48}$$

where the limit of the sum in (3.48) is k, the number of principal components which are included. Thus inclusion of another component cannot increase the residual sum of squares. However, if a component with a very small singular value is introduced, it will contribute a very large amount to the corresponding element of $\boldsymbol{w}$, and $\boldsymbol{x}$ will acquire large elements also. From (3.48), however, it is the interaction between the normalised component $\boldsymbol{u}_j$ and $\boldsymbol{b}$ which determines how much a given component reduces the sum of squares. A least-squares problem will therefore be ill conditioned if $\boldsymbol{b}$ is best approximated by a column of $\mathbf{U}$ which is associated with a small singular value and thus may be computed inaccurately.

On the other hand, if the components corresponding to 'large' singular values are the ones which are responsible for reducing the sum of squares, then the problem has a solution which can be safely computed by leaving out the components which make the elements of $\boldsymbol{w}$ and $\boldsymbol{x}$ large without appreciably reducing the sum of squares. Unless the unwanted components have no part in reducing the sum of squares, that is unless

$$\boldsymbol{u}_i^{\mathrm{T}}\boldsymbol{b}=0 \qquad \text{for } i>k \tag{3.49}$$

under the same condition (3.46) for k, then solutions which omit these components are not properly termed least-squares solutions but principal-components solutions.

In many least-squares problems, poorly determined components will not arise, all singular values being of approximately the same magnitude. As a rule of thumb for my clients, I suggest they look very carefully at their data, and in particular the matrix $\mathbf{A}$, if the ratio of the largest singular value to the smallest exceeds 1000. Such a distribution of singular values suggests that the columns of $\mathbf{A}$ are not truly independent and, regardless of the conditioning of the problem as discussed above, one may wish to redefine the problem by leaving out certain variables (columns of $\mathbf{A}$) from the set used to approximate $\boldsymbol{b}$.

Algorithm 2. Least-squares solution via singular-value decomposition

```
procedure svdlss(nRow, nCol: integer; {order of problem}
                 W : wmatrix; {working array with decomposition}
                 Y: rvector; {right hand side vector}
                 Z : rvector; {squares of singular values}
                 A : rmatrix; {coefficient matrix (for residuals)}
                 var Bvec: rvector); {solution vector}
{alg02.pas ==
   least squares solution via singular value decomposition.
   On entry, W must have the working matrix resulting from the operation of
   NashSVD on a real matrix A in alg1.pas. Z will have the squares of the
   singular values. Y will have the vector to be approximated. Bvec will be
   the vector of parameters (estimates) returned. Note that A could be
   omitted if residuals were not wanted. However, the user would then lose
   the ability to interact with the problem by changing the tolerance q.
   Because this uses a slightly different decomposition from that in the
   first edition of Compact Numerical Methods, the step numbers are not
   given.
                     Copyright 1988 J. C. Nash
}
var
   i, j, k : integer;
   q, s : real;
```

Algorithm 2. Least-squares solution via singular-value decomposition (cont.)

```
begin
    writeln('alg02.pas == svdlss');
    repeat
        writeln;
        writeln('Singular values');
        for j := 1 to nCol do
        begin
            write(sqrt(Z[j]):18,' ');
            if j = 4 * (j div 4) then writeln;
        end;
        writeln;
        write('Enter a tolerance for zero singular value (<0 to quit) ');
        readln(infile,q);
        if length(infname)>0 then writeln(q);
        if q>=0.0 then
        begin
            q := q*q; {we will work with the Z vector directly}
            for i := 1 to nCol do {loop to generate each element of Bvec }
            begin
                s := 0.0;
                for j := 1 to nCol do {loop over columns of V }
                begin
                    for k := 1 to nRow do {loop over elements of Y }
                    begin
                        if Z[j]>q then
                            s := s + W[i+nRow,j]*W[k,j]*Y[k]/Z[j];
                            {this is V * S+ * U-transpose * Y = A+ * Y }
                            {NOTE: we must use the > sign and not >= in case
                            the user enters a value of zero for q, which would
                            result in zero-divide.}
                    end;
                end;
                Bvec[i] := s;
            end;
            writeln('Least squares solution');
            for j := 1 to nCol do
            begin
                write(Bvec[j]:12,' ');
                if j = 5 * (j div 5) then writeln;
            end;
            writeln;
            s := resids(nRow, nCol, A, Y, Bvec, true);
        end; {if q>=0.0 }
    until q<0.0; {this is how we exit from the procedure}
end {alg02.pas == svdlss};
```

In the above code the residual sum of squares is computed in the separate procedure resids.pas. *In* alg02.pas, *I have not included step numbers because the present code is quite different from the original algorithm.*

Example 3.1. The generalised inverse of a rectangular matrix via the singular-value decomposition

Given the matrices **U**, **V** and **S** of the singular-value decomposition (2.53), then by the product

$$\mathbf{A}^{+} = \mathbf{V}\mathbf{S}^{+}\mathbf{U}^{\mathrm{T}} \tag{2.56}$$

the generalised (Moore–Penrose) inverse can be computed directly. Consider the matrix

$$\mathbf{A} = \begin{bmatrix} 5 & 1\cdot0\mathrm{E}{-6} & 1 \\ 6 & 0\cdot999999 & 1 \\ 7 & 2\cdot00001 & 1 \\ 8 & 2\cdot9999 & 1 \end{bmatrix} = \mathbf{U}\mathbf{S}\mathbf{V}^{\mathrm{T}}.$$

A Hewlett–Packard 9830 operating in 12 decimal digit arithmetic computes the singular values of this matrix via algorithm 1 to six figures as

$$13\cdot7530,\ 1\cdot68961 \text{ and } 1\cdot18853\mathrm{E}{-5}$$

with

$$\mathbf{U} = \begin{bmatrix} 0\cdot358943 & -0\cdot755762 & -0\cdot328687 \\ 0\cdot446526 & -0\cdot317194 & 0\cdot111741 \\ 0\cdot534110 & 0\cdot121383 & 0\cdot762674 \\ 0\cdot621692 & 0\cdot559891 & -0\cdot545716 \end{bmatrix}$$

and

$$\mathbf{V} = \begin{bmatrix} 0\cdot958786 & -0\cdot209025 & -0\cdot192451 \\ 0\cdot245748 & 0\cdot950036 & 0\cdot192456 \\ 0\cdot142607 & -0\cdot231819 & 0\cdot962249 \end{bmatrix}.$$

The generalised inverse using the definition (2.57) of $\mathbf{S}^{+}$ is then (to six figures)

$$\mathbf{A}_1^{+} = \begin{bmatrix} 5322\cdot32 & -1809\cdot27 & -12349\cdot4 & 8836\cdot37 \\ -5322\cdot78 & 1809\cdot22 & 12349\cdot9 & -8836\cdot33 \\ -26610\cdot8 & 9046\cdot70 & 61747\cdot0 & -44181\cdot9 \end{bmatrix}.$$

However, we might wonder whether the third singular value is merely an approximation to zero, that is, that the small value computed is a result of rounding errors. Using a new definition (3.37) for $\mathbf{S}^{+}$, assuming this singular value is really zero gives

$$\mathbf{A}_2^{+} = \begin{bmatrix} 0\cdot118521 & 0\cdot070370 & 0\cdot022219 & -0\cdot025924 \\ -0\cdot418538 & -0\cdot170373 & 0\cdot077795 & 0\cdot325925 \\ 0\cdot107415 & 0\cdot048150 & -0\cdot011116 & -0\cdot070372 \end{bmatrix}.$$

If these generalised inverses are used to solve least-squares problems with

$$\boldsymbol{b} = (1, 2, 3, 4)^{\mathrm{T}}$$

as the right-hand sides, the solutions are

$$\boldsymbol{x}_1 = (1{\cdot}000000048, -4{\cdot}79830\text{E}-8, -4{\cdot}00000024)^{\text{T}}$$

with a residual sum of squares of 3·75892E−20 and

$$\boldsymbol{x}_2 = (0{\cdot}222220924, 0{\cdot}777801787, -0{\cdot}111121188)^{\text{T}}$$

with a residual sum of squares of 2·30726E−9. Both of these solutions are probably acceptable in a majority of applications. Note, however, that the first generalised inverse gives

$$\mathbf{A}_1^+\mathbf{A} = \begin{bmatrix} 1{\cdot}0000004 & 1{\cdot}00000\text{E}-7 & 6{\cdot}00000\text{E}-8 \\ -6{\cdot}00000\text{E}-7 & 0{\cdot}9999998 & -6{\cdot}00000\text{E}-8 \\ 1{\cdot}00000\text{E}-6 & -1{\cdot}00000\text{E}-6 & 1{\cdot}0000003 \end{bmatrix}$$

while the second gives

$$\mathbf{A}_2^+\mathbf{A} = \begin{bmatrix} 0{\cdot}962962751 & 0{\cdot}037038331 & 0{\cdot}185185457 \\ 0{\cdot}037038331 & 0{\cdot}962960588 & -0{\cdot}185190864 \\ 0{\cdot}185185457 & -0{\cdot}185190864 & 0{\cdot}074076661 \end{bmatrix}$$

in place of

$$\begin{bmatrix} 1 & 0 & 0 \\ 0 & 1 & 0 \\ 0 & 0 & 0 \end{bmatrix}.$$

In the above solutions and products, all figures printed by the HP 9830 have been given rather than the six-figure approximations used earlier in the example.

Example 3.2. Illustration of the use of algorithm 2

The estimation of the coefficients x_i, $i = 1, 2, 3, 4, 5$, in example 2.3 (p. 23), provides an excellent illustration of the worth of the singular-value decomposition for solving least-squares problems when the data are nearly collinear. The data for the problem are given in table 3.1.

To evaluate the various solutions, the statistic

$$R^2 = 1 - (\boldsymbol{r}^{\text{T}}\boldsymbol{r})\left(\sum_{i=1}^{m} (b_i - \bar{b})^2\right)^{-1} \tag{3.50}$$

will be used, where

$$\boldsymbol{r} = \boldsymbol{b} - \mathbf{A}\boldsymbol{x} \tag{2.15}$$

is the residual vector and $\bar{b}$ is the mean of the elements of $\boldsymbol{b}$, the dependent variable. The denominator in the second term of (3.50) is often called the *total sum of squares* since it is the value of the residual sum of squares for the model

$$\mathrm{y} = \text{constant} = \bar{\mathrm{y}}. \tag{3.51}$$

The statistic R^2 can be corrected for the number of *degrees of freedom* in the least-squares problem. Thus if there are m observations and k fitted parameters,

TABLE 3.1. Index numbers (1940 = 100) for farm money income and agricultural use of nitrogen, phosphate, potash and petroleum in the United States (courtesy Dr S Chin).

Income	Nitrogen	Phosphate	Potash	Petroleum
305	563	262	461	221
342	658	291	473	222
331	676	294	513	221
339	749	302	516	218
354	834	320	540	217
369	973	350	596	218
378	1079	386	650	218
368	1151	401	676	225
405	1324	446	769	228
438	1499	492	870	230
438	1690	510	907	237
451	1735	534	932	235
485	1778	559	956	236

there are $(m-k)$ degrees of freedom and the corrected R^2 is

$$\tilde{R}^2 = R^2 - \frac{(k-1)(1-R^2)}{m-k}. \tag{3.52}$$

R^2 and $\tilde{R}^2$ provide measures of the goodness of fit of our model which are not dependent on the scale of the data.

Using the last four columns of table 3.1 together with a column of ones for the matrix **A** in algorithm 2, with the first column of the table as the dependent variable ***b***, a Data General NOVA operating in 23-bit binary floating-point arithmetic computes the singular values:

$$5298{\cdot}55,\ 345{\cdot}511,\ 36{\cdot}1125,\ 21{\cdot}4208 \text{ and } 5{\cdot}13828\text{E}-2.$$

The ratio of the smallest of these to the largest is only very slightly larger than the machine precision, 2^{-22}, and we may therefore expect that a great number of extremely different models may give very similar degees of approximation to the data. Solutions (a), (b), (c) and (d) in table 3.2 therefore present the solutions corresponding to all, four, three and two principal components, respectively. Note that these have 8, 9, 10 and 11 degrees of freedom because we estimate the coefficients of the principal components, then transform these to give solutions in terms of our original variables. The solution given by only three principal components is almost as good as that for all components, that is, a conventional least-squares solution. However, the coefficients in solutions (a), (b) and (c) are very different.

Neither the algorithms in this book nor those anywhere else can make a clear and final statement as to which solution is 'best'. Here questions of statistical significance will not be addressed, though they would probably enter into consideration if we were trying to identify and estimate a model intended for use in

Table 3.2. Solutions for various principal-component regressions using the data in table 3.1.

	Tolerance for zero	R^2	$X_{constant}$	$X_{nitrogen}$	$X_{phosphate}$	X_{potash}	$X_{petroleum}$
(*a*)	0	0·972586 (0·958879)	207·782	−0·046191	1·0194	−0·15983	−0·290373
(*b*)	1	0·969348 (0·959131)	4·33368E−3	−5·85314E−2	1·1757	−0·252296	0·699621
(*c*)	22	0·959506 (0·951407)	5·14267E−3	4·34851E−2	0·392026	−6·93389E−2	1·0115
(*d*)	40	0·93839 (0·932789)	2·54597E−3	−0·15299	0·300127	0·469294	0·528881
Regression involving a constant and index numbers for phosphate							
(*e*)	0	0·968104 (0·965204)	179·375	—	0·518966	—	—
(*f*)	1	0·273448	2·24851E−3	—	0·945525	—	—

The values in parentheses below each R^2 are the corrected $\tilde{R}^2$ statistic given by formula (3.52).

some analysis or prediction. To underline the difficulty of this task, merely consider the alternative model

$$\text{income} = x_1 + x_3(\text{phosphate}) \tag{3.53}$$

for which the singular values are computed as 1471·19 and 0·87188, again quite collinear. The solutions are (*e*) and (*f*) in table 3.2 and the values of R^2 speak for themselves.

A sample driver program DR0102.PAS is included on the program diskette. Appendix 4 describes the sample driver programs and supporting procedures and functions.

Chapter 4

HANDLING LARGER PROBLEMS

4.1. INTRODUCTION

The previous chapter used plane rotations multiplying a matrix from the right to orthogonalise its columns. By the essential symmetry of the singular-value decomposition, there is nothing to stop us multiplying a matrix by plane rotations from the left to achieve an orthogonalisation of its rows. The amount of work involved is of order m^2n operations per sweep compared to mn^2 for the columnwise orthogonalisation ($\mathbf{A}$ is m by n), and as there are normally more rows than columns it may seem unprofitable to do this. However, by a judicious combination of row orthogonalisation with Givens' reduction, an algorithm can be devised which will handle a theoretically unlimited number of rows.

4.2. THE GIVENS' REDUCTION

The above approach to the computation of a singular-value decomposition and least-squares solution works very well on a small computer for relatively small matrices. For instance, it can handle least-squares regression calculations involving up to 15 economic time series with 25 years of data in less than 2000 words of main memory (where one matrix element occupies two words of storage). While such problems are fairly common in economics, biological and sociological phenomena are likely to have associated with them very large numbers of observations, m, even though the number of variables, n, may not be large. Again, the formation of the sum-of-squares and cross-products matrix $\mathbf{A}^T\mathbf{A}$ and solution via the normal equations (2.22) should be avoided. In order to circumvent the difficulties associated with the storage of the whole of the matrix $\mathbf{A}$, the Givens' reduction can be used. A Givens' transformation is simply a plane rotation which transforms two vectors so that an element of one of them becomes zero. Consider two row vectors $\mathbf{z}^T$ and $\mathbf{y}^T$ and a pre-multiplying rotation:

$$\begin{pmatrix} c & s \\ -s & c \end{pmatrix} \begin{pmatrix} \mathbf{z}^T \\ \mathbf{y}^T \end{pmatrix} = \begin{pmatrix} \mathbf{Z}^T \\ \mathbf{Y}^T \end{pmatrix} \tag{4.1}$$

where

$$c = \cos\phi \qquad s = \sin\phi \tag{4.2}$$

and ϕ is the angle of rotation. If Y_1 is to be zero, then

$$-sz_1 + cy_1 = 0 \tag{4.3}$$

so that the angle of rotation in this case is given by

$$\tan\phi = s/c = y_1/z_1. \tag{4.4}$$

This is a simpler angle calculation than that of §3.3 for the orthogonalisation process, since it involves only one square root per rotation instead of two. That is, if

$$p=(y_1^2+z_1^2)^{1/2} \tag{4.5}$$

then we have

$$c=z_1/p \tag{4.6}$$

and

$$s=y_1/p. \tag{4.7}$$

It is possible, in fact, to perform such transformations with no square roots at all (Gentleman 1973, Hammarling 1974, Golub and Van Loan 1983) but no way has so far come to light for incorporating similar ideas into the orthogonalising rotation of §3.3. Also, it now appears that the extra overhead required in avoiding the square root offsets the expected gain in efficiency, and early reports of gains in speed now appear to be due principally to better coding practices in the square-root-free programs compared to their conventional counterparts.

The Givens' transformations are assembled in algorithm 3 to triangularise a real m by n matrix $\mathbf{A}$. Note that the ordering of the rotations is crucial, since an element set to zero by one rotation must not be made non-zero by another. Several orderings are possible; algorithm 3 acts column by column, so that rotations placing zeros in column k act on zeros in columns $1, 2, \ldots, (k-1)$ and leave these elements unchanged. Algorithm 3 leaves the matrix $\mathbf{A}$ triangular, that is

$$\mathbf{A}[i, j]=0 \qquad \text{for } i>j \tag{4.8}$$

which will be denoted $\mathbf{R}$. The matrix $\mathbf{Q}$ contains the transformations, so that the original m by n matrix is

$$\mathbf{A}=\mathbf{QR}. \tag{4.9}$$

In words, this procedure simply zeros the last $(m-1)$ elements of column 1, then the last $(m-2)$ elements of column $2, \ldots,$ and finally the last $(m-n)$ elements of column n.

Since the objective in considering the Givens' reduction was to avoid storing a large matrix, it may seem like a step backwards to discuss an algorithm which introduces an m by m matrix $\mathbf{Q}$. However, this matrix is not needed for the solution of least-squares problems except in its product $\mathbf{Q}^{\mathrm{T}}\boldsymbol{b}$ with the right-hand side vector $\boldsymbol{b}$. Furthermore, the ordering of the rotations can be altered so that they act on one row at a time, requiring only storage for this one row and for the resulting triangular n by n matrix which will again be denoted $\mathbf{R}$, that is

$$\mathbf{Q}^{\mathrm{T}}\mathbf{A}=\mathbf{Q}^{\mathrm{T}}\mathbf{Q}\begin{pmatrix}\mathbf{R}\\ \mathbf{0}\end{pmatrix}=\begin{pmatrix}\mathbf{R}\\ \mathbf{0}\end{pmatrix}. \tag{4.10}$$

In the context of this decomposition, the normal equations (2.22) become

$$\mathbf{A}^{\mathrm{T}}\mathbf{A}\boldsymbol{x}=\begin{pmatrix}\mathbf{R}\\ \mathbf{0}\end{pmatrix}^{\mathrm{T}}\mathbf{Q}^{\mathrm{T}}\mathbf{Q}\begin{pmatrix}\mathbf{R}\\ \mathbf{0}\end{pmatrix}\boldsymbol{x}=\begin{pmatrix}\mathbf{R}\\ \mathbf{0}\end{pmatrix}^{\mathrm{T}}\mathbf{Q}^{\mathrm{T}}\boldsymbol{b}=\mathbf{A}^{\mathrm{T}}\boldsymbol{b}. \tag{4.11}$$

Thus the zeros below **R** multiply the last $(m-n)$ elements of

$$\mathbf{Q}^{\mathrm{T}}\boldsymbol{b} = \begin{pmatrix} \boldsymbol{d}_1 \\ \boldsymbol{d}_2 \end{pmatrix} \tag{4.12}$$

where $\boldsymbol{d}_1$ is of order n and $\boldsymbol{d}_2$ of order $(m-n)$. Thus

$$\begin{aligned} \mathbf{A}^{\mathrm{T}}\mathbf{A}\boldsymbol{x} &= \mathbf{R}^{\mathrm{T}}\mathbf{R}\boldsymbol{x} \\ &= \mathbf{R}^{\mathrm{T}}\boldsymbol{d}_1 + \mathbf{0}\boldsymbol{d}_2 = \mathbf{A}^{\mathrm{T}}\boldsymbol{b}. \end{aligned} \tag{4.13}$$

These equations are satisfied regardless of the values in the vector $\boldsymbol{d}_2$ by solutions $\boldsymbol{x}$ to the triangular system

$$\mathbf{R}\boldsymbol{x} = \boldsymbol{d}_1. \tag{4.14}$$

This system is trivial to solve if there are no zero elements on the diagonal of **R**. Such zero elements imply that the columns of the original matrix are not linearly independent. Even if no zero or 'small' element appears on the diagonal, the original data may be linearly dependent and the solution $\boldsymbol{x}$ to (4.14) in some way 'unstable'.

Algorithm 3. Givens' reduction of a real rectangular matrix

```
procedure givens( nRow,nCol : integer; {size of matrix to be decomposed}
                  var A, Q: rmatrix); {the matrix to be decomposed
                        which with Q holds the resulting decomposition}
{alg03.pas ==
        Givens' reduction of a real rectangular matrix to Q * R form.
        This is a conventional version, which works one column at a time.

                  Copyright 1988 J. C. Nash
}
var
   i, j, k, mn: integer; {loop counters and array indices}
   b, c, eps, p, s : real; {angle parameters in rotations}
begin
   writeln('alg03.pas -- Givens',chr(39),' reduction -- column-wise');
   {STEP 0 -- partly in the procedure call}
   mn := nRow; if nRow>nCol then mn := nCol; {mn is the minimum of nRow
                  and nCol and gives the maximum size of the triangular matrix
                  resulting from the reduction. Note that the decomposition is
                  still valid when nRow<nCol, but the R matrix is then
                  trapezoidal, i.e. an upper trianglular matrix with added
                  columns on the right.}
   for i := 1 to nRow do
   begin {set Q to a unit matrix of size nRow}
      for j := 1 to nRow do Q[i,j] := 0.0;
      Q[i,i] := 1.0;
   end; {loop on i -- Q now a unit matrix}
   eps := calceps; {the machine precision}
   for j := 1 to (mn-1) do {main loop on diagonals of triangle} {STEP 1}
   begin {STEP 2}
      for k := (j+1) to nRow do {loop on column elements}
      begin {STEP 3}
         c := A[j,j]; s := A[k,j]; {the two elements in the rotation}
```

Algorithm 3. Givens' reduction of a real rectangular matrix (cont.)

```
          b := abs(c); if abs(s)>b then b := abs(s);
          if b>0 then
          begin {rotation is needed}
            c := c/b; s := s/b; {normalise elements to avoid over- or under-flow}
            p := sqrt(c*c+s*s); {STEP 4}
            s := s/p;
            if abs(s)>=eps then {STEP 5}
            begin {need to carry out rotations} {STEP 6}
              c := c/p;
              for i := 1 to nCol do {STEP 7 -- rotation of A}
              begin
                p := A[j,i]; A[j,i] := c*p+s*A[k,i]; A[k,i] := -s*p+c*A[k,i];
              end; {loop on i for rotation of A}
              for i := 1 to nRow do {STEP 8 -- rotation of Q. Note: nRow not nCol.}
              begin
                p := Q[i,j]; Q[i,j] := c*p+s*Q[i,k]; Q[i,k] := -s*p+c*Q[i,k];
              end; {loop on i for rotation of Q}
            end; {if abs(s)>=eps}
          end; {if b>0}
        end; {loop on k -- the end-loop is STEP 9}
      end; {loop on j -- the end-loop is STEP 10}
    end; {alg03.pas == Givens' reduction}
```

After the Givens' procedure is complete, the array A *contains the triangular factor* R *in rows* 1 *to* mn. *If* nRow *is less than* nCol, *then the right-hand (*nCol – nRow*) columns of array* A *contain the transformed columns of the original matrix so that the product* Q*R = A, *in which* R *is now trapezoidal. The decomposition can be used together with a back-substitution algorithm such as algorithm 6 to solve systems of linear equations.*

The order in which non-zero elements of the working array are transformed to zero is not unique. In particular, it may be important in some applications to zero elements row by row instead of column by column. The file alg03a.pas *on the software disk presents such a row-wise variant. Appendix 4 documents driver programs* DR03.PAS *and* DR03A.PAS *which illustrate how the two Givens' reduction procedures may be used.*

Example 4.1. The operation of Givens' reduction

The following output of a Data General ECLIPSE operating in six hexadecimal digit arithmetic shows the effect of Givens' reduction on a rectangular matrix. At each stage of the loops of steps 1 and 2 of algorithm 3 the entire **Q** and **A** matrices are printed so that the changes are easily seen. The loop parameters j and k as well as the matrix elements $c = \mathbf{A}[j, j]$ and $s = \mathbf{A}[k, j]$ are printed also. In this example, the normalisation at step 3 of the reduction is not necessary, and the sine and cosine of the angle of rotation could have been determined directly from the formulae (4.5), (4.6) and (4.7).

The matrix chosen for this example has only rank 2. Thus the last row of the FINAL A MATRIX is essentially null. In fact, small diagonal elements of the triangular matrix **R** will imply that the matrix **A** is 'nearly' rank-deficient. However, the absence of small diagonal elements in **R**, that is, in the final array **A**, do not indicate that the original **A** is of full rank. Note that the recombination of the factors **Q** and **R** gives back the original matrix apart from very small errors which are of the order of the machine precision multiplied by the magnitude of the elements in question.

```
*RUN
TEST GIVENS - GIFT - ALG 3 DEC 12 77
SIZE -- M= ? 3   N= ? 4
MTIN - INPUT M BY N MATRIX
ROW 1 : ? 1 ? 2 ? 3 ? 4
ROW 2 : ? 5 ? 6 ? 7 ? 8
ROW 3 : ? 9 ? 10 ? 11 ? 12

ORIGINAL  A MATRIX
ROW 1 :         1               2               3               4
ROW 2 :         5               6               7               8
ROW 3 :         9               10              11              12

GIVENS TRIANGULARIZATION DEC 12 77
  Q MATRIX
ROW 1 :         1               0               0
ROW 2 :         0               1               0
ROW 3 :         0               0               1

J= 1    K= 2    A[J,J]= 1    A[K,J]= 5
  A MATRIX
ROW 1 :         5.09902         6.27572         7.45242         8.62912
ROW 2 :        -1.19209E-07    -.784466        -1.56893        -2.3534
ROW 3 :         9               10              11              12

  Q MATRIX
ROW 1 :         .196116        -.980581         0
ROW 2 :         .980581         .196116         0
ROW 3 :         0               0               1

J= 1    K= 3    A[J,J]= 5.09902    A[K,J]= 9
  A MATRIX
ROW 1 :         10.3441         11.7942         13.2443         14.6944
ROW 2 :        -1.19209E-07    -.784466        -1.56893        -2.3534
ROW 3 :         0              -.530862        -1.06172        -1.59258

  Q MATRIX
ROW 1 :         9.66738E-02    -.980581        -.170634
ROW 2 :         .483369         .196116        -.853168
ROW 3 :         .870063         0               .492941

J= 2    K= 3    A[J,J]=-.784466    A[K,J]=-.530862
FINAL  A MATRIX
ROW 1 :         10.3441         11.7942         13.2443         14.6944
ROW 2 :         9.87278E-08     .947208         1.89441         2.84162
ROW 3 :        -6.68109E-08     0              -9.53674E-07    -1.90735E-06


FINAL  Q MATRIX
ROW 1 :         9.66738E-02     .907738        -.40825
ROW 2 :         .483369         .315737         .816498
ROW 3 :         .870063        -.276269        -.408249

RECOMBINATION
ROW 1 :         1               2.00001         3.00001         4.00001
ROW 2 :         5.00001         6.00002         7.00002         8.00002
ROW 3 :         9.00001         10              11              12
```

4.3. EXTENSION TO A SINGULAR-VALUE DECOMPOSITION

In order to determine if linear dependencies are present, it is possible to extend the Givens' reduction and compute a singular-value decomposition by orthogonalising the *rows* of **R** by plane rotations which act from the left. It is not necessary to accumulate the transformations in a matrix **U**; instead they can be applied directly to $\mathbf{Q}^{\mathrm{T}}\boldsymbol{b}$, in fact, to the first n elements of this vector which form $\boldsymbol{d}_1$. The rotations can be thought of as acting all at once as the orthogonal matrix $\mathbf{P}^{\mathrm{T}}$. Applying this to equation (4.14) gives

$$\mathbf{P}^{\mathrm{T}}\mathbf{R}\boldsymbol{x}=\mathbf{P}^{\mathrm{T}}\boldsymbol{d}_1=\boldsymbol{f}. \tag{4.15}$$

However, the rows of $\mathbf{P}^{\mathrm{T}}\mathbf{R}$ are orthogonal, that is

$$\mathbf{P}^{\mathrm{T}}\mathbf{R}=\mathbf{S}\mathbf{V}^{\mathrm{T}} \tag{4.16}$$

with

$$\mathbf{S}\mathbf{V}^{\mathrm{T}}\mathbf{V}\mathbf{S}=\mathbf{S}^2. \tag{4.17}$$

Combining the Givens' reduction and orthogonalisation steps gives

$$\mathbf{P}^{\mathrm{T}}\mathbf{Q}^{\mathrm{T}}\mathbf{A}=\mathbf{P}^{\mathrm{T}}\begin{pmatrix}\mathbf{R}\\ \mathbf{0}\end{pmatrix}=\begin{pmatrix}\mathbf{S}\mathbf{V}^{\mathrm{T}}\\ \mathbf{0}\end{pmatrix} \tag{4.18}$$

or

$$\mathbf{A}=\mathbf{Q}\mathbf{P}\begin{pmatrix}\mathbf{S}\\ \mathbf{0}\end{pmatrix}\mathbf{V}^{\mathrm{T}} \tag{4.19}$$

which is a singular-value decomposition.

As in the case of the columnwise orthogonalisation, small singular values (i.e. rows of $\mathbf{P}^{\mathrm{T}}\mathbf{R}$ having small norm) will cause **V** to possess some unnormalised rows having essentially zero elements. In this case (4.17) will not be correct, since

$$\mathbf{V}^{\mathrm{T}}\mathbf{V}\simeq\begin{pmatrix}\mathbf{1}_k & \\ & \mathbf{0}_{n-k}\end{pmatrix} \tag{4.20}$$

where k is the number of singular values larger than some pre-assigned tolerance for zero. Since in the solution of least-squares problems these rows always act only in products with **S** or $\mathbf{S}^{+}$, this presents no great difficulty to programming an algorithm using the above Givens' reduction/row orthogonalisation method.

4.4. SOME LABOUR-SAVING DEVICES

The above method is not nearly so complicated to implement as it may appear. Firstly, all the plane rotations are row-wise for both the Givens' reduction and the orthogonalisation. Moreover, one or more (say g) vectors $\boldsymbol{b}$ can be concatenated with the matrix **A** so that rotations do not have to be applied separately to these, but appear to act on a single matrix.

The second observation which reduces the programming effort is that the rows of this matrix $(\mathbf{A}, \boldsymbol{b})$ are needed only one at a time. Consider a working array

$(n+1)$ by $(n+g)$ which initially has all elements in the first n rows equal to zero. Each of the m observations or rows of $(\mathbf{A}, \boldsymbol{b})$ can be loaded in succession into the $(n+1)$th row of the working array. The Givens' rotations will suitably fill up the workspace and create the triangular matrix $\mathbf{R}$. The same workspace suffices for the orthogonalisation. Note that the elements of the vectors $\boldsymbol{d}_2$ are automatically left in the last g elements of row $(n+1)$ of the working array when the first n have been reduced to zero. Since there are only $(m-n)$ components in each $\boldsymbol{d}_2$, but m rows to process, at least n of the values left in these positions will be zero. These will not necessarily be the first n values.

A further feature of this method is that the residual sum of squares, $\boldsymbol{r}^{\mathrm{T}}\boldsymbol{r}$, is equal to the sum of squared terms $\boldsymbol{d}_2^{\mathrm{T}}\boldsymbol{d}_2$. This can be shown quite easily since

$$\begin{aligned} \boldsymbol{r}^{\mathrm{T}}\boldsymbol{r} &= (\boldsymbol{b}-\mathbf{A}\boldsymbol{x})^{\mathrm{T}}(\boldsymbol{b}-\mathbf{A}\boldsymbol{x}) \\ &= \boldsymbol{b}^{\mathrm{T}}\boldsymbol{b} - \boldsymbol{b}^{\mathrm{T}}\mathbf{A}\boldsymbol{x} - \boldsymbol{x}^{\mathrm{T}}\mathbf{A}^{\mathrm{T}}\boldsymbol{b} + \boldsymbol{x}^{\mathrm{T}}\mathbf{A}^{\mathrm{T}}\mathbf{A}\boldsymbol{x}. \end{aligned} \tag{4.21}$$

By using the normal equations (2.22) the last two terms of this expression cancel leaving

$$\boldsymbol{r}^{\mathrm{T}}\boldsymbol{r} = \boldsymbol{b}^{\mathrm{T}}\boldsymbol{b} - \boldsymbol{b}^{\mathrm{T}}\mathbf{A}\boldsymbol{x}. \tag{4.22}$$

If least-squares problems with large numbers of observations are being solved via the normal equations, expression (4.22) is commonly used to compute the residual sum of squares by accumulating $\boldsymbol{b}^{\mathrm{T}}\boldsymbol{b}$, $\mathbf{A}^{\mathrm{T}}\mathbf{A}$ and $\mathbf{A}^{\mathrm{T}}\boldsymbol{b}$ with a single pass through the data. In this case, however, (4.22) almost always involves the subtraction of nearly equal numbers. For instance, when it is possible to approximate $\boldsymbol{b}$ very closely with $\mathbf{A}\boldsymbol{x}$, then nearly all the digits in $\boldsymbol{b}^{\mathrm{T}}\boldsymbol{b}$ will be cancelled by those in $\boldsymbol{b}^{\mathrm{T}}\mathbf{A}\boldsymbol{x}$, leaving a value for $\boldsymbol{r}^{\mathrm{T}}\boldsymbol{r}$ with very few correct digits.

For the method using rotations, on the other hand, we have

$$\boldsymbol{b}^{\mathrm{T}}\boldsymbol{b} = \boldsymbol{b}^{\mathrm{T}}\mathbf{Q}\mathbf{Q}^{\mathrm{T}}\boldsymbol{b} = \boldsymbol{d}_1^{\mathrm{T}}\boldsymbol{d}_1 + \boldsymbol{d}_2^{\mathrm{T}}\boldsymbol{d}_2 \tag{4.23}$$

and

$$\boldsymbol{b}^{\mathrm{T}}\mathbf{A}\boldsymbol{x} = \boldsymbol{b}^{\mathrm{T}}\mathbf{Q}\begin{pmatrix}\mathbf{R}\\ \mathbf{0}\end{pmatrix}\boldsymbol{x} = \boldsymbol{d}_1^{\mathrm{T}}\boldsymbol{d}_1 \tag{4.24}$$

by equation (4.12). Hence, by substitution of (4.23) and (4.24) into (4.22) we obtain

$$\boldsymbol{r}^{\mathrm{T}}\boldsymbol{r} = \boldsymbol{d}_2^{\mathrm{T}}\boldsymbol{d}_2. \tag{4.25}$$

The cancellation is now accomplished theoretically with the residual sum of squares computed as a sum of positive terms, avoiding the digit cancellation.

The result (4.25) is derived on the assumption that (4.14) holds. In the rank-deficient case, as shown by k zero or 'small' singular values, the vector $\boldsymbol{f}$ in equation (4.15) can be decomposed so that

$$\mathbf{P}^{\mathrm{T}}\boldsymbol{d}_1 = \boldsymbol{f} = \begin{pmatrix}\boldsymbol{f}_1\\ \boldsymbol{f}_2\end{pmatrix} \tag{4.26}$$

where $\boldsymbol{f}_1$ is of order $(n-k)$ and $\boldsymbol{f}_2$ of order k. Now equation (4.24) will have the form

$$\boldsymbol{b}^{\mathrm{T}}\mathbf{A}\boldsymbol{x} = \boldsymbol{b}^{\mathrm{T}}\mathbf{Q}\mathbf{P}\mathbf{S}\mathbf{V}^{\mathrm{T}}\boldsymbol{x} = \boldsymbol{f}_1^{\mathrm{T}}\boldsymbol{f}_1 \tag{4.27}$$

by application of equation (4.16) and the condition that $S_{k+1}, S_{k+2}, \ldots, S_n$ are all 'zero'. Thus, using

$$\boldsymbol{d}_1^{\mathrm{T}}\boldsymbol{d}_1 = \boldsymbol{f}_1^{\mathrm{T}}\boldsymbol{f}_1 + \boldsymbol{f}_2^{\mathrm{T}}\boldsymbol{f}_2 \tag{4.28}$$

and (4.22) with (4.27) and (4.23), the residual sum of squares in the rank-deficient case is

$$\boldsymbol{r}^{\mathrm{T}}\boldsymbol{r} = \boldsymbol{d}_2^{\mathrm{T}}\boldsymbol{d}_2 + \boldsymbol{f}_2^{\mathrm{T}}\boldsymbol{f}_2. \tag{4.29}$$

From a practical point of view (4.29) is very convenient, since the computation of the residual sum of squares is now clearly linked to those singular values which are chosen to be effectively zero by the user of the method. The calculation is once again as a sum of squared terms, so there are no difficulties of digit cancellation.

The vector

$$\begin{pmatrix} \boldsymbol{f}_2 \\ \boldsymbol{d}_2 \end{pmatrix} \tag{4.30}$$

in the context of statistical calculations is referred to as a set of uncorrelated residuals (Golub and Styan 1973).

Nash and Lefkovitch (1976) report other experience with algorithm 4. In particular, the largest problem I have solved using it involved 25 independent variables (including a constant) and two dependent variables, for which there were 196 observations. This problem was initially run on a Hewlett–Packard 9830 calculator, where approximately four hours elapsed in the computation. Later the same data were presented to a FORTRAN version of the algorithm on both Univac 1108 and IBM 370/168 equipment, which each required about 12 seconds of processor time. Despite the order-of-magnitude differences in the timings between the computers and the calculator, they are in fact roughly proportional to the cycle times of the machines. Moreover, as the problem has a singularity of order 2, conventional least-squares regression programs were unable to solve it, and when it was first brought to me the program on the HP 9830 was the only one on hand which could handle it.

Algorithm 4. Givens' reductions, singular-value decomposition and least-squares solution

```
procedure GivSVD( n : integer; {order of problem}
                  nRHS: integer; {number of right hand sides}
                  var B: rmatrix; {matrix of solution vectors}
                  var rss: rvector; {residual sums of squares}
                  var svs: rvector; {singular values}
                  var W: rmatrix; {returns V-transpose}
                  var nobs : integer); {number of observations}

{alg04.pas ==
      Givens' reduction, singular value decomposition and least squares
      solution.

In this program, which is designed to use a very small working array yet
solve least squares problems with large numbers of observations, we do not
explicitly calculate the U matrix of the singular value decomposition.
```

Algorithm 4. Givens' reductions, singular-value decomposition and least-squares solution (cont.)

```
One could save the rotations and carefully combine them to produce the U
matrix. However, this algorithm uses plane rotations not only to zero
elements in the data in the Givens' reduction and to orthogonalize rows of
the work array in the svd portion of the code, but also to move the data
into place from the (n+1)st row of the working array into which the data
is read. These movements i.e. of the observation number nobs,
would normally move the data to row number nobs of the original
matrix A to be decomposed. However, it is possible, as in the array given
by data file ex04.cnm
          3 1 <--- there are 3 columns in the matrix A
                    and 1 right hand side
          -999 <--- end of data flag
          1 2 3 1 <--- the last column is the RHS vector
          2 4 7 1
          2 2 2 1
          5 3 1 1
          -999 0 0 0 <--- end of data row
that this movement does not take place. This is because we use a complete
cycle of Givens' rotations using the diagonal elements W[j,j], j := 1 to
n, of the work array to zero the first n elements of row nobs of the
(implicit) matrix A. In the example, row 1 is rotated from row 4 to row 1
since W is originally null. Observation 2 is loaded into row 4 of W, but
the first Givens' rotation for this observation will zero the first TWO
elements because they are the same scalar multiple of the corresponding
elements of observation 1. Since W[2,2] is zero, as is W[4,2], the second
Givens' rotation for observation 2 is omitted, whereas we should move the
data to row 2 of W. Instead, the third and last Givens' rotation for
observation 2 zeros element W[4,3] and moves the data to row 3 of W.
In the least squares problem such permutations are irrelevant to the final
solution or sum of squared residuals. However, we do not want the
rotations which are only used to move data to be incorporated into U.
Unfortunately, as shown in the example above, the exact form in which such
rotations arise is not easy to predict. Therefore, we do not recommend
that this algorithm be used via the rotations to compute the svd unless
the process is restructured as in Algorithm 3. Note that in any event a
large data array is needed.
The main working matrix W must be n+1 by n+nRHS in size.
                Copyright 1988 J. C. Nash
}
var
   count, EstRowRank, i, j, k, m, slimit, sweep, tcol : integer;
   bb, c, e2, eps, p, q, r, s, tol, trss, vt : real;
   enddata : boolean;
   endflag : real;
procedure rotnsub; {to allow for rotations using variables local
                    to Givsvd. c and s are cosine and sine of
                    angle of rotation.}
var
   i: integer;
   r: real;
```

Algorithm 4. Givens' reductions, singular-value decomposition and least-squares solution (cont.)

```
begin
   for i := m to tcol do {Note: starts at column m, not column 1.}
   begin
      r := W[j,i];
      W[j,i] := r*c+s*W[k,i];
      W[k,i] := -r*s+c*W[k,i];
   end;
end; {rotnsub}
begin {Givsvd}
   writeln('alg04.pas -- Givens',chr(39),
         ' reduction, svd, and least squares solution');
   Write('Order of ls problem and no. of right hand sides: ');
   readln(infile,n,nRHS); {STEP 0}
   if infname<>'con' then writeln(n,' ',nRHS);
   write('Enter a number to indicate end of data ');
   readln(infile,endflag);
   if infname<>'con' then writeln(endflag);
   tcol := n+nRHS; {total columns in the work matrix}
   k := n+1; {current row of interest in the work array during Givens' phase}
   for i := 1 to n do
      for j := 1 to tcol do
      W[i,j] := 0.0; {initialize the work array}
   for i := 1 to nRHS do rss[i] := 0.0; {initialize the residual sums of squares}
   {Note that other quantities, in particular the means and the total
   sums of squares will have to be calculated separately if other
   statistics are desired.}
   eps := calceps; {the machine precision}
   tol := n*n*eps*eps; {a tolerance for zero}
   nobs := 0; {initially there are no observations}
   {STEP 1 -- start of Givens' reduction}
   enddata := false; {set TRUE when there is no more data. Initially FALSE.}
   while (not enddata) do
   begin {main loop for data acquisition and Givens' reduction}
      getobsn( n, nRHS, W, k, endflag, enddata); {STEP 2}
      if (not enddata) then
      begin {We have data, so can proceed.} {STEP 3}
      nobs := nobs+1; {to count the number of observations}
      write('Obsn ',nobs,' ');
      for j := 1 to (n+nRHS) do
      begin
         write(W[k,j]:10:5,' ');
         if (7 * (j div 7) = j) and (j<n+nRHS) then writeln;
      end;
      writeln;
      for j := 1 to (n+nRHS) do
      begin {write to console file}
      end;
      for j := 1 to n do {loop over the rows of the work array to
                  move information into the triangular part of the
                  Givens' reduction} {STEP 4}
      begin
         m := j; s := W[k,j]; c := W[j,j]; {select elements in rotation}
```

Algorithm 4. Givens' reductions, singular-value decomposition and least-squares solution (cont.)

```
            bb := abs(c); if abs(s)>bb then bb := abs(s);
            if bb>0.0 then
            begin {can proceed with rotation as at least one non-zero element}
                c := c/bb; s := s/bb; p := sqrt(c*c+s*s); {STEP 7}
                s := s/p; {sin of angle of rotation}
                if abs(s)>=tol then
                begin {not a very small angle} {STEP8}
                c := c/p; {cosine of angle of rotation}
                rotnsub; {to perform the rotation}
                end; {if abs(s)>=tol}
            end; {if bb>0.0}
        end; {main loop on j for Givens' reduction of one observation} {STEP 9}
        {STEP 10 -- accumulate the residual sums of squares}
        write(' Uncorrelated residual(s):');
        for j := 1 to nRHS do
        begin
            rss[j] := rss[j]+sqr(W[k,n+j]); write(W[k,n+j]:10,' ');
            if (7 * (j div 7) = j) and (j < nRHS) then
            begin
                writeln;
            end;
        end;
        writeln;
        {NOTE: use of sqr function which is NOT sqrt.}
        end; {if (not enddata)}
    end; {while (not enddata)}
    {This is the end of the Givens' reduction part of the program.
        The residual sums of squares are now in place. We could find the
        least squares solution by back-substitution if the problem is of full
        rank. However, to determine the approximate rank, we will continue
        with a row-orthogonalisation.}
    {STEP 11} {Beginning of svd portion of program.}
    m := 1; {Starting column for the rotation subprogram}
    slimit := n div 4; if slimit<6 then slimit := 6; {STEP 12}
    {This sets slimit, a limit on the number of sweeps allowed.
        A suggested limit is max([n/4], 6).}
    sweep := 0; {initialize sweep counter}
    e2 := 10.0*n*eps*eps; {a tolerance for very small numbers}
    tol := eps*0.1; {a convergence tolerance}
    EstRowRank := n; {current estimate of rank};
    repeat
        count := 0;{to initialize the count of rotations performed}
        for j := 1 to (EstRowRank-1) do {STEP 13}
        begin {STEP 14}
            for k := (j+1) to EstRowRank do
            begin {STEP 15}
                p := 0.0; q := 0.0; r := 0.0;
                for i := 1 to n do
                begin
                    p := p+W[j,i]*W[k,i]; q := q+sqr(W[j,i]); r := r+sqr(W[k,i]);
                end; {accumulation loop}
                svs[j] := q; svs[k] := r;
```

Algorithm 4. Givens' reductions, singular-value decomposition and least-squares solution (cont.)

```
            {Now come important convergence test considerations.
            First we will decide if rotation will exchange order of rows.}
            {STEP 16 If q<r then goto step 19}
            {Check if the rows are ordered.}
            if q >= r then
            begin {STEP 17 Rows are ordered, so try convergence test.}
              if not ((q<=e2*svs[1]) or (abs(p)<=tol*q)) then
                {First condition checks for very small row norms in BOTH rows,
                for which no rotation makes sense. The second condition
                determines if the inner product is small with respect to the
                larger of the rows, which implies a very small rotation angle.}
                begin {revised STEP 18}
                {columns are in order, but not converged to smallinner product.
                  Calculate angle and rotate.}
                  p := p/q; r := 1-r/q; vt := sqrt(4*p*p + r*r);
                  c := sqrt(0.5*(1+r/vt)); s := p/(vt*c);
                  rotnsub; {STEP 19 in original algorithm}
                  count := count+1;
                end;
              end
              else { q<r, columns out of order -- must rotate}
              {revised STEP 16. Note: r > q, and cannot be zero since both are
                sums of squares for the svd. In the case of a real symmetric
                matrix, this assumption must be questioned.}
              begin
                p := p/r; q := q/r-1; vt := sqrt(4*p*p + q*q);
                s := sqrt(0.5*(1-q/vt));
                if p<0 then s := -s;
                c := p/(vt*s);
                rotnsub; {STEP 19 in original algorithm}
                count := count+1;
              end;
              {Both angle calculations have been set up so that large numbers
                do not occur in intermediate quantities. This is easy in the svd
                case, since quantities q and r cannot be negative.}
              {STEP 20 has been removed, since we now put the number of
                rotations in count, and do not count down to zero.}
            end; {loop on k -- end-loop is STEP 21}
          end; {loop on j}
          sweep := sweep +1;
          writeln('Sweep ',sweep,' ',count,' rotations performed');
          {Set EstColRank to largest column index for which
            svs[column index] > (svs[1]*tol + tol*tol)
            Note how Pascal expresses this more precisely.}
          while (EstRowRank >= 3) and (svs[EstRowRank] <= svs[1]*tol+tol*tol)
            do EstRowRank := EstRowRank-1;
        until (sweep>slimit) or (count=0); {STEP 22}
          {Singular value decomposition now ready for extraction of information
            and formation of least squares solution.}
        writeln('Singular values and principal components');
        for j := 1 to n do {STEP 23}
        begin
          s := svs[j];
```

Algorithm 4. Givens' reductions, singular-value decomposition and least-squares solution (cont.)

```
            s := sqrt(s); svs[j] := s; {to save the singular value}
            writeln('Singular value [',j,']= ',s);
            if s>=tol then
            begin
                for i := 1 to n do W[j,i] := W[j,i]/s;
                for i := 1 to n do
                begin
                    if (8 * (i div 8) = i) and (i<n) then
                    begin
                        writeln;
                    end;
                end; {for i=1...}
                {principal component is a column of V or a row of V-transpose. W
                    stores V-transpose at the moment.}
                writeln;
            end; {if s>=tol}
            {Principal components are not defined for very small singular values.}
        end; {loop on j over the singular values}
        {STEP 24 -- start least squares solution}
        q := 0.0; {to ensure one pass of least squares solution}
        while q>=0.0 do
        begin
            write('Enter a tolerance for zero (<0 to exit) ');
            readln(infile,q);
            if infname<>'con' then writeln(q);
            if q>=0.0 then
            begin
                {For each value of the tolerance for zero entered as q we must
                    calculate the least squares solution and residual sum of squares
                    for each right hand side vector. The current elements in columns
                n+1 to n+nRHS of the work array W give the contribution of each
                principal coordinate to the least squares solution. However, we do
                not here compute the actual principal coordinates for reasons
                outlined earlier.}
            for i := 1 to nRHS do {STEP 25}
            begin
                trss := rss[i]; {get current sum of squared residuals}
                for j := 1 to n do {loop over the singular values -- STEP 26}
                begin {STEP 27}
                    p := 0.0;
                    for k := 1 to n do
                    begin
                        if svs[k]>q then p := p+W[k,j]*W[k,n+i]/svs[k];
                    end; {loop over singular values}
                    B[j,i] := p; {to save current solution -- STEP 28}
                    writeln('Solution component [',j,']= ',p);
                    if svs[j]<=q then trss := trss+sqr(W[j,n+i]); {to adjust the
                            residual sum of squares in the rank-deficient case}
                end; {loop on j -- end-loop is STEP 29}
                writeln('Residual sum of squares =',trss);
            end; {loop on i -- end-loop is STEP 30}
        end; {if q>=0.0}
    end; {while q>=0.0}
end;{alg04.pas -- Givens' reduction, svd, and least squares solution}
```

Example 4.2. The use of algorithm 4

In the first edition, a Hewlett–Packard 9830A desk computer was used to solve a particular linear least-squares regression problem. This problem is defined by the data in the file EX04.CNM on the software diskette. Using the driver program DR04.PAS, which is also on the diskette according to the documentation in appendix 4, gives rise to the following output.

```
dr04.pas -- run Algorithm 4 problems -- Givens' reduction,
1989/06/03   16:09:47
File for input of control data ([cr] for keyboard) ex04.cnm
File for console image ([cr] = nul) out04.
alg04.pas -- Givens' reduction, svd, and least squares solution
Order of ls problem and no. of right hand sides = 5 1
Enter a number to indicate end of data  -9.9900000000E+02
Obsn 1  563.00000  262.00000  461.00000  221.00000    1.00000  305.00000
    Uncorrelated residual(s):0.0000E+00
Obsn 2  658.00000  291.00000  473.00000  222.00000    1.00000  342.00000
    Uncorrelated residual(s):0.0000E+00
Obsn 3  676.00000  294.00000  513.00000  221.00000    1.00000  331.00000
    Uncorrelated residual(s):0.0000E+00
Obsn 4  749.00000  302.00000  516.00000  218.00000    1.00000  339.00000
    Uncorrelated residual(s):0.0000E+00
Obsn 5  834.00000  320.00000  540.00000  217.00000    1.00000  354.00000
    Uncorrelated residual(s):0.0000E+00
Obsn 6  973.00000  350.00000  596.00000  218.00000    1.00000  369.00000
    Uncorrelated residual(s):-6.563E-02
Obsn 7 1079.00000  386.00000  650.00000  218.00000    1.00000  378.00000
    Uncorrelated residual(s):-9.733E+00
Obsn 8 1151.00000  401.00000  676.00000  225.00000    1.00000  368.00000
    Uncorrelated residual(s):-6.206E+00
Obsn 9 1324.00000  446.00000  769.00000  228.00000    1.00000  405.00000
    Uncorrelated residual(s):1.7473E+01
Obsn 10 1499.00000  492.00000  870.00000  230.00000    1.00000  438.00000
    Uncorrelated residual(s):1.5054E+01
Obsn 11 1690.00000  510.00000  907.00000  237.00000    1.00000  438.00000
    Uncorrelated residual(s):7.4959E+00
Obsn 12 1735.00000  534.00000  932.00000  235.00000    1.00000  451.00000
    Uncorrelated residual(s):1.0754E+00
Obsn 13 1778.00000  559.00000  956.00000  236.00000    1.00000  485.00000
    Uncorrelated residual(s):1.5580E+01
Sweep 1 10 rotations performed
Sweep 2 10 rotations performed
Sweep 3 2 rotations performed
Sweep 4 0 rotations performed
Singular values and principal components
```

```
Singular value [1]=   5.2985598853E+03
 0.82043  0.27690  0.47815  0.14692  0.00065
Singular value [2]=   3.4551146213E+02
-0.49538  0.30886  0.46707  0.66411  0.00322
Singular value [3]=   3.6112521703E+01
-0.26021 -0.12171  0.71337 -0.63919 -0.00344
Singular value [4]=   2.1420869565E+01
 0.11739 -0.90173  0.21052  0.35886  0.00093
Singular value [5]=   5.1382810120E-02
 0.00006 -0.00075  0.00045 -0.00476  0.99999
Enter a tolerance for zero (<0 to exit)   0.0000000000E+00
Solution component [1]=  -4.6192433678E-02
Solution component [2]=   1.0193865559E+00
Solution component [3]=  -1.5982291948E-01
Solution component [4]=  -2.9037627732E-01
Solution component [5]=   2.0778262574E+02
Residual sum of squares =  9.6524564856E+02
Enter a tolerance for zero (<0 to exit)   1.0000000000E+00
Solution component [1]=  -5.8532203918E-02
Solution component [2]=   1.1756920631E+00
Solution component [3]=  -2.5228971048E-01
Solution component [4]=   6.9962158969E-01
Solution component [5]=   4.3336659982E-03
Residual sum of squares =  1.0792302647E+03
Enter a tolerance for zero (<0 to exit)  -1.0000000000E+00
```

4.5. RELATED CALCULATIONS

It sometimes happens that a least-squares solution has to be updated as new data are collected or become available. It is preferable to achieve this by means of a stable method such as the singular-value decomposition. Chambers (1971) discusses the general problem of updating regression solutions, while Businger (1970) has proposed a method for updating a singular-value decomposition. However, the idea suggested in the opening paragraph of this chapter, in particular to orthogonalise $(n+1)$ rows each of n elements by means of plane rotations, works quite well. Moreover, it can be incorporated quite easily into algorithm 4, though a little caution is needed to ensure the correct adjustment of quantities needed to compute statistics such as R^2. Nash and Lefkovitch (1977) present both FORTRAN and BASIC programs which do this. These programs are sub-optimal in the sense that they perform the normal sweep strategy through the rows of **W**, whereas when a new observation is appended the first n rows are already mutually orthogonal. Because the saving only applies during the first sweep, no special steps have been taken to employ this knowledge. Unfortunately, each new orthogonalisation of the rows may take as long as the first, that is, the one that follows the Givens' reduction. Perhaps this is not surprising since new observations may profoundly change the nature of a least-squares problem.

The method suggested is mainly useful for adding single observations, and other approaches are better if more than a very few observations are to be included. For instance, one could update the triangular form which results from the Givens' reduction if this had been saved, then proceed to the singular-value decomposition as in algorithm 4.

No methods will be discussed for removing observations, since while methods exist to accomplish this (see Lawson and Hanson 1974, pp 225–31), the operation is potentially unstable. See also Bunch and Nielsen (1978).

For instance, suppose we have a Givens' **QR** decomposition of a matrix **A** (or any other **QR** decomposition with **Q** orthogonal and **R** upper-triangular), then add and delete a row (observation) denoted $\mathbf{y}^T$. Then after the addition of this row, the (1, 1) elements of the matrices are related by

$$(\tilde{\mathbf{A}}^T\tilde{\mathbf{A}})_{11} = (\tilde{\mathbf{R}}^T\tilde{\mathbf{R}})_{11} = \tilde{R}_{11}^2 = (\mathbf{A}^T\mathbf{A})_{11} + y_1^2 = R_{11}^2 + y_1^2 \tag{4.31}$$

where the tilde is used to indicate matrices which have been updated. Deletion of $\mathbf{y}^T$ now requires the subtraction

$$R_{11}^2 = \tilde{R}_{11}^2 - y_1^2 \tag{4.32}$$

to be performed in some way or another, an operation which will involve digit cancellation if y_1 and $\tilde{R}_{11}$ are close in magnitude. The same difficulty may of course occur in other columns—the first is simply easier to illustrate. Such cases imply that an element of y^T dominates the column in which it occurs and as such should arouse suspicions about the data. Chambers' (1971) subroutine to delete rows from a **QR** decomposition contains a check designed to catch such occurrences.

Of interest to those users performing regression calculations are the estimates of standard errors of the regression coefficients (the least-squares solution elements). The traditional standard error formula is

$$\text{SE}(b_i) = (\sigma^2\,(\mathbf{A}^T\mathbf{A})_{ii}^{-1})^{1/2} \tag{4.33}$$

where σ^2 is an estimate of the variance of data about the fitted model calculated by dividing the sum of squared residuals by the number of degrees of freedom (nRow − nCol) = (nRow − n). The sum of squared residuals has already been computed in algorithm 4, and has been adjusted for rank deficiency within the solution phase of the code.

The diagonal elements of the inverse of the sum of squares and cross-products matrix may seem to pose a bigger task. However, the singular-value decomposition leads easily to the expression

$$(\mathbf{A}^T\,\mathbf{A})^{-1} = \mathbf{V}\,\mathbf{S}^+\,\mathbf{S}^+\,\mathbf{V}^T. \tag{4.34}$$

In particular, diagonal elements of the inverse of the sum of squares and cross-products matrix are

$$(\mathbf{A}^{\mathrm{T}}\,\mathbf{A})_{ii}^{-1} = \sum_{j=1}^{n} (V_{ij} S_{jj}^{+} S_{jj}^{+} V_{ji}^{\mathrm{T}}) \tag{4.35}$$

$$= \sum_{j=1}^{n} (V_{ij} S_{jj}^{+})^2.$$

Thus, the relevant information for the standard errors is obtained by quite simple row sums over the **V** matrix from a singular-value decomposition. When the original **A** matrix is rank deficient, and we decide (via the tolerance for zero used to select 'non-zero' singular values) that the rank is r, the summation above reduces to

$$(\mathbf{A}^{\mathrm{T}}\,\mathbf{A})_{ii}^{-1} = \sum_{j=1}^{r} (V_{ij}/S_{jj})^2. \tag{4.36}$$

However, the meaning of a standard error in the rank-deficient case requires careful consideration, since the standard error will increase very sharply as small singular values are included in the summation given in (4.36). I usually refer to the dispersion measures computed via equations (4.33) through (4.36) for rank $r < n$ cases as 'standard errors under the condition that the rank is 5 (or whatever value r currently has)'. More discussion of these issues is presented in Searle (1971) under the topic 'estimable functions', and in various sections of Belsley, Kuh and Welsch (1980).

Chapter 5

SOME COMMENTS ON THE FORMATION OF THE CROSS-PRODUCTS MATRIX $\mathbf{A}^T\mathbf{A}$

Commonly in statistical computations the diagonal elements of the matrix

$$(\mathbf{A}^T\mathbf{A})^{-1} \tag{5.1}$$

are required, since they are central to the calculation of variances for parameters estimated by least-squares regression. The cross-products matrix $\mathbf{A}^T\mathbf{A}$ from the singular-value decomposition (2.53) is given by

$$\mathbf{A}^T\mathbf{A} = \mathbf{VSU}^T\mathbf{USV}^T = \mathbf{VS}^2\mathbf{V}^T. \tag{5.2}$$

This is a singular-value decomposition of $\mathbf{A}^T\mathbf{A}$, so that

$$(\mathbf{A}^T\mathbf{A})^+ = \mathbf{V}(\mathbf{S}^+)^2\mathbf{V}^T. \tag{5.3}$$

If the cross-products matrix is of full rank, the generalised inverse is identical to the inverse (5.1) and, further,

$$\mathbf{S}^+ = \mathbf{S}^{-1}. \tag{5.4}$$

Thus we have

$$(\mathbf{A}^T\mathbf{A})^{-1} = \mathbf{VS}^{-2}\mathbf{V}^T. \tag{5.5}$$

The diagonal elements of this inverse are therefore computed as simple row norms of the matrix

$$\mathbf{VS}^{-1}. \tag{5.6}$$

In the above manner the singular-value decomposition can be used to compute the required elements of the inverse of the cross-products matrix. This means that the explicit computation of the cross-products matrix is unnecessary.

Indeed there are two basic problems with computation of $\mathbf{A}^T\mathbf{A}$. One is induced by sloppy programming practice, the other is inherent in the formation of $\mathbf{A}^T\mathbf{A}$. The former of these occurs in any problem where one of the columns of $\mathbf{A}$ is constant and the mean of each column is not subtracted from its elements. For instance, let one of the columns of $\mathbf{A}$ (let it be the last) have all its elements equal to 1. The normal equations (2.22) then yield a cross-products matrix with last row (and column), say the nth,

$$(\boldsymbol{a}_n^T\boldsymbol{a}_1, \boldsymbol{a}_n^T\boldsymbol{a}_2, \ldots, \boldsymbol{a}_n^T\boldsymbol{a}_n). \tag{5.7}$$

But

$$\boldsymbol{a}_n^T\boldsymbol{a}_j = \sum_{i=1}^{m} a_{ij} = m\bar{a}_j \tag{5.8}$$

where $\bar{a}_j$ is the mean of the jth column of the m by n matrix $\mathbf{A}$. Furthermore, the right-hand side of the nth normal equation is

$$\boldsymbol{a}_n^{\mathrm{T}}\boldsymbol{b} = m\bar{b}. \tag{5.9}$$

This permits x_n to be eliminated by using the nth normal equation

$$m\sum_{j=1}^{n-1} \bar{a}_j x_j + m x_n = m\bar{b} \tag{5.10}$$

or

$$x_n = \bar{b} - \sum_{j=1}^{n-1} \bar{a}_j x_j. \tag{5.11}$$

When this expression is substituted into the normal equations, the kth equation (note carefully the bars above the symbols) becomes

$$\sum_{j=1}^{n-1} \boldsymbol{a}_k^{\mathrm{T}}(\boldsymbol{a}_j - \bar{\boldsymbol{a}}_j)x_j = \boldsymbol{a}_k^{\mathrm{T}}(\boldsymbol{b} - \bar{\boldsymbol{b}}). \tag{5.12}$$

But since

$$\bar{\boldsymbol{a}}_k^{\mathrm{T}}(\boldsymbol{a}_j - \bar{\boldsymbol{a}}_j) = 0 \tag{5.13}$$

and

$$\bar{\boldsymbol{a}}_k^{\mathrm{T}}(\boldsymbol{b} - \bar{\boldsymbol{b}}) = 0 \tag{5.14}$$

equation (5.12) becomes

$$\sum_{j=1}^{n-1} (\boldsymbol{a}_k - \bar{\boldsymbol{a}}_k)^{\mathrm{T}}(\boldsymbol{a}_j - \bar{\boldsymbol{a}}_j)x_j = (\boldsymbol{a}_k - \bar{\boldsymbol{a}}_k)^{\mathrm{T}}(\boldsymbol{b} - \bar{\boldsymbol{b}}) \tag{5.15}$$

which defines a set of normal equations of order $(n-1)$

$$(\mathbf{A}')^{\mathrm{T}}\mathbf{A}'\boldsymbol{x}' = (\mathbf{A}')^{\mathrm{T}}\boldsymbol{b}' \tag{5.16}$$

where $\mathbf{A}'$ is formed from the $(n-1)$ non-constant columns of $\mathbf{A}$ each adjusted by subtraction of the mean and where $\boldsymbol{b}'$ is formed from $\boldsymbol{b}$ by subtraction of $\bar{\boldsymbol{b}}$. $\boldsymbol{x}'$ is simply $\boldsymbol{x}$ without $\boldsymbol{x}_n$.

Besides reducing the order of the problem, less information is lost in the formation of $(\mathbf{A}')^{\mathrm{T}}\mathbf{A}'$ than $\mathbf{A}^{\mathrm{T}}\mathbf{A}$, since the possible addition of large numbers to the matrix is avoided. These large numbers have subsequently to be subtracted from each other in the solution process, and this subtraction leads to digit cancellation which one should always seek to avoid.

As an example, consider the calculation of the variance of the column vector

$$\boldsymbol{a} = \begin{pmatrix} 1000 \\ 1002 \\ 1004 \\ 1008 \end{pmatrix} \tag{5.17}$$

which has mean $\bar{a} = 1003{\cdot}5$ so that

$$\boldsymbol{a} - \bar{\boldsymbol{a}} = \begin{pmatrix} -3{\cdot}5 \\ -1{\cdot}5 \\ 0{\cdot}5 \\ 4{\cdot}5 \end{pmatrix}. \tag{5.18}$$

The variance is computed via either

$$\text{var}(\boldsymbol{a}) = (\boldsymbol{a} - \bar{\boldsymbol{a}})^{\mathrm{T}}(\boldsymbol{a} - \bar{\boldsymbol{a}})/m \tag{5.19}$$

where $m = 4$ is the number of elements in the vector, or since

$$(\boldsymbol{a} - \bar{\boldsymbol{a}})^{\mathrm{T}}(\boldsymbol{a} - \bar{\boldsymbol{a}}) = \boldsymbol{a}^{\mathrm{T}}\boldsymbol{a} - 2\bar{\boldsymbol{a}}^{\mathrm{T}}\boldsymbol{a} + \bar{\boldsymbol{a}}^{\mathrm{T}}\bar{\boldsymbol{a}} = \boldsymbol{a}^{\mathrm{T}}\boldsymbol{a} - m\bar{a}^2 \tag{5.20}$$

by

$$\text{var}(\boldsymbol{a}) = \boldsymbol{a}^{\mathrm{T}}\boldsymbol{a}/m - \bar{a}^2. \tag{5.21}$$

Note that statisticians prefer to divide by $(m-1)$ which makes a change necessary in (5.21). Equation (5.19) when applied to the example on a six decimal digit computer gives

$$\text{var}(\boldsymbol{a}) = (12{\cdot}25 + 2{\cdot}25 + 0{\cdot}25 + 20{\cdot}25)/4 = 35/4 = 8{\cdot}75. \tag{5.22}$$

By comparison, formula (5.21) produces the results in table 5.1 depending on whether the computer truncates (chops) or rounds. The computation in exact arithmetic is given for comparison.

By using data in deviation form, this difficulty is avoided. However, there is a second difficulty in using the cross-products matrix which is inherent in its formation. Consider the least-squares problem defined by

$$\mathbf{A} = \begin{pmatrix} 1 & 1 & 1 \\ 1\text{E}-5 & 0 & 1 \\ 0 & 1\text{E}-5 & 1 \\ 0 & 0 & 1 \end{pmatrix} \qquad \boldsymbol{b} = \begin{pmatrix} 1 \\ 0 \\ 0 \\ 0 \end{pmatrix}.$$

TABLE 5.1. Results from formula (5.21).

	Exact	Truncated	Rounded
$(a_1)^2$	1000000	100000 * 10	100000 * 10
$(a_2)^2$	1004004	100400 * 10	100400 * 10
$(a_3)^2$	1008016	100801 * 10	100802 * 10
$(a_4)^2$	1016064	101606 * 10	101606 * 10
sum	4028084	402807 * 10	402808 * 10
sum/4	1007021	100701 * 10	100702 * 10
$-\bar{a}^2$	−1007012·25	−100701 * 10	−100701 * 10
var($\boldsymbol{a}$)	8·75	0	1 * 10 = 10

An added note to caution. In this computation all the operations are correctly rounded or truncated. Many computers are not so fastidious with their arithmetic.

In six-digit rounded computation this produces

$$\mathbf{A}^{\mathrm{T}}\mathbf{A} = \begin{pmatrix} 1 & 1 & 1{\cdot}00001 \\ 1 & 1 & 1{\cdot}00001 \\ 1{\cdot}00001 & 1{\cdot}00001 & 4 \end{pmatrix}$$

which is singular since the first two columns or rows are identical. If we use deviations from means (and drop the constant column) a singular matrix still results. For instance, on a Data General NOVA minicomputer using a 23-bit binary mantissa (between six and seven decimal digits), the **A** matrix using deviation from mean data printed by the machine is

$$\mathbf{A}' = \begin{pmatrix} 0{\cdot}749998 & 0{\cdot}749998 \\ -0{\cdot}249992 & -0{\cdot}250002 \\ -0{\cdot}250002 & -0{\cdot}249992 \\ -0{\cdot}250002 & -0{\cdot}250002 \end{pmatrix}$$

and the cross-products matrix as printed is

$$(\mathbf{A}')^{\mathrm{T}}\mathbf{A}' = \begin{pmatrix} 0{\cdot}749995 & 0{\cdot}749995 \\ 0{\cdot}749995 & 0{\cdot}749995 \end{pmatrix}$$

which is singular.

However, by means of the singular-value decomposition given by algorithm 1, the same machine computes the singular values of **A** (not **A**′) as

$$2{\cdot}17533,\ 1{\cdot}12603 \text{ and } 1\mathrm{E}-5.$$

Since the ratio of the smallest to the largest of the singular values is only slightly larger than the machine precision ($2^{-22} \simeq 2{\cdot}38419\mathrm{E}-7$), it is reasonable to presume that the tolerance q in the equation (3.37) should be set to some value between 1E−5 and 1·12603. This leads to a computed least-squares solution

$$\boldsymbol{x} = \begin{pmatrix} 0{\cdot}500002 \\ 0{\cdot}500002 \\ -3{\cdot}51667\mathrm{E}-6 \end{pmatrix}$$

with a residual sum of squares

$$\boldsymbol{r}^{\mathrm{T}}\boldsymbol{r} = 1{\cdot}68955\mathrm{E}-5.$$

With the tolerance of $q = 0$, the computed solution is

$$\boldsymbol{x} = \begin{pmatrix} 0{\cdot}499248 \\ 0{\cdot}500755 \\ -3{\cdot}51667\mathrm{E}-6 \end{pmatrix}$$

with

$$\boldsymbol{r}^{\mathrm{T}}\boldsymbol{r} = 1{\cdot}68956\mathrm{E}-4.$$

(In exact arithmetic it is not possible for the sum of squares with $q = 0$ to exceed that for a larger tolerance.)

When using the singular-value decomposition one could choose to work with deviations from means or to scale the data in some way, perhaps using columns which are deviations from means scaled to have unit variance. This will then prevent 'large' data from swamping 'small' data. Scaling of equations has proved a difficult and somewhat subjective issue in the literature (see, for instance, Dahlquist and Björck 1974, p 181ff).

Despite these cautions, I have found the solutions to least-squares problems obtained by the singular-value decomposition approach to be remarkably resilient to the omission of scaling and the subtraction of means.

As a final example of the importance of using decomposition methods for least-squares problems, consider the data (Nash and Lefkovitch 1976)

$$\mathbf{A} = \begin{pmatrix} \alpha & \alpha & \alpha \\ \alpha & \alpha & \alpha \\ 0 & 1 & 0 \\ 0 & 0 & 1 \end{pmatrix} \qquad \boldsymbol{b} = \begin{pmatrix} 1 \\ 1 \\ 1 \\ 1 \end{pmatrix}.$$

This is a regression through the origin and can be shown to have the exact solution

$$\boldsymbol{x} = \begin{pmatrix} 1/\alpha - 2 \\ 1 \\ 1 \end{pmatrix}$$

with a zero residual sum of squares. If we wish to use a method which only scans the data once, that is, explicit residuals are not computed, then solution of the normal equations allows the residual sum of squares to be computed via

$$\boldsymbol{r}^{\mathrm{T}}\boldsymbol{r} = \boldsymbol{b}^{\mathrm{T}}\boldsymbol{b} - \boldsymbol{b}^{\mathrm{T}}\mathbf{A}\boldsymbol{x}. \tag{5.23}$$

Alternatively, algorithm 4 can be used, to form the sum of squares by means of the uncorrelated residuals (4.30).

The following solutions were found using a Hewlett–Packard 9830 desk calculator (machine precision equal to 1E−11, but all arrays in the examples stored in split precision equal to 1E−5):

(i) Conventional regression performed by using the Choleski decomposition (§7.1) to solve the normal equations gave

(*a*) for $\alpha = 8$

$$\boldsymbol{x} = \begin{pmatrix} -1{\cdot}87434 \\ 0{\cdot}999677 \\ 0{\cdot}999677 \end{pmatrix} \qquad \text{and} \qquad \boldsymbol{r}^{\mathrm{T}}\boldsymbol{r} = 4{\cdot}22\mathrm{E}-4$$

(*b*) for $\alpha = 64$

$$\boldsymbol{x} = \begin{pmatrix} -1{\cdot}93715 \\ 0{\cdot}976384 \\ 0{\cdot}976395 \end{pmatrix} \qquad \text{and} \qquad \boldsymbol{r}^{\mathrm{T}}\boldsymbol{r} = 0{\cdot}046709.$$

(ii) Algorithm 4 gave

(*a*) for $\alpha = 8$

$$x = \begin{pmatrix} -1{\cdot}87498 \\ 0{\cdot}999992 \\ 0{\cdot}999995 \end{pmatrix} \qquad \text{and} \qquad r^T r = 0$$

(*b*) for $\alpha = 64$,

$$x = \begin{pmatrix} -1{\cdot}984378 \\ 0{\cdot}999980 \\ 0{\cdot}999980 \end{pmatrix} \qquad \text{and} \qquad r^T r = 0.$$

Since the first edition of this book appeared, several authors have considered problems associated with the formation of the sum of squares and cross-products matrix, in particular the question of collinearity. See, for example, Nash (1979b) and Stewart (1987).

Chapter 6

LINEAR EQUATIONS—A DIRECT APPROACH

6.1. INTRODUCTION

So far we have been concerned with solving linear least-squares problems. Now the usually simpler problem of linear equations will be considered. Note that a program designed to solve least-squares problems will give solutions to linear equations. The residual sum of squares must be zero if the equations are consistent. While this is a useful way to attack sets of equations which are suspected to possess singular coefficient matrices, since the singular-value decomposition permits such to be identified, in general the computational cost will be too high. Therefore this chapter will examine a direct approach to solving systems of linear equations. This is a variant of the elimination method taught to students in secondary schools, but the advent of automatic computation has changed only its form, showing its substance to be solid.

6.2. GAUSS ELIMINATION

Let us now look at this approach to the solution of equations (2.2). First note that if $\mathbf{A}$ is upper-triangular so that $A_{ij}=0$ if $i>j$, then it is easy to solve for $\boldsymbol{x}$ by a *back-substitution*, that is

$$x_n = b_n/A_{nn} \tag{6.1}$$

$$x_{n-1}=(b_{n-1}-A_{n-1,n}x_n)/A_{n-1,n-1} \tag{6.2}$$

and generally

$$x_j=\left(b_j-\sum_{k=j+1}^{n}A_{jk}x_k\right)\Big/A_{jj}. \tag{6.3}$$

These equations follow directly from equations (2.2) and the supposed upper- or right-triangular structure of $\mathbf{A}$. The Gauss elimination scheme uses this idea to find solutions to simultaneous linear equations by constructing the triangular form

$$\mathbf{R}\boldsymbol{x}=\boldsymbol{f} \tag{6.4}$$

from the original equations.

Note that each of the equations (2.2), that is

$$\sum_{k=1}^{n}A_{ik}x_k=b_i \qquad \text{for } i=1,2,\ldots,n \tag{6.5}$$

can be multiplied by an arbitrary quantity without altering the validity of the equation; if we are not to lose any information this quantity must be non-zero.

Furthermore, the sum of any two equations of (6.5) is also an equation of the set (6.5). Multiplying the first equation (i.e. that for $i=1$) by

$$m_{i1} = A_{i1}/A_{11} \tag{6.6}$$

and subtracting it from the ith equation gives new equations

$$\sum_{k=1}^{n} A'_{ik}x_k = b'_i \qquad \text{for } i=2, 3, \ldots, n \tag{6.7}$$

where

$$\begin{aligned} A'_{ik} &= A_{ik} - A_{1k}A_{i1}/A_{11} \\ &= A_{ik} - m_{i1}A_{1k} \end{aligned} \tag{6.8}$$

and

$$b'_i = b_i - b_1 A_{i1}/A_{11} = b_i - m_{i1}b_1. \tag{6.9}$$

But

$$A'_{i1} = A_{i1} - A_{11}A_{i1}/A_{11} = 0 \tag{6.10}$$

so that we have eliminated all but the first element of column 1 of $\mathbf{A}$. This process can now be repeated with new equations 2, 3, . . . , n to eliminate all but the first two elements of column 2. The element A_{12} is unchanged because equation 1 is not a participant in this set of eliminations. By performing $(n-1)$ such sets of eliminations we arrive at an upper-triangular matrix $\mathbf{R}$. This procedure can be thought of as an ordered sequence of multiplications by elementary matrices. The elementary matrix which eliminates A_{ij} will be denoted $\mathbf{M}_{ij}$ and is defined by

$$\mathbf{M}_{ij} = \mathbf{1}_n - m_{ij}\mathbf{\Delta}_{ij} \tag{6.11}$$

where

$$m_{ij} = A_{ij}/A_{jj} \tag{6.12}$$

(the elements in $\mathbf{A}$ are all current, not original, values) and where $\mathbf{\Delta}_{ij}$ is the matrix having 1 in the position ij and zeros elsewhere, that is

$$\Delta_{rs} = \delta_{ir}\delta_{js} \tag{6.13}$$

which uses the Kronecker delta, $\delta_{ir}=1$ for $i=r$ and $\delta_{ir}=0$ otherwise. The effect on $\mathbf{M}_{ij}$ when pre-multiplying a matrix $\mathbf{A}$ is to replace the ith row with the difference between the ith row and m_{ij} times the jth row, that is, if

$$\mathbf{A}' = \mathbf{M}_{ij}\mathbf{A} \tag{6.14}$$

then

$$A'_{rk} = A_{rk} \qquad \text{for } r \neq i \tag{6.15}$$

$$A'_{ik} = A_{ik} - m_{ij}A_{jk} \tag{6.16}$$

with $k=1, 2, \ldots, n$. Since $A_{jk}=0$ for $k<j$, for computational purposes one need only use $k=j, (j+1), \ldots, n$. Thus

$$\mathbf{R} = \left(\prod_{j=1}^{n-1}\prod_{i=j+1}^{n} \mathbf{M}_{ij}\right)\mathbf{A} \tag{6.17}$$

$$= \mathbf{L}^{-1}\mathbf{A} \tag{6.18}$$

gives the triangular matrix in question. The choice of symbol

$$\mathbf{L}^{-1} = \prod_{j=1}^{n-1} \prod_{i=j+1}^{n} \mathbf{M}_{ij} \tag{6.19}$$

is deliberate, since the matrix product is lower-triangular by virtue of the lower-triangular nature of each $\mathbf{M}_{ij}$ and the lemmas below which will be stated without proof (see, for instance, Mostow and Sampson 1969, p 226).

Lemma 1. The product of two lower-/upper-triangular matrices is also lower-/upper-triangular.

Lemma 2. The inverse of a lower-/upper-triangular matrix is also lower-/upper-triangular.

By virtue of lemma 2, the matrix $\mathbf{L}$ (the inverse of $\mathbf{L}^{-1}$) is lower-triangular. The elements of $\mathbf{L}$ are given by

$$L_{ij} = \begin{cases} 0 & \text{for } j > i \\ 1 & \text{for } j = i \\ m_{ij} & \text{for } j < i. \end{cases} \tag{6.20}$$

This is proved simply by forming

$$\mathbf{L}^{-1}\mathbf{L} = \mathbf{1} \tag{6.21}$$

using (6.19) and (6.20). Note that (6.18) implies that

$$\mathbf{A} = \mathbf{LR} \tag{6.22}$$

which is a *triangular decomposition* of the matrix $\mathbf{A}$. It permits us to rewrite the original equations

$$\mathbf{A}\boldsymbol{x} = \mathbf{LR}\boldsymbol{x} = \boldsymbol{b} \tag{6.23}$$

as

$$\mathbf{R}\boldsymbol{x} = \mathbf{L}^{-1}\mathbf{A}\boldsymbol{x} = \mathbf{L}^{-1}\boldsymbol{b} = \boldsymbol{f}. \tag{6.24}$$

Because we can retain the triangular structure by writing the unit matrix

$$\mathbf{1} = \mathbf{DD}^{-1} \tag{6.25}$$

where $\mathbf{D}$ is a non-singular diagonal matrix so that

$$\mathbf{A} = \mathbf{LR} = \mathbf{LDD}^{-1}\mathbf{R} = \mathbf{L}'\mathbf{R}' \tag{6.26}$$

the Gauss elimination is not the only means to obtain a triangular decomposition of $\mathbf{A}$.

In fact, the Gauss elimination procedure as it has been explained so far is unsatisfactory for computation in finite arithmetic because the m_{ij} are computed from *current* elements of the matrix, that is, those elements which have been computed as a result of eliminating earlier elements in the scheme. Recall that

$$m_{ij} = A'_{ij}/A'_{jj} \tag{6.12}$$

using the prime to denote current values of the matrix elements, that is, those values which have resulted from eliminating elements in columns 1, 2, . . . , $(j-1)$.

If A'_{jj} is zero, we cannot proceed, and 'small' A'_{jj} are quite likely to occur during subtractions involving digit cancellations, so that multipliers m_{ij} that are large and inaccurate are possible. However, we can ensure that multipliers m_{ij} are all less than one in magnitude by permuting the rows of $\mathbf{A}'$ (and hence $\mathbf{A}$) so that the largest of A'_{ij}, for $i=j, (j+1), \ldots, n$, is in the diagonal or *pivot* position. This modified procedure, called Gauss elimination with *partial pivoting*, has a large literature (see Wilkinson (1965) for a discussion with error analysis). Since the rows of $\mathbf{A}$ have been exchanged, this procedure gives the triangular decomposition of a transformed matrix

$$\mathbf{PA}=\mathbf{LR} \tag{6.27}$$

where $\mathbf{P}$ is the permutation matrix, simply a re-ordering of the columns of a unit matrix appropriate to the re-ordering of the rows of $\mathbf{A}$.

Particular methods exist for further minimising error propagation in the Gauss elimination procedure by using double-precision accumulation of vector inner products. These go by the names of Crout and Doolittle and are discussed, for instance, by Dahlquist and Björck (1974) as well as by Wilkinson (1965). Since the accumulation of inner products in several computer programming languages on a variety of machines is a non-trivial operation (though on a few machines it is simpler to accumulate in double than in single precision), these will not be discussed here. Nor will the Gauss elimination with complete pivoting, which chooses as pivot the largest element in the current matrix, thereby requiring both column and row permutations. The consensus of opinion in the literature appears to be that this involves too much work for the very few occasions when complete pivoting is distinctly more accurate than partial pivoting.

The Gauss elimination with partial pivoting and back-substitution are now stated explicitly. All work is done in an array $\mathbf{A}$ of dimension n by $n+p$, where p is the number of right-hand sides $\boldsymbol{b}$ to be solved.

Algorithm 5. Gauss elimination with partial pivoting

```
Procedure gelim( n : integer; {order of equations}
                 p : integer; {number of right hand sides}
                 var A : rmatrix; {equation coefficients in row order with right
                     hand sides built intothe matrix columns n+1, . . ,n+p}
                 tol : real); {pivot tolerance}

{alg05.pas == Gauss elimination with partial pivoting.
   This form does not save the pivot ordering, nor does it keep
   the matrix decomposition which results from the calculations.
              Copyright 1988 J. C. Nash
}
var
   det, s : real;
   h,i,j,k: integer;
begin {STEP 0}
   det := 1.0; {to initialise determinant value}
   writeln('alg05.pas -- Gauss elimination with partial pivoting');
   for j := 1 to (n-1) do {STEP 1}
```

Algorithm 5. Gauss elimination with partial pivoting (cont.)

```
      begin {STEP 2a}
        s := abs(A[j,j]); k := j;
        for h := (j+1) to n do {STEP 2b}
        begin
          if abs(A[h,j])>s then
          begin
            s := abs(A[h,j]); k := h;
          end;
        end; {loop on h}
        if k<>j then {STEP 3 -- perform row interchange. Here we do this
                        explicitly and trade time for the space to store indices
                        and the more complicated program code.}
        begin
          writeln('Interchanging rows ',k,' and ',j);
          for i  :=  j to (n+p) do
          begin
            s  := A[k,i];  A[k,i] := A[j,i];  A[j,i] := s;
          end; {loop on i}
          det := -det; {to change sign on determinant because of interchange}
        end; {interchange}
        det := det*A[j,j]; {STEP 4}
        if abs(A[j,j])<tol then
        begin
          writeln('Matrix computationally singular -- pivot < ',tol);
          halt;
        end;
        for k := (j+1) to n do {STEP 5}
        begin
          A[k,j] := A[k,j]/A[j,j]; {to form multiplier m[k,j]}
          for i := (j+1) to (n+p) do
            A[k,i] := A[k,i]-A[k,j]*A[j,i]; {main elimination step}
        end; {loop on k -- STEP 6}
        det := det*A[n,n]; {STEP 7}
        if abs(A[n,n])<tol then
        begin
          writeln('Matrix computationally singular -- pivot < ',tol);
        halt;
        end;
      end; {loop on j -- this ends the elimination steps}
      writeln('Gauss elimination complete -- determinant = ',det);
  end; {alg05.pas}
```

The array A *now contains the information*

$$m_{ij} = A[i, j] \qquad \textit{for } i > j$$

$$R_{ij} = A[i, j] \qquad \textit{for } i \leqslant j \leqslant n$$

$$f_i^{(j)} = A[i, j] \qquad \textit{for } j = n+1, n+2, \ldots, n+p$$

that is, the elements of the p *right-hand sides.*

Algorithm 6. Gauss elimination back-substitution

This algorithm is designed to follow Gauss elimination (algorithm 5), but can be applied also to systems of equations which are already triangular or which have been brought to triangular form by other methods, for instance, the Givens' reduction of §4.2 (algorithm 3).

```
procedure gebacksub(n, p:integer; {size of problem n=nRow, p=nRHS}
                    var A : rmatrix); {work array containing
                        Gauss elimination reduced coefficient
                        matrix and transformed right hand sides}

{alg06.pas == Gauss elimination back-substitution.
Places solutions to linear equation systems in columns n+1,..,n+p
of matrix A. Alg05.pas (Gauss elimination) must be executed first with
right hand sides in the columns n+1,..,n+p of matrix A in order
to triangularize the system of equations.
                    Copyright 1988 J. C. Nash
}
var
   s : real; {accumulator}
   i, j, k: integer;
begin
   writeln('alg06.pas -- Gauss elimination back-substitution');
   for i:=(n+1) to (n+p) do {STEP 1}
   begin
      A[n,i]:=A[n,i]/A[n,n]; {STEP 2}
      for j:=(n-1) downto 1 do {STEP 3}
      begin
         s:=A[j,i]; {STEP 4}
         for k:=(j+1) to n do {STEP 5}
         begin
            s:=s-A[j,k]*A[k,i]; {to subtract contributions from solution
                                 elements which have already been determined}
         end; {loop on k}
         A[j,i]:=s/A[j,j]; {STEP 6 -- to fix solution element j}
      end; {loop on j -- STEP 7}
   end; {loop on i -- STEP 8}
end; {alg06.pas}
```

The solutions to the triangular system(s) of equations and hence to the original equations (2.2) are contained in columns n+1, n+2, . . . , n+p, *of the working array.*

Example 6.1. The use of linear equations and linear least-squares problems

Organisations which publish statistics frequently use indices to summarise the change in some set of measurable quantities. Already in example 3.2 we have used indices of the use of various chemicals in agriculture and an index for farm income. The consumer price index, and the Dow Jones and Financial Times indices provide other examples. Such indices are computed by dividing the average value of the quantity for period t by the average for some *base period* $t = 0$ which is usually given the index value 100. Thus, if the quantity is called P, then

$$I_t = 100(\bar{P}_t/\bar{P}_0) \tag{6.28}$$

where

$$\bar{P}_t = \left(\sum_{j=1}^{n} W_j P_{tj}\right)\left(\sum_{j=1}^{n} W_j\right)^{-1} \tag{6.29}$$

given n classes or types of quantity P, of which the jth has value P_{tj} in period t and is assigned weight W_j in the average. Note that it is assumed that the weighting W_j is independent of the period, that is, of time. However, the weightings or 'shopping basket' may in fact change from time to time to reflect changing patterns of product composition, industrial processes causing pollution, stocks or securities in a portfolio, or consumer spending.

Substitution of (6.29) into (6.28) gives

$$I_t = 100\left(\sum_{j=1}^{n} W_j P_{tj}\right)\left(\sum_{j=1}^{n} W_j P_{0j}\right)^{-1}.$$

Finally, letting

$$K = 100\left(\sum_{j=1}^{n} W_j P_{0j}\right)^{-1} \tag{6.30}$$

gives

$$I_t = \sum_{j=1}^{n} (KW_j)P_{tj}. \tag{6.31}$$

Thus, if n periods of data I_t, P_{tj}, $j = 1, \ldots, n$, are available, we can compute the weightings KW_j. Hence, by assuming

$$\sum_{j=1}^{n} W_j = 1 \tag{6.32}$$

that is, that the weights are fractional contributions of each component, we can find the value of K and each of the W_j. This involves no more nor less than the solution of a set of linear equations. The work of solving these is, of course, unnecessary if the person who computes the index publishes his set of weights—as indeed is the case for several indices published in the *Monthly Digest of Statistics*†. Unfortunately, many workers do not deem this a useful or courteous practice towards their colleagues, and I have on two occasions had to attempt to discover the weightings. In both cases it was not possible to find a consistent set of weights over more than n periods, indicating that these were being adjusted over time. This created some difficulties for my colleagues who brought me the problems, since they were being asked to use current price data to generate a provisional estimate of a price index considerably in advance of the publication of the indices by the agency which normally performed the task. Without the weights, or even approximate values from the latest period for which they were available, it was not possible to construct such estimates. In one case the calculation was to have

† *Monthly Digest of Statistics* UK Central Statistical Office (London: HMSO).

TABLE 6.1. Prices and indices.

P_1	P_2	P_3	P_4	I_1	I_2
1	0·5	1·3	3·6	100	100
1·1	0·5	1·36	3·6	103·718	103·718
1·1	0·5	1·4	3·6	104·487	104·487
1·25	0·6	1·41	3·6	109·167	109·167
1·3	0·6	1·412	3·95	114·974	114·974
1·28	0·6	1·52	3·9	115·897	98·4615
1·31	0·6	1·6	3·95	118·846	101·506

used various proposed oil price levels to ascertain an index of agricultural costs. When it proved impossible to construct a set of consistent weights, it was necessary to try to track down the author of the earlier index values.

As an example of such calculations, consider the set of prices shown in table 6.1 and two indices I_1 and I_2 calculated from them. I_1 is computed using proportions 0·4, 0·1, 0·3 and 0·2 respectively of P_1, P_2, P_3 and P_4. I_2 uses the same weights except for the last two periods where the values 0·35, 0·15, 0·4 and 0·1 are used.

Suppose now that these weights are unknown. Then the data for the first four periods give a set of four equations (6.31) which can be solved to give

$$\boldsymbol{KW} = \begin{bmatrix} 25\cdot6448 \\ 6\cdot41001 \\ 19\cdot2253 \\ 12\cdot8215 \end{bmatrix}$$

using Gauss elimination (Data General NOVA, 23-bit binary mantissa). Applying the normalisation (6.32) gives

$$\boldsymbol{W} = \begin{bmatrix} 0\cdot400065 \\ 0\cdot0999976 \\ 0\cdot29992 \\ 0\cdot200018 \end{bmatrix}.$$

If these weights are used to generate index numbers for the last three periods, the values I_1 will be essentially reproduced, and we would detect a change in the weighting pattern if the values I_2 were expected.

An alternative method is to use a least-squares formulation, since if the set of weights is consistent, the residual sum of squares will be zero. Note that there is no constant term (column of ones) in the equations. Again on the NOVA in 23-bit arithmetic, I_1 gives

$$\boldsymbol{KW} = \begin{bmatrix} 25\cdot6362 \\ 6\cdot41743 \\ 19\cdot2329 \\ 12\cdot8201 \end{bmatrix} \qquad \boldsymbol{W} = \begin{bmatrix} 0\cdot3999 \\ 0\cdot100106 \\ 0\cdot300014 \\ 0\cdot199881 \end{bmatrix}$$

with a residual sum of squares (using KW) over the seven periods of $4{\cdot}15777E-7$. The same calculation with I_2 gives a residual sum of squares of $241{\cdot}112$, showing that there is not a consistent set of weights. It is, of course, possible to find a consistent set of weights even though index numbers have been computed using a varying set; for instance, if our price data had two elements identical in one period, any pair of weights for these prices whose sum was fixed would generate the same index number.

6.3. VARIATIONS ON THE THEME OF GAUSS ELIMINATION

Gauss elimination really presents only one type of difficulty to the programmer—which of the many possible variations to implement. We have already touched upon the existence of two of the better known ones, those of Crout and Doolittle (see Dahlquist and Björck 1974, pp 157–8). While these methods are useful and important, they require double-precision arithmetic to be used to full advantage, so cannot be used effectively if the computing system at hand lacks this capability.

Bowdler *et al* (1966) present ALGOL versions of the Crout algorithm which implicitly scale the rows of the coefficient matrix. These algorithms are complicated by ALGOL's lack of double-length arithmetic, necessitating calls to machine code procedures. (This comment applies to ALGOL-60, not ALGOL-68.)

By and large I have avoided scaling within my programs because of the great difficulty of making any reliable general recommendations. Indeed, given any two non-singular diagonal matrices **D** and **E**, the system of equations

$$\mathbf{DAEE}^{-1}x = \mathbf{D}b \tag{6.33}$$

has the same solution x as the equations

$$\mathbf{A}x = b. \tag{2.2}$$

In scaling the equations by row multiplication we are adjusting **D**, which adjusts the pivot selection. It is often recommended that the scaling factors in **D** be chosen to *equilibrate* the matrix **A**, that is, so that

$$\max_j |(\mathbf{DA})_{ij}| = 1 \qquad \text{for } i = 1, 2, \ldots, n \tag{6.34}$$

where for the moment **E** is assumed to be a unit matrix. This is simply a dose of common sense which attempts to avoid arithmetic involving numbers widely different in magnitude. However, as Dahlquist and Björck (1974, pp 181–3) point out, the scaling $\mathbf{E}^{-1}$ of the solution x can frustrate our efforts to stabilise the computation. Furthermore, optimal scaling depends on knowledge of the matrix $\mathbf{A}^{-1}$, which is not known. They therefore suggest **E** be chosen to reflect 'the importance of the unknowns'. This statement is suitably amorphous to cover whatever situations arise, so I shall venture the opinion that the magnitudes of the solution elements

$$\mathbf{y} = \mathbf{E}^{-1}x \tag{6.35}$$

should be roughly equivalent. That is to say, the variables in the problem at hand should be measured in units which give the expected solution elements approximately the same size. Is this worth the bother? I can only add that I rarely scale sets of equations unless there is some very obvious and natural way to do it.

Similar comments apply to iterative improvement of a computed solution (Dahlquist and Björck 1974, pp 183–5, Bowdler *et al* 1966). Given a computed solution $\tilde{\boldsymbol{x}}$,

$$\mathbf{A}(\boldsymbol{x}-\tilde{\boldsymbol{x}})=\boldsymbol{b}-\mathbf{A}\tilde{\boldsymbol{x}}=\boldsymbol{r} \tag{6.36}$$

if

$$\boldsymbol{c}=\boldsymbol{x}-\tilde{\boldsymbol{x}} \tag{6.37}$$

then a triangular decomposition of $\mathbf{A}$ permits solution of

$$\mathbf{A}\boldsymbol{c}=\mathbf{LR}\boldsymbol{c}=\boldsymbol{r} \tag{6.38}$$

for $\boldsymbol{c}$ and computation of a new solution

$$\tilde{\boldsymbol{x}}+\boldsymbol{c}. \tag{6.39}$$

The process can be repeated until the computed solution has converged, but in virtually all cases the improvement of the solution occurs in the first application of (6.36) to (6.39). A similar algorithm can be used to improve least-squares solutions (Dahlquist and Björck 1974, pp 204–5). Unfortunately these improvement procedures are dependent on accurate computation of residuals, and double-precision arithmetic must be called into play. As the systems for which this book is designed often lack this feature, one may fall into the habit of not using iterative improvement of linear-equation or least-squares solutions.

Even when computing in an environment where double-length arithmetic is available, I generally do not bother to employ it. Personally, very little of my work has concerned numbers which have arisen out of precise measurement. In fact, my clients are often only sure of the first one or two digits of their data, so that it is unnecessary to provide an extremely accurate solution, though it is important in many cases to identify near-linear dependencies (hence singularities) by means of a technique such as the singular-value decomposition.

So far in this section I have mentioned only techniques of which I do *not* generally make use. To finish, then, consider the one excursion from algorithms 5 and 6 which has diverted me from time to time. This is the purely organisational question of how the information in the working array $\mathbf{A}$ should be organised and accessed. In the algorithms as presented, I have chosen to perform interchanges explicitly and store the coefficient matrix and right-hand sides together in a single two-dimensional array. The choice of a single working array with solutions overwriting the right-hand sides $\boldsymbol{b}$ I feel to be the sensible one for small-computer implementations. The choice of method for accessing the elements of this array is less simple. Besides the direct, two-dimensional method which has been used, it is possible to perform pivot interchanges implicitly if the pivot positions are saved, for instance in an integer vector $\boldsymbol{q}$ so that the ith pivot is stored in $\mathbf{A}[q[i], i]$. Thus if the algorithm is started so that

$$q[i]=i \qquad \text{for } i=1,2,\ldots,n \tag{6.40}$$

then Gauss elimination and back-substitution can be carried out exactly as in algorithms 5 and 6 if every array reference is made with **A**[,] replaced by **A**[q[ ],]. However, a simplification occurs in the interchange step 3, which can be replaced by a simple interchange of the row indices. That is, at step j, if the pivot is in row $q[k] \neq q[j]$, or $k \neq j$, then the indices are simply interchanged rather than the entire rows. However, all array access operations are complicated.

Some overall increases in efficiency may be obtained if we take over the compiler or interpreter function in accessing two-dimensional arrays. That is, we store the working array **A** which is $m = (n+p)$ by n in a single vector $\boldsymbol{a}$ of mn elements. We can do this columnwise, so that

$$\mathbf{A}[i, j] = a[n * (j-1) + i] \tag{6.41}$$

or row-wise, so that

$$\mathbf{A}[i, j] = a[m * (i-1) + j]. \tag{6.42}$$

These translations offer some simplifications of the elimination and back-substitution algorithms. In fact, the row-wise form (6.41) is more useful for elimination where the index of an element is simply incremented to proceed across a row of the coefficient matrix. For back-substitution, we need to form matrix–vector products which oblige us to access array elements by marching simultaneously across rows and down columns. Implicit pivoting is also possible with a one-dimensional storage scheme. This adds just one more item to those from which a method must be selected.

It is probably clear to my readers that I have already decided that simplest is best and intend to stick with algorithms 5 and 6. My reasons are as follows.

(i) Despite the elegance of implicit pivoting, the extra index vector and the program code needed to make it work are counter to the spirit of a compact algorithm.
(ii) The implicit interchange only gains in efficiency relative to the direct method if an interchange is needed; this is without counting the overhead which array access via $\boldsymbol{q}$ implies. But in many instances very few interchanges are required and the whole discussion then boils down to an argument over the likely number of interchanges in the problem set to be solved.
(iii) In coding Gauss elimination with back-substitution and the Gauss–Jordan reduction with various of the above choices, S G Nash and I (unpublished work) found that the implicit pivoting methods were surprisingly prone to 'bugs' which were difficult to discover. This applied particularly to the one-dimensional storage forms. Most of these errors were simple typographical errors in entry of the code. Since it is hoped the algorithms in this book will prove straightforward to implement, only a direct method has been included.

6.4. COMPLEX SYSTEMS OF EQUATIONS

Consider the system of equations (where $\mathrm{i} = (-1)^{1/2}$)

$$(\mathbf{Y} + \mathrm{i}\mathbf{Z})(\boldsymbol{u} + \mathrm{i}\boldsymbol{v}) = \boldsymbol{g} + \mathrm{i}\boldsymbol{h}. \tag{6.43}$$

Separating these into real and imaginary components gives the real equations

$$\mathbf{Y}\boldsymbol{u} - \mathbf{Z}\boldsymbol{v} = \boldsymbol{g} \tag{6.44}$$

$$\mathbf{Y}\boldsymbol{v} + \mathbf{Z}\boldsymbol{u} = \boldsymbol{h} \tag{6.45}$$

which is a set of linear equations (2.22) with

$$\mathbf{A} = \begin{bmatrix} \mathbf{Y} & -\mathbf{Z} \\ \mathbf{Z} & \mathbf{Y} \end{bmatrix} \tag{6.46}$$

$$\boldsymbol{b} = \begin{bmatrix} \boldsymbol{g} \\ \boldsymbol{h} \end{bmatrix} \tag{6.47}$$

and

$$\boldsymbol{x} = \begin{bmatrix} \boldsymbol{u} \\ \boldsymbol{v} \end{bmatrix}. \tag{6.48}$$

This is how complex systems of linear equations can be solved using real arithmetic only. Unfortunately the repetition of the matrices $\mathbf{Y}$ and $\mathbf{Z}$ in (6.46) means that for a set of equations of order n, $2n^2$ storage locations are used unnecessarily. However, the alternative is to recode algorithms 5 and 6 to take account of the complex arithmetic in (6.43). Bowdler *et al* (1966) give ALGOL procedures to perform the Crout variant of the elimination for such systems of equations, unfortunately again requiring double-length accumulation.

6.5. METHODS FOR SPECIAL MATRICES

The literature contains a number of methods for solving special systems of equations. For instance, several contributions in Wilkinson and Reinsch (1971) deal with band matrices, that is, those for which

$$A_{ij} = 0 \qquad \text{if} \quad |i-j| > k \tag{6.49}$$

for some k. Thus if $k = 1$, the matrix is tridiagonal. While these methods are undoubtedly useful and save memory, I have not included them in this monograph because I feel any serious user with enough special problems to warrant a method tailored to the task is likely to find and implement one. Others may only find too many special cases tedious or bewildering. Thus no discussion of banded or other special forms is given, though the user should be alert to triangular forms since it is very wasteful of effort to apply Gauss elimination to a lower-triangular matrix when simple forward-substitution will suffice. Likewise, no treatment is included of the various iteration methods for the systems of equations arising from partial differential equations (see Varga 1962). It should be pointed out, however, that the Givens' reduction can often be organised to take advantage of patterns of zeros in matrices. Even as it stands, algorithm 3 is quite efficient for such problems, since very little work is done when zero elements are encountered and no pivot interchanges are made.

The only special form which will be considered is a symmetric positive definite matrix. Chapter 7 deals with a decomposition of such a matrix useful for solving special sets of linear equations. Chapter 8 discusses a very compact algorithm for inverting such a matrix *in situ*, that is, on top of itself.

Chapter 7

THE CHOLESKI DECOMPOSITION

7.1. THE CHOLESKI DECOMPOSITION

When the matrix $\mathbf{A}$ is symmetric and positive definite (see discussion below for definition), it is possible to perform (*without pivoting*) the symmetric decomposition

$$\mathbf{A}=\mathbf{L}\mathbf{L}^{\mathrm{T}}=\mathbf{R}^{\mathrm{T}}\mathbf{R} \tag{7.1}$$

where

$$\mathbf{L}=\mathbf{R}^{\mathrm{T}} \tag{7.2}$$

is a lower-triangular matrix. In fact, it is also possible to perform Gauss elimination in a symmetric fashion for symmetric positive definite matrices without pivoting for stability (see Dahlquist and Björck 1974, pp 162–4).

The Choleski algorithm (Wilkinson 1965) is derived directly from (7.1), that is

$$A_{ij}=\sum_{k=1}^{\min(i,j)} L_{ik}L_{kj}^{\mathrm{T}}=\sum_{k=1}^{\min(i,j)} L_{ik}L_{jk}. \tag{7.3}$$

Note that the summation runs only from 1 to the minimum of i and j due to the triangular nature of $\mathbf{L}$. Thus we have

$$A_{11}=L_{11}^2 \tag{7.4}$$

so that

$$L_{11}=(A_{11})^{1/2}. \tag{7.5}$$

Furthermore

$$A_{i1}=L_{i1}L_{11} \tag{7.6}$$

so that we obtain

$$L_{i1}=A_{i1}/L_{11}. \tag{7.7}$$

Consider now the mth column of $\mathbf{L}$ which is defined for $i>m$ by

$$L_{mm}L_{im}=\left(A_{im}-\sum_{k=1}^{m-1} L_{ik}L_{mk}\right) \tag{7.8}$$

with the diagonal element determined first by setting $i=m$. It is straightforward to see that every element in the right-hand side of equation (7.8) comes from columns 1, 2, . . . , $(m-1)$ of $\mathbf{L}$ or from column m of $\mathbf{A}$. Since (7.5) and (7.7) define the first column of $\mathbf{L}$, we have a stepwise procedure for computing its remaining columns, and furthermore this can be arranged so that $\mathbf{L}$ overwrites $\mathbf{A}$

within the computer. It remains to be shown that the procedure is stable and that for $i=m$ the right-hand side of (7.8) is positive, so no square roots of negative numbers are required.

Firstly, $\mathbf{A}$ is positive definite if

$$\boldsymbol{x}^{\mathrm{T}}\mathbf{A}\boldsymbol{x}>0 \qquad \text{for all } \boldsymbol{x}\neq\mathbf{0}. \tag{7.9}$$

An equivalent statement is that all the eigenvalues of $\mathbf{A}$ are positive. From (7.9) it follows by setting $\boldsymbol{x}$ to any column of the unit matrix that no diagonal element of $\mathbf{A}$ can be non-positive. Likewise, by taking only x_i and x_j non-zero

$$x_i^2A_{ii}+x_j^2A_{jj}+2x_ix_jA_{ij}>0 \tag{7.10}$$

which requires that the quadratic equation

$$z^2A_{ii}+2zA_{ij}+A_{jj}=0 \tag{7.11}$$

has only complex roots. This occurs if

$$A_{ij}^2<A_{ii}A_{jj}. \tag{7.12}$$

Consider now the ith step of the Choleski decomposition. For the moment suppose that only rows $1, 2, \ldots, (i-1)$ of $\mathbf{L}$ have been computed, giving a submatrix $\mathbf{L}_{i-1}$ which is a decomposition of the submatrix $\mathbf{A}_{i-1}$ of $\mathbf{A}$; hence

$$\begin{pmatrix}\mathbf{L}_{i-1} & \mathbf{0}\\ \boldsymbol{c}^{\mathrm{T}} & L_{ii}\end{pmatrix}\begin{pmatrix}\mathbf{L}_{i-1}^{\mathrm{T}} & \boldsymbol{c}\\ \mathbf{0} & L_{ii}\end{pmatrix}=\mathbf{A}_i=\begin{pmatrix}\mathbf{A}_{i-1} & \boldsymbol{a}\\ \boldsymbol{a}^{\mathrm{T}} & A_{ii}\end{pmatrix}. \tag{7.13}$$

Following Kowalik and Osborne (1968), we have

$$\mathbf{L}_{i-1}\boldsymbol{c}=\boldsymbol{a} \tag{7.14}$$

or

$$\boldsymbol{c}=\mathbf{L}_{i-1}^{-1}\boldsymbol{a} \tag{7.15}$$

where $\mathbf{L}_{i-1}$ is assumed non-singular. In fact, it is positive definite providing the positive square root is chosen in the computation of each of its diagonal elements via (7.8). Consider now the choice in (7.9) of an $\boldsymbol{x}$ such that the first $(i-1)$ elements of $\boldsymbol{x}$ are given by $\mathbf{L}_{i-1}^{-1}\boldsymbol{c}$, $x_i=-1$, and $x_j=0$ for $j>i$. This choice, using (7.13) gives

$$(\boldsymbol{c}^{\mathrm{T}}\mathbf{L}_{i-1}^{-1},-1)\begin{pmatrix}\mathbf{L}_{i-1}\mathbf{L}_{i-1}^{\mathrm{T}} & \mathbf{L}_{i-1}\boldsymbol{c}\\ \boldsymbol{c}^{\mathrm{T}}\mathbf{L}_{i-1}^{\mathrm{T}} & A_{ii}\end{pmatrix}\begin{pmatrix}\mathbf{L}_{i-1}^{-1}\boldsymbol{c}\\ -1\end{pmatrix}>0 \tag{7.16}$$

which reduces to

$$A_{ii}-\boldsymbol{c}^{\mathrm{T}}\boldsymbol{c}>0. \tag{7.17}$$

But a comparison of this with (7.8) shows that it implies the square of each diagonal element of $\mathbf{L}$ is positive, so that all the elements of $\mathbf{L}$ are real providing $\mathbf{A}$ is positive definite. Furthermore, an analysis similar to that used in (7.10), (7.11) and (7.12) demands that

$$L_{ij}^2<4L_{ii}L_{jj}. \tag{7.18}$$

(Again, the diagonal elements must be chosen to be positive in the decomposition.) Equations (7.17) and (7.18) give bounds to the size of the subdiagonal elements of $\mathbf{L}$, which suggests the algorithm is stable. A much more complete analysis which confirms this conjecture is given by Wilkinson (1961) who shows the matrix $\mathbf{LL}^{\mathrm{T}}$ as computed is always close in some norm to $\mathbf{A}$.

Once the Choleski decomposition has been performed, linear equations

$$\mathbf{A}\boldsymbol{x} = \mathbf{LL}^{\mathrm{T}}\boldsymbol{x} = \boldsymbol{b} \tag{7.19}$$

can be solved by a combination of a forward- and a back-substitution, that is

$$\mathbf{L}\boldsymbol{v} = \boldsymbol{b} \tag{7.20}$$

followed by

$$\mathbf{R}\boldsymbol{x} = \mathbf{L}^{\mathrm{T}}\boldsymbol{x} = \boldsymbol{v} \tag{7.21}$$

where we have used $\mathbf{R}$ to emphasise the fact that $\mathbf{L}^{\mathrm{T}}$ is upper-triangular. In a computer program, $\boldsymbol{b}$, $\boldsymbol{v}$ and $\boldsymbol{x}$ can all occupy the same storage vector, that is, $\boldsymbol{v}$ overwrites $\boldsymbol{b}$, and $\boldsymbol{x}$ overwrites $\boldsymbol{v}$. The solution of (7.20) is termed forward-substitution because the triangular structure defines the elements v_j in the order $1, 2, \ldots, n$, that is

$$v_1 = b_1/L_{11} \tag{7.22}$$

and

$$v_j = \left(b_j - \sum_{k=1}^{j-1} L_{jk}v_k\right) \Bigg/ L_{jj} \qquad \text{for } j = 2, 3, \ldots, n. \tag{7.23}$$

Likewise, the solution elements x_j of (7.21) are obtained in the backward order n, $(n-1), \ldots, 1$ from

$$x_n = v_n/L_{nn} \tag{7.24}$$

$$x_j = \left(v_j - \sum_{k=j+1}^{n} R_{jk}x_k\right) \Bigg/ R_{jj} \tag{7.25}$$

$$= \left(v_j - \sum_{k=j+1}^{n} L_{kj}x_k\right) \Bigg/ L_{jj}. \tag{7.26}$$

7.2. EXTENSION OF THE CHOLESKI DECOMPOSITION TO NON-NEGATIVE DEFINITE MATRICES

When $\mathbf{A}$ is non-negative definite, the inequality (7.9) becomes

$$\boldsymbol{x}^{\mathrm{T}}\mathbf{A}\boldsymbol{x} \geqslant 0 \qquad \text{for all } \boldsymbol{x} \neq \mathbf{0} \tag{7.27}$$

and inequalities (7.10), (7.12), (7.17) and (7.18) must be amended similarly. There is no difficulty in performing the decomposition unless a zero diagonal element appears in $\mathbf{L}$ before the decomposition is complete. For $L_{mm} = 0$, equations (7.3) and (7.8) are satisfied by any values of L_{im} for $i > m$. However, if we desire $\mathbf{L}$ to be non-negative definite and wish to satisfy the amended form of (7.18), that is

$$L_{im}^2 \leqslant 4L_{ii}L_{mm} \tag{7.28}$$

we should set $L_{im}=0$ for $i \geqslant m$. This is a relatively trivial modification to the decomposition algorithm. L_{mm} is, of course, found as the square root of a quantity which arises by subtraction, and in practice this quantity may be slightly negative due to rounding error. One policy, adopted here, is to proceed with the decomposition, setting all $L_{im}=0$ for $i \geqslant m$ even if this quantity is negative, thus assuming the matrix $\mathbf{A}$ is non-negative definite. Since there may be very good reasons for presupposing $\mathbf{A}$ to be non-negative definite, for instance if it has been formed as a sum-of-squares and cross-products matrix in a least-squares regression calculation, this is not as dangerous as it appears. Furthermore the decision to continue the decomposition in §7.1 when the computed square of L_{mm} is positive, rather than greater than some tolerance for zero, has been made after the following considerations.

(i) The decomposition is valid to the precision available in the arithmetic being used. When zero diagonal elements arise in $\mathbf{L}$ they reflect linear dependencies in the set of equations for which a solution is to be found, and any of the infinity of solutions which exist is taken to be acceptable. However, in recognition of the possibility that there may only be a near-linear dependence, it does not seem wise to presume a small number is zero, since the Choleski decomposition, unlike the singular-value decomposition (algorithm 1), does not allow the user to decide at the time solutions are computed which small numbers are to be assumed zero.
(ii) The size of the computed square of L_{mm} is dependent on the scale of the matrix $\mathbf{A}$. Unless the tolerance for zero is proportional to a norm of $\mathbf{A}$, its application has an effect which is not consistent from one problem to another.

If the computed square of L_{mm} is non-positive, the mth column of $\mathbf{L}$ is therefore set to zero

$$L_{im}=0 \qquad \text{for } i=m,(m+1),\ldots,n. \tag{7.29}$$

The forward- and back-substitutions to solve the linear equations

$$\mathbf{A}\boldsymbol{x}=\boldsymbol{b} \tag{2.2}$$

are unfortunately not now possible since division by zero occurs. If, however, the equations are *consistent*, that is, if $\boldsymbol{b}$ belongs to the column space of $\mathbf{A}$, at least one solution $\boldsymbol{x}$ exists (see, for example, Finkbeiner 1966, p 98).

Consider the forward-substitution to solve

$$\mathbf{L}\boldsymbol{v}=\boldsymbol{b} \tag{7.30}$$

for $\boldsymbol{v}$. If v_k is set to zero whenever $L_{kk}=0$, then solutions

$$\mathbf{L}^{\mathrm{T}}\boldsymbol{x}=\boldsymbol{v} \tag{7.31}$$

are solutions to (2.2) for arbitrary values of those x_k for which $L_{kk}=0$. This is, of course, only possible if

$$\sum_{j=1}^{k-1} L_{kj}v_j=b_k \tag{7.32}$$

which is another way of stating the requirement of consistency for the equations.

For a specific example, consider that $L_{mm}=0$ as above. Thus, in the forward-substitution v_m is always multiplied by zero and could have arbitrary value, except that in the back-substitution the mth row of $\mathbf{L}^{\mathrm{T}}$ is null. Denoting this by the vector

$$\boldsymbol{c}_m^{\mathrm{T}}=\mathbf{0} \tag{7.33}$$

it is easily seen that

$$\boldsymbol{c}_m^{\mathrm{T}}\boldsymbol{x}=v_m \tag{7.34}$$

so that v_m must be zero or the equation (7.34) is not satisfied. From (7.30) and (7.31) one has

$$\mathbf{L}\boldsymbol{v}=\mathbf{L}\mathbf{L}^{\mathrm{T}}\boldsymbol{x}=\boldsymbol{b}=\mathbf{A}\boldsymbol{x}. \tag{7.35}$$

Since x_m is arbitrary by virtue of (7.34), the value chosen will influence the values of all x_i, $i<m$, so some standard choice here is useful when, for instance, an implementation of the algorithm is being tested. I have always chosen to set $x_m=0$ (as below in step 14 of algorithm 8).

The importance of the above ideas is that the solution of linear least-squares problems by the normal equations

$$\mathbf{B}^{\mathrm{T}}\mathbf{B}\boldsymbol{x}=\mathbf{B}^{\mathrm{T}}\mathbf{y} \tag{7.36}$$

provides a set of *consistent* linear equations with a symmetric non-negative definite matrix $\mathbf{A}=\mathbf{B}^{\mathrm{T}}\mathbf{B}$, that is

$$\boldsymbol{x}^{\mathrm{T}}\mathbf{B}^{\mathrm{T}}\mathbf{B}\boldsymbol{x}=\boldsymbol{z}^{\mathrm{T}}\boldsymbol{z}=\sum_{k=1}^{m} z_k^2 \tag{7.37}$$

($\mathbf{B}$ is presumed to be m by n). The equations (7.36) are always consistent since the vector $\mathbf{B}^{\mathrm{T}}\mathbf{y}$ belongs to the row space of $\mathbf{B}$, which is identical to the column space of $\mathbf{B}^{\mathrm{T}}\mathbf{B}$.

There remain two organisational details: (*a*) in any program designed to save storage it is useful to keep only one triangle of a symmetric matrix and further to store only the non-zero parts of triangular matrices, and (*b*) the forward- and back-substitutions can be organised to overwrite $\boldsymbol{v}$ on $\boldsymbol{b}$ and then $\boldsymbol{x}$ on $\boldsymbol{v}$.

These ideas are incorporated into the following algorithms for the Choleski decomposition of a non-negative definite symmetric matrix (see Healy (1968) for a FORTRAN implementation) and solution of consistent linear equations by the substitutions described above.

Algorithm 7. Choleski decomposition in compact storage

```
procedure choldcmp(n: integer; {order of problem}
                   var a: smatvec; {matrix to decompose}
                   var singmat: boolean); {singularity flag}

{alg07.pas ==
   Choleski decomposition of symmetric positive definite matrix stored in
   compact row-order form. a[i*(i-1)/2+j] = A[i,j]

              Copyright 1988 J.C.Nash
```

Algorithm 7. Choleski decomposition in compact storage (cont.)

```
}
var
    i,j,k,m,q: integer;
    s : real; {accumulator}
begin
    singmat := false; {singmat will be set true if matrix found
                        computationally singular}

    for j := 1 to n do {STEP 1}
    begin {STEP 2}
        q := j*(j+1) div 2; {index of the diagonal element of row j}
        if j>1 then {STEP 3}
        begin {prepare for the subtraction in Eqn. (7.8). This is not needed
                for the first column of the matrix.}
            for i := j to n do {STEP 4}
            begin
                m := (i*(i-1) div 2)+j; s := a[m];
                for k := 1 to (j-1) do s := s-a[m-k]*a[q-k];
                a[m] := s;
            end; {loop on i}
        end; {of STEP 4}
        if a[q]<=0.0 then {STEP 5}
        begin {matrix singular}
            singmat := true;
            a[q] := 0.0; {since we shall assume matrix is non-negative definite}
        end;
        s := sqrt(a[q]); {STEP 7}
        for i := j to n do {STEP 8}
        begin
            m := (i*(i-1) div 2)+j;
            if s=0.0 then a[m] := 0 {to zero column elements in singular case}
                else a[m] := a[m]/s; {to perform the scaling}
        end; {loop on i}
    end; {loop on j -- end-loop is STEP 9}
end; {alg07.pas == Choleski decomposition choldcmp}
```

This completes the decomposition. The lower-triangular factor L *is left in the vector* **a** *in row-wise storage mode.*

Algorithm 8. Choleski back-substitution

```
procedure cholback(n: integer; {order of problem}
                    a: smatvec; {the decomposed matrix}
                    var x: rvector); {the right hand side}

{alg08.pas ==
    Choleski back substitution for the solution of consistent sets of
    linear equations with symmetric coefficient matrices.

                    Copyright 1988 J.C.Nash
```

Algorithm 8. Choleski back-substitution (cont.)

```
}
var
   i,j,q : integer;
begin {Forward substitution phase -- STEP 1}
   if a[1]=0.0 then x[1]:=0.0 {to take care of singular case}
                  else x[1]:=x[1]/a[1];
   {STEP 2}
   if n>1 then {do Steps 3 to 8; otherwise problem is trivial}
   begin
      {STEP 3}
      q:=1; {to initialize the index of matrix elements}
      for i:=2 to n do {STEP 4}
      begin {STEP 5}
         for j:=1 to (i-1) do
         begin
            q:=q+1; x[i]:=x[i]-a[q]*x[j];
         end; {loop on j}
         q:=q+1; {STEP 6 -- to give index of diagonal element of row i}
         if a[q]=0.0 then x[i]:=0.0 {to handle singular case}
                        else x[i]:=x[i]/a[q]; {STEP 7}
      end; {loop on i -- STEP 8}
   end; {non-trivial case. This completes the forward substitution}
   {STEP 9 -- Back substitution phase}
   if a[n*(n+1) div 2]=0.0 then x[n]:=0.0 {for singular case}
                              else x[n]:=x[n]/a[n*(n+1) div 2];
   if n>1 then {STEP 10}{test for trivial case; otherwise do steps 11 to 15}
   begin {STEP 11}
      for i:=n downto 2 do
      begin {STEP 12}
         q:=i*(i-1) div 2; {to give base index for row i}
         for j:=1 to (i-1) do x[j]:=x[j]-x[i]*a[q+j]; {STEP 13}
         if a[q]=0.0 then x[i-1]:=0.0 {for singular case}
                   else x[i-1]:=x[i-1]/a[q]; {STEP 14}
      end; {loop on i -- STEP 15}
   end; {non-trivial case -- STEP 16}
end; {alg08.pas == Choleski back-substitution cholback}
```

Note that this algorithm will solve consistent sets of equations whose coefficient matrices are symmetric and non-negative definite. It will not detect the cases where the equations are not consistent or the matrix of coefficients is indefinite.

7.3. SOME ORGANISATIONAL DETAILS

The algorithm given for the Choleski decomposition uses the most direct form for computing the array indices. In practice, however, this may not be the most efficient way to program the algorithm. Indeed, by running a FORTRAN version of algorithm 7 against the subroutine of Healy (1968) and that of Kevin Price (Nash 1984a, pp 97–101) on an IBM 370/168 it became quite clear that the latter two codes were markedly faster in execution (using the FORTRAN G compiler) than my own. On the Data General NOVA and Hewlett–Packard 9830, however, this finding seemed to be reversed, probably because these latter machines are run as *interpreters*, that is, they evaluate the code as they encounter it in a form of simultaneous translation into machine instructions whereas a compiler makes the translation first then discards the source code. The Healy and Price routines most likely are slowed down by the manner in which their looping is organised, requiring a backward jump in the program. On a system which interprets, this is likely to require a scan of the program until the appropriate label is found in the code and it is my opinion that this is the reason for the more rapid execution of algorithm 7 in interpreter environments. Such examples as these serve as a warning that programs do show very large changes in their relative, as well as absolute, performance depending on the hardware and software environment in which they are run.

Algorithm 7 computes the triangular factor $\mathbf{L}$ column by column. Row-by-row development is also possible, as are a variety of sequential schemes which perform the central subtraction in STEP 4 in piecemeal fashion. If double-precision arithmetic is possible, the forms of the Choleski decomposition which compute the right-hand side of (7.8) via a single accumulation as in algorithm 7 have the advantage of incurring a much smaller rounding error.

Example 7.1. The Choleski decomposition of the Moler matrix

The Moler matrix, given by

$$\begin{aligned} A_{ij} &= \min(i, j) - 2 \qquad &\text{for } i \neq j \\ A_{ii} &= i \end{aligned}$$

has the decomposition

$$\mathbf{A} = \mathbf{L}\mathbf{L}^{\mathrm{T}}$$

where

$$L_{ij} = \begin{cases} 0 & \text{for } j > i \\ 1 & \text{for } i = j \\ -1 & \text{for } j < i. \end{cases}$$

Note that this is easily verified since

$$A_{ij} = \sum_{k=1}^{\min(i,j)} L_{ik}L_{jk} = \begin{cases} \min(i, j) + 2(-1)(1) & \text{for } i \neq j \\ i & \text{for } i = j. \end{cases}$$

On a Data General NOVA operating in arithmetic having six hexadecimal digits (that is, a machine precision of 16^{-5}) the correct decomposition was observed for Moler matrices of order 5, 10, 15, 20 and 40. Thus for order 5, the Moler matrix has a lower triangle

$$\begin{array}{ccccc} 1 & & & & \\ -1 & 2 & & & \\ -1 & 0 & 3 & & \\ -1 & 0 & 1 & 4 & \\ -1 & 0 & 1 & 2 & 5 \end{array}$$

which gives the Choleski factor

$$\begin{array}{ccccc} 1 & & & & \\ -1 & 1 & & & \\ -1 & -1 & 1 & & \\ -1 & -1 & -1 & 1 & \\ -1 & -1 & -1 & -1 & 1. \end{array}$$

Example 7.2. Solving least-squares problems via the normal equations

Using the data in example 3.2, it is straightforward to form the matrix **B** from the last four columns of table 3.1 together with a column of ones. The lower triangle of $\mathbf{B}^\mathrm{T}\mathbf{B}$ is then (via a Hewlett–Packard 9830 in 12 decimal digit arithmetic)

$$\begin{bmatrix} 18926823 & & & & \\ 6359705 & 2164379 & & & \\ 10985647 & 3734131 & 6445437 & & \\ 3344971 & 1166559 & 2008683 & 659226 & \\ 14709 & 5147 & 8859 & 2926 & 13 \end{bmatrix}.$$

Note that, despite the warnings of chapter 5, means have not been subtracted, since the program is designed to perform least-squares computations when a constant (column of ones) is not included. This is usually called regression through the origin in a statistical context. The Choleski factor **L** is computed as

$$\begin{bmatrix} 4350{\cdot}496868 & & & & \\ 1461{\cdot}834175 & 165{\cdot}5893864 & & & \\ 2525{\cdot}147663 & 258{\cdot}3731371 & 48{\cdot}05831416 & & \\ 768{\cdot}8710282 & 257{\cdot}2450797 & 14{\cdot}66763457 & 40{\cdot}90441964 & \\ 3{\cdot}380993125 & 1{\cdot}235276666 & 0{\cdot}048499519 & 0{\cdot}194896363 & 0{\cdot}051383414 \end{bmatrix}$$

Using the right-hand side

$$\mathbf{B}^{\mathrm{T}}\mathbf{y} = \begin{bmatrix} 5937938 \\ 2046485 \\ 3526413 \\ 1130177 \\ 5003 \end{bmatrix}$$

the forward- and back-substitution algorithm 8 computes a solution

$$\boldsymbol{x} = \begin{bmatrix} -0{\cdot}046192435 \\ 1{\cdot}019386565 \\ -0{\cdot}159822924 \\ -0{\cdot}290376225 \\ 207{\cdot}7826146 \end{bmatrix}.$$

This is to be compared with solution (*a*) of table 3.2 or the first solution of example 4.2 (which is on pp 62 and 63), which shows that the various methods all give essentially the same solution under the assumption that none of the singular values is zero. This is despite the fact that precautions such as subtracting means have been ignored. This is one of the most annoying aspects of numerical computation—the foolhardy often get the right answer! To underline, let us use the above data (that is $\mathbf{B}^{\mathrm{T}}\mathbf{B}$ and $\mathbf{B}^{\mathrm{T}}\mathbf{y}$) in the Gauss elimination method, algorithms 5 and 6. If a Data General NOVA operating in 23-bit binary arithmetic is used, the largest integer which can be represented exactly is

$$2^{23} - 1 = 8388607$$

so that the original matrix of coefficients cannot be represented exactly. However, the solution found by this method, which ignores the symmetry of $\mathbf{B}^{\mathrm{T}}\mathbf{B}$, is

$$\boldsymbol{x} = \begin{bmatrix} -4{\cdot}62306\mathrm{E}-2 \\ 1{\cdot}01966 \\ -0{\cdot}159942 \\ -0{\cdot}288716 \\ 207{\cdot}426 \end{bmatrix}.$$

While this is not as close as solution (*a*) of table 3.2 to the solutions computed in comparatively double-length arithmetic on the Hewlett–Packard 9830, it retains the character of these solutions and would probably be adequate for many practitioners. The real advantage of caution in computation is not, in my opinion, that one gets better answers but that the answers obtained are known not to be unnecessarily in error.

Chapter 8

THE SYMMETRIC POSITIVE DEFINITE MATRIX AGAIN

8.1. THE GAUSS–JORDAN REDUCTION

Another approach to the solution of linear equations (and in fact nonlinear ones) is the *method of substitution*. Here, one of the unknowns x_k is chosen and one of the equations in which it has a non-zero coefficient is used to generate an expression for it in terms of the other x_j, $j \neq k$, and $\boldsymbol{b}$. For instance, given

$$\mathbf{A}\boldsymbol{x} = \boldsymbol{b} \tag{2.2}$$

and $A_{11} \neq 0$, we might use

$$x_1 = \left(b_1 - \sum_{j=2}^{n} A_{1j}x_j\right) \Big/ A_{11}. \tag{8.1}$$

The substitution of this into the other equations gives a new set of $(n-1)$ equations in $(n-1)$ unknowns which we shall write

$$\mathbf{A}'\boldsymbol{x}' = \boldsymbol{b}' \tag{8.2}$$

in which the indices will run from 2 to n. In fact $\boldsymbol{x}'$ will consist of the last $(n-1)$ elements of $\boldsymbol{x}$. By means of (8.1) it is simple to show that

$$b'_k = b_k - b_1 A_{k1}/A_{11} \tag{8.3}$$

and

$$A'_{kj} = A_{kj} - A_{k1}A_{1j}/A_{11} \tag{8.4}$$

for $k, j = 2, \ldots, n$. Notice that if $\boldsymbol{b}$ is included as the $(n+1)$th column of $\mathbf{A}$, (8.4) is the only formula needed, though j must now run to $(n+1)$.

We now have a set of equations of order $(n-1)$, and can continue the process until only a set of equations of order 1 remains. Then, using the trivial solution of this, a set of substitutions gives the desired solution to the original equations. This is entirely equivalent to Gauss elimination (without pivoting) and back-substitution, and all the arithmetic is the same.

Consider, however, that the second substitution is made not only of x_2 into the remaining $(n-2)$ equations, but also into the formula (8.1) for x_1. Then the final order-1 equations yield all the x_j at once. From the viewpoint of elimination it corresponds to eliminating upper-triangular elements in the matrix $\mathbf{R}$ in the system

$$\mathbf{R}\boldsymbol{x} = \boldsymbol{f} \tag{6.4}$$

then dividing through by diagonal elements. This leaves

$$\mathbf{1}x = f' \tag{8.5}$$

that is, the solution to the set of equations.

Yet another way to look at this procedure is as a series of elementary row operations (see Wilkinson 1965) designed to replace the pth column of an n by n matrix $\mathbf{A}$ with the pth column of the unit matrix of order n, that is, e_p. To accomplish this, the pth row of $\mathbf{A}$ must be divided by A_{pp}, and A_{ip} times the resulting pth row subtracted from every row i for $i \neq p$. For this to be possible, of course, A_{pp} cannot be zero.

A combination of n such steps can be used to solve sets of linear equations. To avoid build-up of rounding errors, some form of pivoting must be used. This will involve one of a variety of schemes for recording pivot positions or else will use explicit row interchanges. There are naturally some trade-offs here between simplicity of the algorithm and possible efficiency of execution, particularly if the set of equations is presented so that many row exchanges are needed.

By using the Gauss–Jordan elimination we avoid the programming needed to perform the back-substitution required in the Gauss elimination method. The price we pay for this is that the amount of work rises from roughly $n^3/3$ operations for a single equation to $n^3/2$ as n becomes large (see Ralston 1965 p 401). For small n the Gauss–Jordan algorithm may be more efficient depending on factors such as implementation and machine architecture. In particular, it is possible to arrange to overwrite the ith column of the working matrix with the corresponding column of the inverse. That is, the substitution operations of equations (8.1) and (8.4) with 1 replaced by i give elimination formulae

$$\tilde{A}_{ij} = A_{ij}/A_{ii} \tag{8.1a}$$

$$\tilde{A}_{kj} = A_{kj} - A_{ki}(A_{ij}/A_{ii}) \tag{8.4a}$$

for $j = 1, 2, \ldots, n$, $k = 1, 2, \ldots, n$, but $k \neq i$, with the tilde representing the transformed matrix. However, these operations replace column i with e_i, the ith column of the unit matrix $\mathbf{1}_n$, information which need not be stored. The right-hand side b is transformed according to

$$\tilde{b}_i = b_i/A_{ii} \tag{8.1b}$$

$$\tilde{b}_k = b_k - b_i(A_{ki}/A_{ii}) \tag{8.3a}$$

for $k = 1, 2, \ldots, n$ with $k \neq i$. To determine the inverse of a matrix, we could solve the linear-equation problems for the successive columns of $\mathbf{1}_n$. But now all columns e_j for $j > i$ will be unaltered by (8.1b) and (8.3a). At the ith stage of the reduction, e_i can be substituted on column i of the matrix by storing the pivot A_{ii}, substituting the value of 1 in this diagonal position, then performing the division implied in (8.1a). Row i of the working matrix now contains the multipliers $\tilde{A}_{ij} = (A_{ij}/A_{ii})$. By performing (8.4$a$) row-wise, each value A_{ki} can be saved, a zero substituted from e_i, and the elements of A_{kj}, $j = 1, 2, \ldots, n$, computed.

This process yields a very compact algorithm for inverting a matrix in the working storage needed to store only a single matrix. Alas, the lack of pivoting may be disastrous. The algorithm will not work, for instance, on the matrix

$$\begin{pmatrix} 0 & 1 \\ 1 & 0 \end{pmatrix}$$

which is its own inverse. Pivoting causes complications. In this example, interchanging rows to obtain a non-zero pivot implies that the columns of the resulting inverse are also interchanged.

The extra work involved in column interchanges which result from partial pivoting is avoided if the matrix is symmetric and positive definite—this special case is treated in detail in the next section. In addition, in this case complete pivoting becomes diagonal pivoting, which does not disorder the inverse. Therefore algorithms such as that discussed above are widely used to perform *stepwise regression*, where the pivots are chosen according to various criteria other than error growth. Typically, since the pivot represents a new independent variable entering the regression, we may choose the variable which most reduces the residual sum of squares at the current stage. The particular combination of (say) m out of n independent variables chosen by such a *forward selection rule* is not necessarily the combination of m variables of the n available which gives the smallest residual sum of squares. Furthermore, the use of a sum-of-squares and cross-products matrix is subject to all the criticisms of such approaches to least-squares problems outlined in chapter 5.

As an illustration, consider the problem given in example 7.2. A Data General ECLIPSE operating in six hexadecimal digit arithmetic gave a solution

$$\boldsymbol{x} = \begin{bmatrix} -4{\cdot}64529\text{E}-2 \\ 1{\cdot}02137 \\ -0{\cdot}160467 \\ -0{\cdot}285955 \\ 206{\cdot}734 \end{bmatrix}$$

when the pivots were chosen according to the residual sum-of-squares reduction criterion. The average relative difference of these solution elements from those of solution (a) of table 3.2 is 0·79%. Complete (diagonal) pivoting for the largest element in the remaining submatrix of the Gauss–Jordan working array gave a solution with an average relative difference (root mean square) of 0·41%. There are, of course, 120 pivot permutations, and the differences measured for each solution ranged from 0·10% to 0·79%. Thus pivot ordering does not appear to be a serious difficulty in this example.

The operations of the Gauss–Jordan algorithm are also of utility in the solution of linear and quadratic programming problems as well as in methods derived from such techniques (for example, minimum absolute deviation fitting). Unfortunately, these topics, while extremely interesting, will not be discussed further in this monograph.

8.2. THE GAUSS–JORDAN ALGORITHM FOR THE INVERSE OF A SYMMETRIC POSITIVE DEFINITE MATRIX

Bauer and Reinsch (in Wilkinson and Reinsch 1971, p 45) present a very compact algorithm for inverting a positive definite symmetric matrix *in situ*, that is, overwriting itself. The principal advantages of this algorithm are as follows.

(i) No pivoting is required. This is a consequence of positive definiteness and symmetry. Peters and Wilkinson (1975) state that this is 'well known', but I believe the full analysis is as yet unpublished.
(ii) Only a triangular portion of the matrix need be stored due to symmetry, though a working vector of length n, where n is the order of the matrix, is needed.

The algorithm is simply the substitution procedure outlined above. The modifications which are possible due to symmetry and positive definiteness, however, cause the computational steps to look completely different.

Consider an intermediate situation in which the first k of the elements $\boldsymbol{x}$ and $\boldsymbol{b}$ have been exchanged in solving

$$\mathbf{A}\boldsymbol{x} = \boldsymbol{b} \tag{8.6}$$

by the Gauss–Jordan algorithm. At this stage the matrix of coefficients will have the form

$$\begin{pmatrix} \mathbf{W} & \mathbf{X} \\ \mathbf{Y} & \mathbf{Z} \end{pmatrix} \tag{8.7}$$

with $\mathbf{W}$, k by k; $\mathbf{X}$, k by $(n-k)$; $\mathbf{Y}$, $(n-k)$ by k; and $\mathbf{Z}$, $(n-k)$ by $(n-k)$. Then $\mathbf{W}$ is the inverse of the corresponding block of $\mathbf{A}$ since the equations (8.6) are now given by their equivalents

$$\begin{pmatrix} \mathbf{W} & \mathbf{0} \\ \mathbf{Y} & \mathbf{1}_{n-k} \end{pmatrix} \boldsymbol{b} = \begin{pmatrix} \mathbf{1}_k & \mathbf{X} \\ \mathbf{0} & \mathbf{Z} \end{pmatrix} \boldsymbol{x}. \tag{8.8}$$

Thus by setting $x_j = 0$, for $j = (k+1), (k+2), \ldots, n$, in both (8.6) and (8.8) the required association of $\mathbf{W}$ and the leading k by k block of $\mathbf{A}^{-1}$ is established. Likewise, by setting $b_j = 0$, for $j = 1, \ldots, k$, in (8.6) and (8.8), $\mathbf{Z}$ is the inverse of the corresponding block of $\mathbf{A}^{-1}$. (Note that $\mathbf{W}$ and $\mathbf{Z}$ are symmetric because $\mathbf{A}$ and $\mathbf{A}^{-1}$ are symmetric.)

From these results, $\mathbf{W}$ and $\mathbf{Z}$ are both positive definite for all k since $\mathbf{A}$ is positive definite. This means that the diagonal elements needed as pivots in the Gauss–Jordan steps are always positive. (This follows directly from the definition of positive definiteness for $\mathbf{A}$, that is, that $\boldsymbol{x}^{\mathrm{T}}\mathbf{A}\boldsymbol{x} > 0$ for all $\boldsymbol{x} \neq \mathbf{0}$.)

In order to avoid retaining elements above or below the diagonal in a program, we need one more result. This is that

$$\mathbf{Y} = -\mathbf{X}^{\mathrm{T}} \tag{8.9}$$

in matrix (8.7). This can be proved by induction using the Gauss–Jordan substitution formulae for step k of the algorithm $(i, j \neq k)$

$$A_{kk}^{(k)} = 1/A_{kk}^{(k-1)} \tag{8.10}$$

$$A_{ik}^{(k)} = A_{ik}^{(k-1)}/A_{kk}^{(k-1)} \tag{8.11}$$

$$A_{kj}^{(k)} = -A_{kj}^{(k-1)}/A_{kk}^{(k-1)} \tag{8.12}$$

$$A_{ij}^{(k)} = A_{ij}^{(k-1)} - A_{ik}^{(k-1)}A_{kj}^{(k-1)}/A_{kk}^{(k-1)}. \tag{8.13}$$

For $k = 1$, therefore, the condition $\mathbf{Y} = -\mathbf{X}^{\mathrm{T}}$ is given as

$$A_{1j}^{(1)} = -A_{j1}^{(1)} \tag{8.14}$$

from equations (8.11) and (8.12).

Now assume

$$A_{hj}^{(k-1)} = -A_{jh}^{(k-1)} \tag{8.15}$$

$1 \leqslant h \leqslant k-1$, $k \leqslant j \leqslant n$, for any of $k = 2, \ldots, n$. We will show that the hypothesis is then true for k. By equation (8.13) we have

$$\begin{aligned} A_{hj}^{(k)} &= A_{hj}^{(k-1)} - A_{hk}^{(k-1)}A_{kj}^{(k-1)}/A_{kk}^{(k-1)} \\ &= -A_{jh}^{(k-1)} - (-A_{kh}^{(k-1)})A_{jk}^{(k-1)}/A_{kk}^{(k-1)} \\ &= -A_{jh}^{(k)} \end{aligned} \tag{8.16}$$

where we use the identity

$$A_{jk}^{(k-1)} = A_{kj}^{(k-1)} \tag{8.17}$$

since these elements belong to a submatrix $\mathbf{Z}$ which is symmetric in accord with the earlier discussion.

It remains to establish that

$$A_{kj}^{(k)} = -A_{jk}^{(k)} \qquad \text{for } j = (k+1), \ldots, n \tag{8.18}$$

but this follows immediately from equations (8.11) and (8.12) and the symmetry of the submatrix $\mathbf{Z}$. This completes the induction.

There is one more trick needed to make the Bauer–Reinsch algorithm extremely compact. This is a sequential cyclic re-ordering of the rows and columns of $\mathbf{A}$ so that the arithmetic is always performed with $k = 1$. This re-numeration relabels $(j+1)$ as j for $j = 1, 2, \ldots, (n-1)$ and relabels 1 as n. Letting

$$p = A_{11}^{(k-1)} \tag{8.19}$$

this gives a new Gauss–Jordan step

$$A_{nn}^{(k)} = 1/p \tag{8.20}$$

$$A_{i-1,n}^{(k)} = A_{i1}^{(k-1)}/p \tag{8.21}$$

$$A_{n,j-1}^{(k)} = -A_{1j}^{(k-1)}/p \tag{8.22}$$

$$A_{i-1,j-1}^{(k)} = A_{ij}^{(k-1)} - A_{i1}^{(k-1)}A_{1j}^{(k-1)}/p \tag{8.23}$$

for $i, j = 2, \ldots, n$.

A difficulty in this is that the quantities $A_{1j}^{(k-1)}/p$ have to be stored or they will be overwritten by $A_{i-1,j-1}$ during the work. This is overcome by using a working vector to store the needed quantities temporarily.

Because of the re-numeration we also have the matrix of coefficients in the form

$$\begin{pmatrix} \mathbf{Z} & \mathbf{Y} \\ \mathbf{X} & \mathbf{W} \end{pmatrix}. \tag{8.24}$$

This permits, via $\mathbf{Z} = \mathbf{Z}^{\mathrm{T}}$ and $\mathbf{Y} = -\mathbf{X}^{\mathrm{T}}$, the computation involved in (8.23), where

$$A_{1j} = \begin{cases} A_{j1} & \text{for } j \leqslant n-k \\ -A_{j1} & \text{for } j > n-k \end{cases} \tag{8.25}$$

without recourse to a full matrix. By using row-wise storage of the lower triangle of **A**, the algorithm below is obtained. Note that after n re-numerations the array elements are in the correct order. Also by counting backwards (step 1) from n to 1 the counter will contain $(n-k+1)$.

Algorithm 9. Bauer–Reinsch inversion of a positive definite symmetric matrix

```
procedure brspdmi(n : integer; {order of problem}
                  var avector : smatvec; {matrix in vector form}
                  var singmat : boolean); {singular matrix flag}
{alg09.pas ==
  Bauer - Reinsch inversion of a symmetric positive definite matrix in
  situ. Lower triangle of matrix is stored as a vector using row
  ordering of the original matrix.

              Copyright 1988 J.C.Nash
}
var
  i,j,k,m,q : integer;
  s,t : real;
  X : rvector;
begin
  writeln('alg09.pas -- Bauer Reinsch inversion');
  singmat := false; {initial setting for singularity flag}
  for k := n downto 1 do {STEP 1}
  begin
    if (not singmat) then
    begin
      s := avector[1]; {STEP 2}
      if s>0.0 then {STEP 3}
      begin
        m := 1; {STEP 4}
        for i := 2 to n do {STEP 5}
        begin {STEP 6}
          q := m; m := m+i; t := avector[q+1]; X[i] := -t/s;
          {Note the use of the temporary vector X[]}
          if i>k then X[i] := -X[i];{take account of Eqn. 8.22 -- STEP 7}
```

Algorithm 9. Bauer–Reinsch inversion of a positive definite symmetric matrix (cont.)

```
                for j := (q+2) to m do {STEP 8}
                begin
                    avector[j-i] := avector[j]+t*X[j-q];
                end; {loop on j}
            end; {loop on i} {STEP 9}
            q := q-1; avector[m] := 1.0/s; {STEP 10}
            for i := 2 to n do avector[q+i] := X[i]; {STEP 11}
        end {if s>0.0} {STEP 12}
        else
            singmat := true; {s<=0.0 and we cannot proceed}
    end; {if (not singmat)}
  end; {loop on k}
end; {alg09.pas == Bauer Reinsch inversion brspdm}
```

This completes the inversion, the original matrix having been overwritten by its inverse.

Example 8.1. The behaviour of the Bauer–Reinsch Gauss–Jordan inversion

Since the result of the algorithm is identical in form to the input, re-entering the procedure will compute the inverse of the inverse, which should be the same as the original matrix. The following output was obtained by applying the Bauer–Reinsch algorithm in this fashion to the Frank and Moler matrices (see appendix 1) on a Data General NOVA having a 23-bit mantissa.

```
NEW
LOAD ENHBRT
LOAD ENHMT4
RUN
ENHBRG AUG 19 75
BAUER REINSCH
ORDER? 5
FRANK MATRIX
1
1  2
1  2  3
1  2  3  4
1  2  3  4  5

INVERSE

ROW 1
2
ROW 2
-1  2
ROW 3
0 -1  2
ROW 4
0  0 -1  2
ROW 5
0  0  0 -1  1
```

```
NEW
LOAD ENHBRT
LOAD ENHMT5
RUN
ENHBRG AUG 19 75
BAUER REINSCH
ORDER? 5
MOLER MATRIX
 1
-1  2
-1  0  3
-1  0  1  4
-1  0  1  2  5

INVERSE

ROW 1
 86
ROW 2
 43  22
ROW 3
 22  11  6
ROW 4
 12  6  3  2
ROW 5
 8  4  2  1  1
```

```
INVERSE OF INVERSE

ROW 1
1
ROW 2
1  2
ROW 3
.999999  2  3
ROW 4
.999999  2  3  4
ROW 5
1  2  3  4  5
```

```
INVERSE OF INVERSE

ROW 1
 .999994
ROW 2
-1  2
ROW 3
-.999987  0  2.99997
ROW 4
-.999989  0  .999976  3.99998
ROW 5
-.999999  0  .999978  1.99998  4.99998
```

Chapter 9

THE ALGEBRAIC EIGENVALUE PROBLEM

9.1. INTRODUCTION

The next three chapters are concerned with the solution of algebraic eigenvalue problems

$$\mathbf{A}x = ex \tag{2.62}$$

and

$$\mathbf{A}x = e\mathbf{B}x. \tag{2.63}$$

The treatment of these problems will be highly selective, and only methods which I feel are particularly suitable for small computers will be considered. The reader interested in other methods is referred to Wilkinson (1965) and Wilkinson and Reinsch (1971) for an introduction to the very large literature on methods for the algebraic eigenproblem. Here I have concentrated on providing methods which are reliable and easy to implement for matrices which can be stored in the computer main memory, except for a short discussion in chapter 19 of two methods suitable for sparse matrices. The methods have also been chosen to fit in with ideas already developed so the reader should not feel on too unfamiliar ground. Thus it is possible, even likely, that better methods exist for most applications and any user should be aware of this. Section 10.5 discusses some of the possible methods for real symmetric methods.

9.2. THE POWER METHOD AND INVERSE ITERATION

One of the simplest methods for finding the eigenvalue of largest magnitude of a matrix $\mathbf{A}$ with its associated eigenvector is to iterate using the scheme

$$\mathbf{y}_i = \mathbf{A}x_i \tag{9.1a}$$

$$x_{i+1} = \mathbf{y}_i/\|\mathbf{y}_i\| \tag{9.1b}$$

from some starting vector x_1 until successive x_i are identical. The eigenvector is then given by x and the magnitude of the eigenvalue by $\|\mathbf{y}\|$ where $\|\ \ \|$ represents any vector norm. The simplicity of this algorithm and its requirement for only a matrix–vector product makes it especially suitable for small computers, though convergence may at times be very slow.

The method works as follows. Consider the expansion of x_1 in terms of the eigenvectors $\boldsymbol{\phi}_j$, $j = 1, 2, \ldots n$, which span the space. This may not be possible for non-symmetric matrices. However, the algorithm is generally less useful for such matrices and will not be considered further in this context. Wilkinson (1965, chap 9) has a good discussion of the difficulties. Therefore, returning to the expansion of $\boldsymbol{x}_1$, we have

$$\boldsymbol{x}_1 = \sum_{j=1}^{n} a_j \boldsymbol{\phi}_j. \tag{9.2}$$

The first iteration of the *power method* then gives

$$\mathbf{y}_1 = \mathbf{A}\boldsymbol{x}_1 = \sum_{j=1}^{n} a_j e_j \boldsymbol{\phi}_j \tag{9.3}$$

where e_j is the eigenvalue of $\mathbf{A}$ corresponding to $\boldsymbol{\phi}_j$. Denoting the reciprocal of the norm of $\mathbf{y}_j$ by N_j, that is

$$N_j = \|\mathbf{y}_j\|^{-1} \tag{9.4}$$

the vector $\boldsymbol{x}_{i+1}$ is given by

$$\boldsymbol{x}_{i+1} = \left(\prod_{j=1}^{i} N_j\right)\left(\sum_{j=1}^{n} a_j e_j^i \boldsymbol{\phi}_j\right). \tag{9.5}$$

The first factor in this expression is simply a normalisation to prevent numbers becoming too large to work with in the computer. Presuming e_1 is the largest eigenvalue in magnitude, $\boldsymbol{x}_{i+1}$ can also be written

$$\boldsymbol{x}_{i+1} = e_1^i \left(\prod_{j=1}^{i} N_j\right)\left(\sum_{j=1}^{n} a_j \boldsymbol{\phi}_j (e_j/e_1)^i\right). \tag{9.6}$$

But since

$$|e_j/e_1| < 1 \tag{9.7}$$

unless $j = 1$ (the case of degenerate eigenvalues is treated below), the coefficients of $\boldsymbol{\phi}_j$, $j \neq 1$, eventually become very small. The ultimate rate of convergence is given by

$$r = |e_2/e_1| \tag{9.8}$$

where e_2 is the eigenvalue having second largest magnitude. By working with the matrix

$$\mathbf{A}' = \mathbf{A} - k\mathbf{1} \tag{9.9}$$

this rate of convergence can be improved if some estimates of e_1 and e_2 are known. Even if such information is not at hand, *ad hoc* shifts may be observed to improve convergence and can be used to advantage. Furthermore, shifts permit

(i) the selection of the most positive eigenvalue or the most negative eigenvalue and, in particular,
(ii) evasion of difficulties when these two eigenvalues are equal in magnitude.

Degenerate eigenvalues present no difficulty to the power method except that it now converges to a vector in the subspace spanned by all eigenvectors corresponding to e_1. Specific symmetry or other requirements on the eigenvector must be imposed separately.

In the above discussion the possibility that $a_1 = 0$ in the expansion of $\boldsymbol{x}_1$ has been conveniently ignored, that is, some component of $\boldsymbol{x}_1$ in the direction of $\boldsymbol{\phi}_1$ is assumed to exist. The usual advice to users is, ‘Don’t worry, rounding errors will eventually introduce a component in the right direction'. However, if the matrix **A** has elements which can be represented exactly within the machine, that is, if **A** can be scaled so that all elements are integers small enough to fit in one machine word, it is quite likely that rounding errors in the 'right direction' will not occur. Certainly such matrices arise often enough to warrant caution in choosing a starting vector. Acton (1970) and Ralston (1965) discuss the power method in more detail.

The power method is a simple, yet highly effective, tool for finding the extreme eigensolutions of a matrix. However, by applying it with the inverse of the shifted matrix $\mathbf{A}'$ (9.9) an algorithm is obtained which permits all distinct eigensolutions to be determined. The iteration does not, of course, use an explicit inverse, but solves the linear equations

$$\mathbf{A}'\mathbf{y}_i = \boldsymbol{x}_i \tag{9.10a}$$

then normalises the solution by

$$\boldsymbol{x}_{i+1} = \mathbf{y}_i / \|\mathbf{y}_i\|. \tag{9.10b}$$

Note that the solution of a set of simultaneous linear equations must be found at each iteration.

While the power method is only applicable to the matrix eigenproblem (2.62), *inverse iteration* is useful for solving the generalised eigenproblem (2.63) using

$$\mathbf{A}' = \mathbf{A} - k\mathbf{B} \tag{9.11}$$

in place of (9.9). The iteration scheme is now

$$\mathbf{A}'\mathbf{y}_i = \mathbf{B}\boldsymbol{x}_i \tag{9.12a}$$

$$\boldsymbol{x}_{i+1} = \mathbf{y}_i / \|\mathbf{y}_i\|. \tag{9.12b}$$

Once again, the purpose of the normalisation of $\mathbf{y}$ in (9.1b), (9.10b) and (9.12b) is simply to prevent overflow in subsequent calculations (9.1a), (9.10a) or (9.12a). The end use of the eigenvector must determine the way in which it is standardised. In particular, for the generalised eigenproblem (2.63), it is likely that $\boldsymbol{x}$ should be normalised so that

$$\boldsymbol{x}^{\mathrm{T}}\mathbf{B}\boldsymbol{x} = 1. \tag{9.13}$$

Such a calculation is quite tedious at each iteration and should not be performed until convergence has been obtained, since a much simpler norm will suffice, for instance the infinity norm

$$\|\mathbf{y}\|_\infty = \max_j |y_j| \tag{9.14}$$

where y_j is the jth element of $\mathbf{y}$. On convergence of the algorithm, the eigenvalue is

$$e = k + x_j / y_j \tag{9.15}$$

(where the absolute value is *not* used).

Inverse iteration works by the following mechanism. Once again expand $\boldsymbol{x}_1$ as in (9.2); then

$$\mathbf{A}'\mathbf{y}_1 = \sum_{j=1}^{n} a_j \mathbf{B}\boldsymbol{\phi}_j$$

$$= \sum_{j=1}^{n} a_j \mathbf{A}'\boldsymbol{\phi}_j/(e_j - k) \tag{9.16}$$

or

$$\mathbf{y}_1 = \sum_{j=1}^{n} a_j \boldsymbol{\phi}_j/(e_j - k). \tag{9.17}$$

Therefore

$$\boldsymbol{x}_{i+1} = \left(\prod_{j=1}^{i} N_j\right)\left(\sum_{j=1}^{n} a_j \boldsymbol{\phi}_j/(e_j - k)^i\right) \tag{9.18}$$

and the eigenvector(s) corresponding to the eigenvalue closest to k very quickly dominate(s) the expansion. Indeed, if k is an eigenvalue, $\mathbf{A}'$ is singular, and after solution of the linear equations (9.12*a*) (this can be forced to override the singularity) the coefficient of the eigenvector $\boldsymbol{\phi}$ corresponding to k should be of the order of 1/eps, where eps is the machine precision. Peters and Wilkinson (1971, pp 418–20) show this 'full growth' to be the only reliable criterion for convergence in the case of non-symmetric matrices. The process then converges in one step and obtaining full growth implies the component of the eigenvector in the expansion (9.2) of $\boldsymbol{x}_1$ is not too small. Wilkinson proposes choosing different vectors $\boldsymbol{x}_1$ until one gives full growth. The program code to accomplish this is quite involved, and for symmetric matrices repetition of the iterative step is simpler and, because of the nature of the symmetric matrix eigenproblem, can also be shown to be safe. The caution concerning the choice of starting vector for matrices which are exactly representable should still be heeded, however. In the case where k is not an eigenvalue, inverse iteration cannot be expected to converge in one step. The algorithm given below therefore iterates until the vector $\boldsymbol{x}$ has converged.

The form of equation (9.12*a*) is amenable to transformation to simplify the iteration. That is, pre-multiplication by a (non-singular) matrix $\mathbf{Q}$ gives

$$\mathbf{QA}\mathbf{y}_i = \mathbf{QB}\boldsymbol{x}_i. \tag{9.19}$$

Note that neither $\boldsymbol{x}_i$ nor $\mathbf{y}_i$ are affected by this transformation. If $\mathbf{Q}$ is taken to be the matrix which accomplishes one of the decompositions of §2.5 then it is straightforward to carry out the iteration. The Gauss elimination, algorithm 5, or the Givens' reduction, algorithm 3, can be used to effect the transformation for this purpose. The matrix $\mathbf{Q}$ never appears explicitly. In practice the two matrices $\mathbf{A}$ and $\mathbf{B}$ will be stored in a single working array $\mathbf{W}$. Each iteration will correspond to a back-substitution similar to algorithm 6. One detail which deserves attention is the handling of zero diagonal elements in $\mathbf{QA}'$, since such zeros imply a division by zero during the back-substitution which solves (9.19). The author has found that replacement of zero (or very small) elements by some small number, say the machine precision multiplied by the norm of $\mathbf{A}'$, permits the process to continue quite satisfactorily. Indeed I have performed a number of successful computations on a non-symmetric generalised eigenvalue problem wherein both **A** and **B** were singular by means of this artifice! However, it must be admitted that this particular problem arose in a context in which a great deal was known about the properties of the solutions.

Algorithm 10. Inverse iteration via Gauss elimination

The algorithm requires a working array W, n *by* 2 ∗ n, *and two vectors* **x** *and* **y** *of order* n.

```
procedure gii(nRow : integer; {order of problem}
                    var A : rmatrix; {matrices of interest packed
                    into array as ( A | B) }
                    var Y : rvector; {the eigenvector}
                    var shift : real; {the shift -- on input, the value
                    nearest which an eigenvalue is wanted. On output, the
                    eigenvalue approximation found.}
                    var itcount: integer); {On input a limit to the number
                    of iterations allowed to find the eigensolution. On
                    output, the number of iterations used. The returned
                    value is negative if the limit is exceeded.}
{alg10.pas == Inverse iteration to find matrix an eigensolution of
                    A Y = ev * B * Y
for real matrices A and B of order n. The solution found corresponds to
one of the eigensolutions having eigenvalue, ev, closest to the value
shift. Y on input contains a starting vector approximation.
                    Copyright 1988 J.C.Nash
}
var
   i, itlimit, j, k, m, msame, nRHS :integer;
   ev, s, t, tol : real; {eigenvalue approximation}
   X : rvector;
begin
   itlimit:=itcount; {to get the iteration limit from the call}
   nRHS:=nRow; {the number of right hand sides is also n since we will
            store matrix B in array A in the right hand nRow columns}
   tol:=Calceps;
   s:=0.0; {to initialize a matrix norm}
   for i:=1 to nRow do
   begin
      X[i]:=Y[i]; {copy initial vector to X for starting iteration}
      Y[i]:=0.0; {to initialize iteration vector Y}
      for j:=1 to nRow do
      begin
         A[i,j]:=A[i,j]-shift*A[i,j+nRow];
         s:=s+abs(A[i,j]);
      end;
   end;
   tol:=tol*s; {to set a reasonable tolerance for zero pivots}
   gelim(nRow, nRHS, A, tol); {Gauss elimination STEPS 2-10}
   itcount:=0;
   msame :=0; {msame counts the number of eigenvector elements which
                    are unchanged since the last iteration}
```

Algorithm 10. Inverse iteration via Gauss elimination (cont.)

```
    while (msame<nRow) and (itcount<itlimit) do
    begin {STEP 11 -- perform the back-substitution first}
        itcount:=itcount+1; {to count the iterations}
        m:=nRow; s:=X[nRow];
        X[nRow]:=Y[nRow]; {save last trial vector -- zeros on iteration 1}
        if abs(A[nRow,nRow])<tol then Y[nRow]:=s/tol
                        else Y[nRow]:=s/A[nRow,nRow];
        t:=abs(Y[nRow]);{to set the first trial value for vector Y}
        for i:=(nRow-1) downto 1 do {STEP 12}
        begin {back-substition}
            s:=X[i]; X[i]:=Y[i];
            for j:=(i+1) to nRow do
            begin
                s:=s-A[i,j]*Y[j];
            end;
            if abs(A[i,i])<tol then Y[i]:=s/tol else Y[i]:=s/A[i,i];
            if abs(Y[i])>t then
            begin
                m:=i; t:=abs(Y[i]);
            end; {to update new norm and its position}
        end; {loop on i}
        ev:=shift+X[m]/Y[m];{current eigenvalue approximation -- STEP 13}
        writeln('Iteration ',itcount,' approx. ev=',ev);
        {Normalisation and convergence tests -- STEP 14}
        t:=Y[m]; msame:=0;
        for i:=1 to nRow do
        begin
            Y[i]:=Y[i]/t;
            if reltest+Y[i] = reltest+X[i] then msame:=msame+1;
            {This test is designed to be machine independent. Mixed mode
                arithmetic is avoided by the use of the constant reltest. The
                variable msame is used in place of m to avoid confusion during the
                scope of the 'while' loop.}
        end; {loop on i}
        {STEP 15 -- now part of while loop control}
        if msame<nRow then
        begin {STEP 16 -- multiplication of vector by transformed B matrix}
            for i:=1 to nRow do
            begin
                s:=0.0;
                for j:=1 to nRow do s:=s+A[i,j+nRow]*Y[j];
                X[i]:=s;
            end;{loop on i}
        end; {if msame < nRow}
    end; {while loop -- STEP 17 is now part of the while loop control}
    if itcount>=itlimit then itcount:=-itcount; {set negative to
                        indicate failure to converge within iteration limit}
    shift:=ev; {to pass eigenvalue to calling program}
end; {alg10.pas == gii}
```

Example 9.1. Inverse iteration

The following output, from a Data General NOVA operating in 23-bit binary arithmetic, shows the application of algorithm 10 to the algebraic eigenproblem of the order-4 Hilbert segment (see appendix 1).

```
RUN
ENHGII OCT 21 76
GAUSS ELIMINATION FOR INVERSE ITERATION
ORDER=? 4
HILBERT SEGMENT
SHIFT=? 0
APPROX EV= 0
APPROX EV= 9.67397E-5
APPROX EV= 9.66973E-5
APPROX EV= 9.66948E-5
APPROX EV= 9.66948E-5
CONVERGED TO EV= 9.66948E-5   IN  5  ITNS
4  EQUAL CPNTS
HILBERT SEGMENT
VECTOR
-2.91938E-2
.328714
-.791412
.514551
RESIDUALS
-2.98023E-8
-7.45058E-8
-2.98023E-8
-2.98023E-8
```

9.3. SOME NOTES ON THE BEHAVIOUR OF INVERSE ITERATION

The algorithm just presented may in some details appear complicated.

(i) The convergence test uses a comparison of all elements in the vector. For many applications the norm of **y** or some similar measure may suffice; however, it is not foolproof, particularly when the starting vector is set to some simple choice such as a column of ones. In this case, inverse iteration with a diagonal matrix $A_{ii} = i$, a unit matrix **B** and a shift of zero will 'converge' at iteration 2, which is its earliest opportunity to stop. However, the vector is left very much in error, though the dominant component has converged.

(ii) The eigenvalue is given by

$$\text{shift} + x_i/y_i = k + x_i/y_i \tag{9.20}$$

from the analysis of equations (9.16)–(9.18). When the element y_i is zero, of course, this is not suitable for determining the eigenvalue. Therefore the program must save the vectors $\boldsymbol{x}$ and **y**, search for the largest element in **y**, then divide it into the corresponding element of $\boldsymbol{x}$ in order to get the eigenvalue. It is tempting to suggest that the expression should be simply

$$\text{shift} + 1/y_i = k + 1/y_i \tag{9.21}$$

since a normalisation is performed at each stage. Alas, too many matrices have

symmetries which cause the dominant 'component' of a vector to be a pair of elements with equal magnitude but opposite sign. A program may therefore choose first one, then the other, as dominant component, thereby calculating a correct eigenvector but an eigenvalue having the wrong sign!
(iii) The limit on the number of iterations is a necessary precaution since the convergence test may be too stringent. Unfortunately, the rate of convergence may be very slow since it is governed by the eigenvalue distribution. If the program can be controlled manually, one may prefer to allow the process to continue until the approximate eigenvalue has settled on a value, then intervene to halt execution if the convergence test does not become satisfied after a few more iterations. In fully automatic computation, some limit, say 100 iterations, does not seem unreasonable.

A comparison of Gauss elimination and Givens' reduction algorithms for inverse iteration using the nine test matrices of appendix 1 suggests that the former is slightly preferable. Using matrices of order 5, 10, 15 and 20 and a shift of $k=0$, I timed the execution of inverse iteration including the triangularisation. Of the 36 problems, two were theoretically impossible since the shift was exactly between two eigenvalues. Both methods failed to converge in 101 iterations for these two problems (the Wilkinson W− matrices of even order) and the vectors at the point of termination had large residuals so these two cases were dropped from the test. On the remaining 34, Gauss elimination yielded a smaller residual than Givens' reduction on 20, had a lower execution time on 21 for a total time (a sum of the separate times) of 3489·3 seconds compared to 4160·9 seconds, and failed to satisfy the convergence requirements in 101 iterations a total of 8 times, as compared to 11 times out of 34 for Givens' reduction. The above tests were performed in BASIC on a Data General NOVA operating interpretively in six hexadecimal (base 16) digit arithmetic.

Together with the overall timing, we should consider the fact that the methods have very different relative performances depending on the matrix in question; in particular, how many interchanges must be performed in Gauss elimination. Furthermore, as we might expect, the Givens' method is faster for sparse matrices.

Note that algorithm 10 is not the most compact form for the ordinary algebraic eigenvalue problem (2.62), since the Gauss elimination algorithm 5 gives

$$\mathbf{PA}'\boldsymbol{x} = \mathbf{LR}\boldsymbol{x} = \mathbf{P}e\boldsymbol{x} \tag{9.22}$$

by virtue of (6.27), where $\mathbf{P}$ is the permutation matrix defined by the interchanges resulting from pivoting. These can be stored, as discussed in §6.3 in a single integer vector of indices q. Then to perform inverse iteration, it is only necessary to store this vector plus a working array n by n instead of the n by $2n$ array used in algorithm 10. Two vectors $\boldsymbol{x}$ and $\mathbf{y}$ are still needed. The elements of the lower-triangular matrix $\mathbf{L}$ are

$$L_{ij} = \begin{cases} 1 & \text{for } i=j \\ 0 & \text{for } j>i \\ m_{ij} & \text{for } j<i. \end{cases} \tag{9.23}$$

The subdiagonal elements m_{ij} are left in place after the Gauss elimination step, and the upper-triangular matrix **R** forms the upper triangle of the working array. Then the inverse iteration step (9.10a) involves the forward-substitution

$$\mathbf{L}\boldsymbol{v} = \mathbf{P}\boldsymbol{x}_i \tag{9.24}$$

and back-substitution

$$\mathbf{R}\mathbf{y}_i = \boldsymbol{v}. \tag{9.25}$$

The latter substitution step has been treated before, but the former is simplified by the ones on the diagonal of **L** so that

$$v_1 = (\mathbf{P}\boldsymbol{x})_1 = x_{q_1} \tag{9.26}$$

$$v_i = (\mathbf{P}\boldsymbol{x})_i - \sum_{j=1}^{i-1} L_{ij}v_j = x_{q_i} - \sum_{j=1}^{i-1} m_{ij}v_j. \tag{9.27}$$

The calculation can be arranged so that $\boldsymbol{v}$ is not needed, that is, so $\boldsymbol{x}$ and $\mathbf{y}$ are the only working vectors needed.

9.4. EIGENSOLUTIONS OF NON-SYMMETRIC AND COMPLEX MATRICES

The algebraic eigenvalue problem is considerably more difficult to solve when the matrix **A** is non-symmetric or complex. If the matrix is Hermitian, that is, if the complex-conjugate transpose of **A** is equal to **A**, the eigenvalues are then real and methods for real symmetric matrices can easily be extended to solve this case (Nash 1974). However, in general, it is possible for the matrix to be defective, that is, have less than n eigenvectors, and the problem may be ill conditioned in that the eigenvalues may be highly sensitive to small changes in the matrix elements. The procedures which have been published for this problem are, by and large, long. They must, after all, contend with the possibility of complex eigenvalues and eigenvectors. On a small machine, the space which must be allocated for these very rapidly exhausts the memory available, since where previously one space was required, two must now be reserved and the corresponding program code is more than doubled in length to handle the arithmetic.

Fortunately, such matrices occur rarely in practice. For both real and complex cases I have used a direct translation of Eberlein's ALGOL program *comeig* (Wilkinson and Reinsch 1971). This uses a generalisation of the Jacobi algorithm and I have found it to function well. One point which is not treated in Eberlein's discussion is that the eigenvectors can all be multiplied by any complex number, c, and still be eigenvectors. In the real symmetric case all arithmetic is in the real domain and by normalisation of the eigenvectors the results of a computation can be standardised with the exception of eigenvectors corresponding to multiple eigenvalues. In the complex case, however, the eigenvector

$$\boldsymbol{x} + \mathrm{i}\mathbf{y} \tag{9.28}$$

will be completely unrecognisable after multiplication by

$$c = re^{i\phi} = r\cos\phi + ir\sin\phi \tag{9.29}$$

that is, we obtain

$$\boldsymbol{x}' + i\boldsymbol{y}' = (\boldsymbol{x}r\cos\phi - \boldsymbol{y}r\sin\phi) + i(\boldsymbol{x}r\sin\phi + \boldsymbol{y}r\cos\phi). \tag{9.30}$$

Therefore, it is useful to *standardise* every computed eigenvector so that the largest component is

$$1 + i0. \tag{9.31}$$

Furthermore it is worthwhile to compute residuals for each vector. While this may be a trivial programming task for those familiar with complex numbers and linear algebra, algorithms for both standardisation and residual calculation follow as an aid to those less familiar with the eigenproblem of general square matrices.

We now present three algorithms which are intended to be used together:

algorithm 11, to standardise a complex (eigen)vector;
algorithm 12, to compute residuals for a purported complex eigenvector; and
algorithm 26, Eberlein's Jacobi-like method for eigensolutions of a complex square matrix.

The driver program DR26.PAS on the software diskette is designed to use these three algorithms in computing and presenting eigensolutions of a general square matrix, that is, a square matrix which may have real or complex elements and need not have any symmetries in its elements.

Algorithm 11. Standardisation of a complex vector

```
procedure stdceigv(n: integer; {number of elements in vector}
                    var T, U: rmatrix); {eigenvector k is given
                    (column k of T) + sqrt(-1) * (column k of U)
                    = T[.,k] + sqrt(-1) * U[.,k] }
{alg11.pas == Standardisation of complex eigensolutions.
                    Copyright 1988 J.C.Nash
}
var
   i, k, m : integer;
   b, e, g, s : real;
begin
   writeln('alg11.pas -- standardized eigensolutions');
   for i := 1 to n do {loop over the eigensolutions}
   begin {the standardization of the solution so that the largest
               element of each solution is set to 1 + 0 sqrt(-1) }
      g := T[1,i]*T[1,i]+U[1,i]*U[1,i]; {STEP 1}
         {the magnitude of element 1 of eigensolution i}
      k := 1; {to set index for the largest element so far}
      if n>1 then
      begin {STEP 2}
         for m := 2 to n do {loop over the other elements of the solution}
```

Algorithm 11. Standardisation of a complex vector (cont.)

```
          begin {STEP 3}
            b := T[m,i]*T[m,i]+U[m,i]*U[m,i]; {the magnitude of element m}
            if b>g then {STEP 4}
            begin {STEP 5}
              k := m; {save the index of the largest element}
              g := b; {and its size}
            end; {if b>g}
          end; { loop on m -- STEP 6}
        end; {if n>1}
        e := T[k,i]/g; {STEP 7}
        s := -U[k,i]/g; {e & s establish the rotation constant in Eq. 9.29}
        for k := 1 to n do {STEP 8}
        begin {the rotation of the elements}
          g := T[k,i]*e-U[k,i]*s; U[k,i] := U[k,i]*e+T[k,i]*s; T[k,i] := g;
        end; { loop on k}
    end; {loop on i -- over the eigensolutions}
end; {alg11.pas == stdceigv}
```

Algorithm 12. Residuals of a complex eigensolution

```
procedure comres( i, n: integer; {eigensolution index for which
                        residuals wanted, and order of problem}
                        A, Z, T, U, Acopy, Zcopy : rmatrix);{ output
                        from comeig (alg26). A and Z store the
                        eigenvalues, T and U the eigenvectors, and
                        Acopy and Zcopy provide a copy of the
                        original complex matrix.}
{alg12.pas == Residuals for complex eigenvalues and eigenvectors.
This is slightly different in form from the step-and-description algorithm
given in the first edition of Compact Numerical Methods; we work with the
i'th eigensolution as produced by comeig.
                        Copyright 1988 J.C.Nash
}
var
  j, k: integer;
  g, s, ss : real;
begin
  writeln('alg12.pas -- complex eigensolution residuals');
  ss := 0.0; {sum of squares accumulator}
  for j := 1 to n do {STEP 1}
  begin {computation of the residuals, noting that the
                    eigenvalue is located on the diagonal of A and Z}
    s := -A[i,i]*T[j,i]+Z[i,i]*U[j,i]; g := -Z[i,i]*T[j,i]-A[i,i]*U[j,i];
    {s + sqrt(-1) g = - eigenvalue * vector_element_j}
    for k := 1 to n do
    begin
      s := s+Acopy[j,k]*T[k,i]-Zcopy[j,k]*U[k,i];
      g := g+Acopy[j,k]*U[k,i]+Zcopy[j,k]*T[k,i];
    end; { loop on k}
    writeln('(',s,',',g,')');
    ss := ss+s*s+g*g;
```

Algorithm 12. Residuals of a complex eigensolution (cont.)

```
    end; { loop on j}
    writeln('Sum of squares = ',ss);
    writeln;
end; {alg12.pas == comres}
```

Algorithm 26. Eigensolutions of a complex matrix by Eberlein's method

```
procedure comeig( n : integer; {order of problem}
                var itcount: integer; {on entry, a limit to the iteration
                    count; on output, the iteration count to convergence
                    which is set negative if the limit is exceeded}
                var A, Z, T, U : rmatrix); {the matrix for which the
                    eigensolutions are to be found is A + sqrt(-1)*Z,
                    and this will be transformed so that the diagonal
                    elements become the eigenvalues; on exit,
                    T + sqrt(-1)*U has the eigenvectors in its columns}
{alg26.pas == Pascal version of Eberlein's comeig.
        Translated from comeig.bas, comeig.for and the original Algol
        version in Eberlein (1971).
Copyright J C Nash 1988
}
var
    Rvec : rvector;
    { Rvec was called en by Eberlein, but we use upper case vector names.}
    i, itlimit, j, k, k1, m, m9, n1 : integer;
    aki, ami, bv, br, bi : real;
    c, c1i, c1r, c2i, c2r, ca, cb, ch, cos2a, cot2x, cotx, cx : real;
    d, de, di, diag, dr, e, ei, er, eps, eta, g, hi, hj, hr : real;
    isw, max, nc, nd, root1, root2, root : real;
    s, s1i, s1r, s2i, s2r, sa, sb, sh, sig, sin2a, sx : real;
    tanh, tau, te, tee, tem, tep ,tse, zki, zmi : real;
    mark : boolean;
begin {comeig}
    {Commentary in this version has been limited to notes on differences
    between this implementation and the Algol original.}
    writeln('alg26.pas -- comeig');
    eps := Calceps; {calculate machine precision}
    {NOTE: we have not scaled eps here, but probably should do so to avoid
    unnecessary computations.}
    mark := false; n1 := n-1;
    for i := 1 to n do
    begin
        for j := 1 to n do
        begin {initialize eigenvector arrays}
            T[i,j] := 0.0; U[i,j] := 0.0; if i=j then T[i,i] := 1.0;
        end; { loop on j}
    end; { loop on i}
    itlimit := itcount; {use value on entry as a limit}
    itcount := 0; {and then set counter to zero}
    while (itcount<=itlimit) and (not mark) do
```

Algorithm 26. Eigensolutions of a complex matrix by Eberlein's method (cont.)

```
    begin
      itcount := itcount+1;{safety loop counter}
      tau := 0.0;{convergence criteria}
      diag := 0.0; {to accumulate diagonal norm}
      for k := 1 to n do
      begin
        tem := 0;
        for i := 1 to n do if i<>k then tem := tem+ABS(A[i,k])+ABS(Z[i,k]);
        tau := tau+tem; tep := abs(A[k,k])+abs(Z[k,k]);
        diag := diag+tep; {rem accumulate diagonal norm}
        Rvec[k] := tem+tep;
      end; { loop on k}
      writeln('TAU=',tau,' AT ITN ',itcount);
      for k := 1 to n1 do {interchange rows and columns}
      begin
        max := Rvec[k]; i := k; k1 := k+1;
        for j := k1 to n do
        begin
          if max<Rvec[j] then
          begin
            max := Rvec[j]; i := j;
          end; {if max<Rvec[j]}
        end; {loop on j}
        if i<>k then
        begin
          Rvec[i] := Rvec[k];
          for j := 1 to n do
          begin
            tep := A[k,j]; A[k,j] := A[i,j]; A[i,j] := tep; tep := Z[k,j];
            Z[k,j] := Z[i,j]; Z[i,j] := tep;
          end; { loop on j}
          for j := 1 to n do
          begin
            tep := A[j,k]; A[j,k] := A[j,i]; A[j,i] := tep; tep := Z[j,k];
            Z[j,k] := Z[j,i]; Z[j,i] := tep; tep := T[j,k]; T[j,k] := T[j,i];
            T[j,i] := tep; tep := U[j,k]; U[j,k] := U[j,i]; U[j,i] := tep;
          end; { loop on j}
        end; {if i<>k}
      end; {loop on k }
      if tau>=100.0*eps then {note possible change in convergence test from
          form of Eberlein to one which uses size of diagonal elements}
      begin {sweep}
        mark := true;
        for k := 1 to n1 do {main outer loop}
        begin
          k1 := k+1;
          for m := k1 to n do {main inner loop}
          begin
            hj := 0.0; hr := 0.0; hi := 0.0; g := 0.0;
            for i := 1 to n do
            begin
              if (i<>k) and (i<>m) then
```

Algorithm 26. Eigensolutions of a complex matrix by Eberlein's method (cont.)

```
          begin
            hr := hr+A[k,i]*A[m,i]+Z[k,i]*Z[m,i];
            hr := hr-A[i,k]*A[i,m]-Z[i,k]*Z[i,m];
            hi := hi+Z[k,i]*A[m,i]-A[k,i]*Z[m,i];
            hi := hi-A[i,k]*Z[i,m]+Z[i,k]*A[i,m];
            te := A[i,k]*A[i,k]+Z[i,k]*Z[i,k]+A[m,i]*A[m,i]+Z[m,i]*Z[m,i];
            tee := A[i,m]*A[i,m]+Z[i,m]*Z[i,m]+A[k,i]*A[k,i]+Z[k,i]*Z[k,i];
            g := g+te+tee; hj := hj-te+tee;
          end; {if i<>k and i<>m}
        end; {loop on i}
        br := A[k,m]+A[m,k]; bi := Z[k,m]+Z[m,k]; er := A[k,m]-A[m,k];
        ei := Z[k,m]-Z[m,k]; dr := A[k,k]-A[m,m]; di := Z[k,k]-Z[m,m];
        te := br*br+ei*ei+dr*dr; tee := bi*bi+er*er+di*di;
        if te>=tee then
        begin
          isw := 1.0; c := br; s := ei; d := dr; de := di;
          root2 := sqrt(te);
        end
        else {te<tee}
        begin
          isw := -1.0; c := bi; s := -er; d := di; de := dr;
          root2 := sqrt(tee);
        end;
        root1 := sqrt(s*s+c*c); sig := -1.0; if d>=0.0 then sig := 1.0;
        sa := 0.0; ca := -1.0; if c>=0.0 then ca := 1.0;
        if root1<=eps then
        begin
          sx := 0.0; sa := 0.0; cx := 1.0; ca := 1.0;
          if isw<=0.0 then
          begin
            e := ei; bv := -br;
          end
          else
          begin
            e := er; bv := bi;
          end; {if isw<=0.0}
          nd := d*d+de*de;
        end
        else {root1>eps}
        begin
          if abs(s)>eps then
          begin
            ca := c/root1; sa := s/root1;
          end; {abs(s)>eps}
          cot2x := d/root1; cotx := cot2x+(sig*sqrt(1.0+cot2x*cot2x));
          sx := sig/sqrt(1.0+cotx*cotx); cx := sx*cotx;
          { find rotated elements }
          eta := (er*br+ei*bi)/root1; tse := (br*bi-er*ei)/root1;
          te := sig*(tse*d-de*root1)/root2; tee := (d*de+root1*tse)/root2;
          nd := root2*root2+tee*tee; tee := hj*cx*sx; cos2a := ca*ca-sa*sa;
          sin2a := 2.0*ca*sa; tem := hr*cos2a+hi*sin2a;
          tep := hi*cos2a-hr*sin2a; hr := hr*cx*cx-tem*sx*sx-ca*tee;
```

Algorithm 26. Eigensolutions of a complex matrix by Eberlein's method (cont.)

```
                    hi := hi*cx*cx+tep*sx*sx-sa*tee;
                    bv := isw*te*ca+eta*sa; e := ca*eta-isw*te*sa;
                end; {root1>eps}
                { label 'enter1' is here in Algol version}
                s := hr-sig*root2*e; c := hi-sig*root2*bv; root := sqrt(c*c+s*s);
                if root<eps then
                begin
                    cb := 1.0; ch := 1.0; sb := 0.0; sh := 0.0;
                end {if root<eps}
                else {root>=eps}
                begin
                    cb := -c/root; sb := s/root; tee := cb*bv-e*sb; nc := tee*tee;
                    tanh := root/(g+2.0*(nc+nd)); ch := 1.0/sqrt(1.0-tanh*tanh);
                    sh := ch*tanh;
                end;{root>=eps}
                tem := sx*sh*(sa*cb-sb*ca); c1r := cx*ch-tem; c2r := cx*ch+tem;
                c1i := -sx*sh*(ca*cb+sa*sb); c2i := c1i; tep := sx*ch*ca;
                tem := cx*sh*sb; s1r := tep-tem; s2r := -tep-tem; tep := sx*ch*sa;
                tem := cx*sh*cb; s1i := tep+tem; s2i := tep-tem;
                tem := sqrt(s1r*s1r+s1i*s1i); tep := sqrt(s2r*s2r+s2i*s2i);
                if tep>eps then mark := false;
                if (tep>eps) and (tem>eps) then
                begin
                    for i := 1 to n do
                    begin
                        aki := A[k,i]; ami := A[m,i]; zki := Z[k,i]; zmi := Z[m,i];
                        A[k,i] := c1r*aki-c1i*zki+s1r*ami-s1i*zmi;
                        Z[k,i] := c1r*zki+c1i*aki+s1r*zmi+s1i*ami;
                        A[m,i] := s2r*aki-s2i*zki+c2r*ami-c2i*zmi;
                        Z[m,i] := s2r*zki+s2i*aki+c2r*zmi+c2i*ami;
                    end; { loop on i}
                    for i := 1 to n do
                    begin
                        aki := A[i,k]; ami := A[i,m]; zki := Z[i,k]; zmi := Z[i,m];
                        A[i,k] := c2r*aki-c2i*zki-s2r*ami+s2i*zmi;
                        Z[i,k] := c2r*zki+c2i*aki-s2r*zmi-s2i*ami;
                        A[i,m] := -s1r*aki+s1i*zki+c1r*ami-c1i*zmi;
                        Z[i,m] := -s1r*zki-s1i*aki+c1r*zmi+c1i*ami;
                        aki := T[i,k]; ami := T[i,m]; zki := U[i,k]; zmi := U[i,m];
                        T[i,k] := c2r*aki-c2i*zki-s2r*ami+s2i*zmi;
                        U[i,k] := c2r*zki+c2i*aki-s2r*zmi-s2i*ami;
                        T[i,m] := -s1r*aki+s1i*zki+c1r*ami-c1i*zmi;
                        U[i,m] := -s1r*zki-s1i*aki+c1r*zmi+c1i*ami;
                    end; { loop on i}
                end; {if tep and tem >eps}
            end; {loop on m}
        end; { loop on k}
    end {if tau>=100*eps}
    else mark := true; {to indicate convergence}
  end; {while itcount<=itlimit}
  if itcount>itlimit then itcount := -itcount; {negative iteration count
                        means process has not converged properly}
end; {alg26.pas == comeig}
```

Example 9.2. Eigensolutions of a complex matrix

The following output from a Data General NOVA operating in 23-bit binary arithmetic shows the computation of the eigensolutions of a complex matrix due to Eberlein from the test set published by Gregory and Karney (1969). The notation (,) is used to indicate a complex number, real part followed by imaginary part. Note that the residuals computed are quite large by comparison with those for a real symmetric matrix. This is due to the increased difficulty of the problem, to the extra operations needed to take account of the complex numbers and to the standardisation of the eigenvectors, which will introduce some additional errors (for instance, in the first eigenvector, 5·96046E−8 for zero). This comment must be tempered by the observation that the norm of the matrix is quite large, so that the residuals divided by this norm are still only a reasonably small multiple of the machine precision.

```
RUN
ENHCMG - COMEIG AT SEPT 3 74
ORDER? 3
ELEMENT( 1 , 1 );REAL=? 1 IMAGINARY? 2
ELEMENT( 1 , 2 );REAL=? 3 IMAGINARY? 4
ELEMENT( 1 , 3 );REAL=? 21 IMAGINARY? 22
ELEMENT( 2 , 1 );REAL=? 43 IMAGINARY? 44
ELEMENT( 2 , 2 );REAL=? 13 IMAGINARY? 14
ELEMENT( 2 , 3 );REAL=? 15 IMAGINARY? 16
ELEMENT( 3 , 1 );REAL=? 5 IMAGINARY? 6
ELEMENT( 3 , 2 );REAL=? 7 IMAGINARY? 8
ELEMENT( 3 , 3 );REAL=? 25 IMAGINARY? 26

TAU= 194 AT ITN 1
TAU= 99.7552  AT ITN 2
TAU= 64.3109  AT ITN 3
TAU= 25.0133  AT ITN 4
TAU= 7.45935  AT ITN 5
TAU= .507665  AT ITN 6
TAU= 6.23797E-4  AT ITN 7
TAU= 1.05392E-7  AT ITN 8
EIGENSOLUTIONS
RAW VECTOR 1
( .371175 ,-.114606 )
( .873341 ,-.29618 )
( .541304 ,-.178142 )

EIGENVALUE 1 =( 39.7761 , 42.9951 )
VECTOR
( .42108 , 1.15757E-2 )
( 1 , 5.96046E-8 )
( .617916 , 5.57855E-3 )
RESIDUALS
( 2.2918E-4 , 2.34604E-4 )
( 5.16415E-4 , 5.11169E-4 )
( 3.70204E-4 , 3.77655E-4 )

RAW VECTOR 2
(-9.52917E-2 ,-.491205 )
( 1.19177 , .98026 )
(-.342159 ,-9.71221E-2 )
```

```
EIGENVALUE 2 =( 6.7008 ,-7.87591 )
VECTOR
(-.249902 ,-.206613 )
( 1 , 1.19209E-7 )
(-.211227 , 9.22453E-2 )
RESIDUALS
(-3.8147E-5 , 3.8147E-6 )
( 7.55787E-5 ,-7.48634E-5 )
(-1.52588E-5 , 2.57492E-5 )

RAW VECTOR 3
( .408368 , .229301 )
(-.547153 ,-1.39186 )
(-4.06002E-2 , .347927 )

EIGENVALUE 3 =(-7.47744 , 6.88024 )
VECTOR
(-.242592 , .198032 )
( 1 , 0 )
(-.206582 ,-.110379 )
RESIDUALS
( 5.24521E-6 ,-4.00543E-5 )
(-7.9155E-5 , 7.82013E-5 )
( 2.81334E-5 ,-1.04904E-5 )
```

Chapter 10

REAL SYMMETRIC MATRICES

10.1. THE EIGENSOLUTIONS OF A REAL SYMMETRIC MATRIX

The most common practical algebraic eigenvalue problem is that of determining all the eigensolutions of a real symmetric matrix. Fortunately, this problem has the most agreeable properties with respect to its solution (see, for instance, Wilkinson 1965).

(i) All the eigenvalues of a real symmetric matrix are real.
(ii) It is possible to find a complete set of n eigenvectors for an order-n real symmetric matrix and these can be made mutually orthogonal. Usually they are normalised so that the Euclidean norm (sum of squares of the elements) is unity. Thus, the total eigenproblem can be expressed

$$\mathbf{AX} = \mathbf{XE} \qquad (10.1)$$

where the matrix $\mathbf{X}$ has column j such that

$$\mathbf{A}\boldsymbol{x}_j = e_j\boldsymbol{x}_j \qquad (10.2)$$

where e_j is the jth eigenvalue. $\mathbf{E}$ is the diagonal matrix

$$E_{ij} = e_j\delta_{ij} \qquad (10.3)$$

with δ_{ij} the familiar Kronecker delta (see §6.2, p 60).

By virtue of the orthogonality, we have

$$\mathbf{X}^{\mathrm{T}}\mathbf{X} = \mathbf{1}_n \qquad (10.4)$$

where $\mathbf{1}_n$ is the unit matrix of order n, but because $\mathbf{X}$ is composed of n orthogonal and non-null vectors, it is of full rank and invertible. Thus from equation (10.4) by left multiplication with $\mathbf{X}$ we get

$$\mathbf{XX}^{\mathrm{T}}\mathbf{X} = \mathbf{X} \qquad (10.5)$$

so that right multiplication with $\mathbf{X}^{-1}$ gives

$$\mathbf{XX}^{\mathrm{T}}\mathbf{XX}^{-1} = \mathbf{XX}^{\mathrm{T}} = \mathbf{XX}^{-1} = \mathbf{1}_n \qquad (10.6)$$

showing that $\mathbf{X}^{\mathrm{T}}$ is the inverse $\mathbf{X}^{-1}$ and that $\mathbf{X}$ is an orthogonal matrix.
(iii) If the matrix $\mathbf{A}$ is not only symmetric, so

$$\mathbf{A}^{\mathrm{T}} = \mathbf{A} \qquad (10.7)$$

but also positive definite (see §7.1, p 71), then from the singular-value decomposition

$$\mathbf{A} = \mathbf{USV}^{\mathrm{T}} \qquad (2.53)$$

the eigenvalues of **A** are found as the diagonal elements S_{ii}, $i = 1, 2, \ldots, n$, and the matrices **U** and **V** both contain complete sets of eigenvectors. These sets (the columns of each matrix) are not necessarily identical since, if any two eigenvalues are equal (also denoted as being degenerate or of multiplicity greater than one), any linear combination of their corresponding eigenvectors is also an eigenvector of the same eigenvalue. If the original two eigenvectors are orthogonal, then orthogonal linear combinations are easily generated by means of orthogonal matrices such as the plane rotations of §3.3. This is an important point to keep in mind; recently a computer service bureau wasted much time and money trying to find the 'bug' in a program taken from a Univac 1108 to an IBM 370 because the eigenvectors corresponding to degenerate eigenvalues were computed differently by the two computers.

Property (iii) above will now be demonstrated. First note that the eigenvalues of a symmetric positive definite matrix are positive. For, from (7.9) and equation (10.1), we have

$$0 < \mathbf{y}^{\mathrm{T}}\mathbf{A}\mathbf{y} = \mathbf{y}^{\mathrm{T}}\mathbf{X}\mathbf{E}\mathbf{X}^{\mathrm{T}}\mathbf{y}$$

$$= \boldsymbol{w}^{\mathrm{T}}\mathbf{E}\boldsymbol{w} = \sum_{j=1}^{n} e_j w_j^2 \tag{10.8}$$

thus implying that all the e_j must be positive or else a vector $\boldsymbol{w} = \mathbf{X}^{\mathrm{T}}\mathbf{y}$ could be devised such that $w_j = 1$, $w_i = 0$ for $i \neq j$ corresponding to the non-positive eigenvalue, thus violating the condition (7.9) for definiteness. Hereafter, **E** and **S** will be ordered so that

$$S_i \geqslant S_{i+1} > 0 \tag{10.9}$$

$$e_i \geqslant e_{i+1}. \tag{10.10}$$

This enables **S** and **E** to be used interchangably once their equivalence has been demonstrated.

Now consider pre-multiplying equation (10.1) by **A**. Thus we obtain

$$\mathbf{A}^2\mathbf{X} = \mathbf{AAX} = \mathbf{AXE} = \mathbf{XEE} = \mathbf{XE}^2 \tag{10.11}$$

while from symmetry (10.7) and the decomposition (2.53)

$$\mathbf{A}^2\mathbf{V} = \mathbf{A}^{\mathrm{T}}\mathbf{AV} = \mathbf{VS}^2. \tag{10.12}$$

Since (10.11) and (10.12) are both eigenvalue equations for $\mathbf{A}^2$, $\mathbf{S}^2$ and $\mathbf{E}^2$ are identical to within ordering, and since all e_i are positive, the orderings (10.9) and (10.10) imply

$$\mathbf{S} = \mathbf{E}. \tag{10.13}$$

Now it is necessary to show that

$$\mathbf{AV} = \mathbf{VS}. \tag{10.14}$$

From (10.1), letting $\mathbf{Q} = \mathbf{X}^{\mathrm{T}}\mathbf{V}$, we obtain

$$\mathbf{AV} = \mathbf{XEX}^{\mathrm{T}}\mathbf{V} = \mathbf{XEQ} = \mathbf{XSQ}. \tag{10.15}$$

However, from (10.11) and (10.12), we get

$$\mathbf{QS}^2 = \mathbf{S}^2\mathbf{Q}. \tag{10.16}$$

Explicit analysis of the elements of equation (10.16) shows that (*a*) if $S_{ii} \neq S_{jj}$, then $Q_{ij} = 0$, and (*b*) the commutation

$$\mathbf{QS} = \mathbf{SQ} \tag{10.17}$$

is true even in the degenerate eigenvalue case; thus,

$$\mathbf{AV} = \mathbf{XSQ} = \mathbf{XQS} = \mathbf{XX}^{\mathrm{T}}\mathbf{VS} = \mathbf{VS}. \tag{10.18}$$

The corresponding result for **U** is shown in the same fashion.

10.2. EXTENSION TO MATRICES WHICH ARE NOT POSITIVE DEFINITE

The above result shows that if **A** is a symmetric positive definite matrix, its eigensolutions can be found via the singular-value decomposition algorithms 1 and 4. Moreover, only one of the matrices **U** or **V** is required, so that the eigenvectors overwrite the original matrix. Furthermore, the algorithm 1, for example, generates the matrix $\mathbf{B} = \mathbf{US}$ in performing an orthogonalisation of the columns of **A**, so that the eigenvalues are kept implicitly as the norms of the columns of the orthogonalised matrix and a separate vector to store the eigenvalues is not required.

What of the case where the matrix **A** is not positive definite? This is hardly any extra trouble, since the matrix

$$\mathbf{A}' = \mathbf{A} - h\mathbf{1}_n \tag{10.19}$$

has the same eigenvectors as **A** (as proved by direct substitution in equation (10.1)) and has eigenvalues

$$E'_{ii} = E_{ii} - h \qquad \text{for } i = 1, 2, \ldots, n \tag{10.20}$$

where E_{ii}, $i = 1, 2, \ldots, n$, are the eigenvalues of **A**. Thus it is always possible to generate a positive definite matrix from **A** by adding an appropriate constant $-h$ to each of its diagonal elements. Furthermore, it is very simple to compute an appropriate value for h from Gerschgorin's theorem (Ralston 1965, p 467), which states that all the eigenvalues of a matrix **A** (a general square matrix, real or complex) are found in the domain which is the union of the circles having centres A_{ii}, $i = 1, 2, \ldots, n$, and respective radii

$$r_i = \sum_{\substack{j=1 \\ j \neq i}}^{n} |A_{ij}|. \tag{10.21}$$

Because a symmetric matrix has real eigenvalues this implies

$$\begin{aligned} E_{nn} \geqslant E &= \min(A_{ii} - r_i) \\ &= \min\left(A_{ii} - \sum_{\substack{j=1 \\ j \neq i}}^{n} |A_{ij}|\right). \end{aligned} \tag{10.22}$$

If $E > 0$, the matrix **A** is positive definite; otherwise a shift equal to E will make it so. Thus we can define

$$h = 0 \qquad \text{for } E > \varepsilon \tag{10.23a}$$

$$h = -(|E| + \varepsilon^{1/2}) = E - \varepsilon^{1/2} \qquad \text{for } E \leqslant \varepsilon \tag{10.23b}$$

to ensure a positive definite matrix $\mathbf{A}'$ results from the shift (10.19). The machine precision ε is used simply to take care of those situations, such as a matrix with a null row (and column), where the lower bound E is in fact a small eigenvalue.

Unfortunately, the accuracy of eigensolutions computed via this procedure is sensitive to the shift. For instance, the largest residual element R, that is, the element of largest magnitude in the matrix

$$\mathbf{AX} - \mathbf{XE} \tag{10.24}$$

and the largest inner product P, that is, the off-diagonal element of largest magnitude in the matrix

$$\mathbf{X}^{\mathrm{T}}\mathbf{X} - \mathbf{1}_n \tag{10.25}$$

for the order-10 Ding Dong matrix (see appendix 1) are: for $h = -3{\cdot}57509$, $R = 5{\cdot}36442\text{E}-6$ and $P = 1{\cdot}24425\text{E}-6$ while for $h = -10{\cdot}7238$, $R = 1{\cdot}49012\text{E}-5$ and $P = 2{\cdot}16812\text{E}-6$. These figures were computed on a Data General NOVA (23-bit binary arithmetic) using single-length arithmetic throughout as no extended precision was available. The latter shift was obtained using

$$h = \begin{cases} 0 & \text{for } E > 0 \qquad (10.26a) \\ 3E - \varepsilon & \text{for } E < 0. \qquad (10.26b) \end{cases}$$

In general, in a test employing all nine test matrices from appendix 1 of order 4 and order 10, the shift defined by formulae (10.23) gave smaller residuals and inner products than the shift (10.26). The eigenvalues used in the above examples were computed via the Rayleigh quotient

$$E_{jj} \simeq Q_j = \boldsymbol{x}_j^{\mathrm{T}}\mathbf{A}\boldsymbol{x}_j / \boldsymbol{x}_j^{\mathrm{T}}\boldsymbol{x}_j \tag{10.27}$$

rather than the singular value, that is, equation (10.20). In the tests mentioned above, eigenvalues computed via the Rayleigh quotient gave smaller residuals than those found merely by adding on the shift. This is hardly surprising if the nature of the problem is considered. Suppose that the true eigenvectors are $\boldsymbol{\phi}_i$, $i = 1, 2, \ldots, n$. Let us add a component $c\boldsymbol{w}$ to $\boldsymbol{\phi}_j$, where $\boldsymbol{w}$ is some normalised combination of the $\boldsymbol{\phi}_i$, $i \neq j$, and c measures the size of the component (error); the normalised approximation to the eigenvector is then

$$\boldsymbol{x}_j = (1 + c^2)^{-1/2}(\boldsymbol{\phi}_j + c\boldsymbol{w}). \tag{10.28}$$

The norm of the deviation $(\boldsymbol{x}_j - \boldsymbol{\phi}_j)$ is found, using the binomial expansion and ignoring terms in c^4 and higher powers relative to those in c^2, to be approximately equal to c. The Rayleigh quotient corresponding to the vector given by (10.28) is

$$Q_j = (E_{jj} + c^2\boldsymbol{w}^{\mathrm{T}}\mathbf{A}\boldsymbol{w})/(1 + c^2) \tag{10.29}$$

since $\boldsymbol{\phi}_j^{\mathrm{T}}\mathbf{A}\boldsymbol{w}$ is zero by virtue of the orthogonality of the eigenvectors. The deviation of Q_j from the eigenvalue is

TABLE 10.1. Maximum absolute residual element R and maximum absolute inner product P between normalised eigenvectors for eigensolutions of order $n = 10$ real symmetric matrices. All programs in BASIC on a Data General NOVA. Machine precision $= 2^{-22}$.

Matrix		Rutishauser Jacobi	Rutishauser with Nash formulae	Algorithm 14 type: Jacobi which orders	Algorithm 14 type: Jacobi using symmetry	Algorithm 13 type: with equation (10.27)	Algorithm 13 type: with equation (10.20)
Hilbert	R	7·26E−7	5·76E−6	4·82E−6	6·29E−6	6·68E−6	7·15E−6
	P	0	8·64E−7	1·13E−6	1·10E−6	2·32E−6	2·32E−6
Ding Dong	R	2·32E−6	2·86E−6	8·86E−6	1·08E−5	5·36E−6	1·54E−5
	P	0	5·36E−7	1·43E−6	1·19E−6	1·24E−6	1·24E−6
Moler	R	1·74E−5	3·62E−5	6·34E−5	1·01E−4	3·91E−5	9·46E−5
	P	1·94E−7	8·64E−7	8·05E−7	8·94E−7	2·21E−6	2·21E−6
Frank	R	2·29E−5	5·53E−5	8·96E−5	1·25E−4	5·72E−5	9·72E−5
	P	2·09E−7	6·85E−7	1·07E−6	8·57E−7	1·66E−6	1·66E−6
Bordered	R	1·79E−6	1·91E−6	6·20E−6	2·05E−5	1·43E−6	1·91E−6
	P	5·34E−9	5·96E−7	9·98E−7	1·40E−6	5·54E−7	5·54E−7
Diagonal	R	0	0	0	0	0	0
	P	0	0	0	0	0	0
W+	R	2·32E−6	4·59E−6	2·45E−5	2·01E−5	9·16E−6	1·43E−5
	P	1·79E−6	1·26E−6	1·88E−6	1·91E−6	1·75E−6	1·75E−6
W−	R	1·94E−6	8·58E−6	1·63E−5	2·86E−5	1·35E−5	2·00E−5
	P	4·77E−7	6·26E−7	7·97E−7	5·41E−7	2·10E−6	2·10E−6
Ones	R	4·65E−6	1·06E−6	1·06E−5	5·05E−5	2·43E−5	1·19E−5
	P	0	3·65E−7	9·92E−7	1·04E−3	9·89E−7	9·89E−7

$$(E_{jj} - Q_j) = c^2(E_{jj} - \mathbf{w}^{\mathrm{T}}\mathbf{A}\mathbf{w})/(1 + c^2). \tag{10.30}$$

Thus the error has been squared in the sense that the deviation of $\boldsymbol{x}_j$ from $\boldsymbol{\phi}_j$ is of order c, while that of Q_j from E_{jj} is of order c^2. Since c is less than unity, this implies that the Rayleigh quotient is in some way 'closer' to the eigenvalue than the vector is to an eigenvector.

Unfortunately, to take advantage of the Rayleigh quotient (and residual calculation) it is necessary to keep a copy of the original matrix in the memory or perform some backing store manipulations. A comparison of results for algorithm 13 using formulae (10.20) and (10.27) are given in table 10.1.

Algorithm 13. Eigensolutions of a real symmetric matrix via the singular-value decomposition

```
Procedure evsvd(n: integer; {order of matrix eigenproblem}
                var A,V : rmatrix; {matrix and eigenvectors}
                initev: boolean; {switch -- if TRUE eigenvectors
                are initialized to a unit matrix of order n}
```

Algorithm 13. Eigensolutions of a real symmetric matrix via the singular-value decomposition (cont.)

```
                              W : wmatrix; {working array}
                              var Z: rvector); {to return the eigenvalues}
{alg13.pas ==
        eigensolutions of a real symmetric matrix via the singular value
        decomposition by shifting eigenvalues to form a positive definite
        matrix.
        This algorithm replaces Algorithm 13 in the first edition of Compact
        Numerical Methods.
                              Copyright 1988 J.C.Nash
}
var
    count, i, j, k, limit, skipped : integer;
    c, p, q, s, shift, t : real ; {rotation angle quantities}
    oki, okj, rotn : boolean;
    ch : char;
begin
    writeln('alg13.pas -- symmetric matrix eigensolutions via svd');
    {Use Gerschgorin disks to approximate the shift. This version
        calculates only a positive shift.}
    shift:=0.0;
    for i:=1 to n do
    begin
        t:=A[i,i];
        for j:=1 to n do
            if i<>j then t:=t-abs(A[i,j]);
        if t<shift then shift:=t; {looking for lowest bound to eigenvalue}
    end; {loop over rows}
    shift:=-shift; {change sign, since current value < 0 if useful}
    if shift<0.0 then shift:=0.0;
    writeln('Adding a shift of ',shift,' to diagonal of matrix.');
    for i:=1 to n do
    begin
        for j:=1 to n do
        begin
            W[i,j]:=A[i,j]; {copy matrix to working array}
            if i=j then W[i,i]:=A[i,i]+shift; {adding shift in process}
            if initev then {initialize eigenvector matrix}
            begin
                if i=j then W[i+n,i]:=0.0
                else
                begin
                    W[i+n,j]:=0.0;
                end;
            end; {eigenvector initialization}
        end; {loop on j}
    end; {loop on i}
    NashSVD(n, n, W, Z); {call alg01 to do the work}
    for i:=1 to n do
    begin
        Z[i]:=sqrt(Z[i])-shift; {to adjust eigenvalues}
        for j:=1 to n do
            V[i,j]:=W[n+i,j]; {extract eivenvectors}
    end; {loop on i}
end; {alg13.pas == evsvd}
```

Example 10.1. Principal axes of a cube

Consider a cube of uniform density, mass m and edge length a situated so that three of its edges which meet at a vertex form the coordinate axes x, y, z. The moments and products of inertia (see, for instance, Synge and Griffith 1959, pp 282–93) for the cube in this coordinate frame are

$$I_{xx} = I_{yy} = I_{zz} = 2ma^2/3$$
$$I_{xy} = I_{xz} = I_{yz} = -ma^2/4.$$

These can be formed into the moment-of-inertia tensor

$$\begin{bmatrix} I_{xx} & -I_{xy} & -I_{xz} \\ -I_{xy} & I_{yy} & -I_{yz} \\ -I_{xz} & -I_{yz} & I_{zz} \end{bmatrix} = \begin{bmatrix} 8 & -3 & -3 \\ -3 & 8 & -3 \\ -3 & -3 & 8 \end{bmatrix}$$

where all the elements have been measured in units $12/(ma^2)$.

Algorithm 13 can be used to find the eigensolutions of this matrix. The eigenvalues then give the principal moments of inertia and the eigenvectors give the principal axes to which these moments apply. Algorithm 13 when used on a Data General NOVA operating in 23-bit binary arithmetic finds

$$I_1 = 2ma^2/12 \qquad \boldsymbol{v}_1 = (0{\cdot}57735,\ 0{\cdot}577351,\ 0{\cdot}57735)^{\mathrm{T}}$$
$$I_2 = 11ma^2/12 \qquad \boldsymbol{v}_2 = (0{\cdot}707107,\ -0{\cdot}707107, -4{\cdot}33488\mathrm{E}-8)^{\mathrm{T}}$$
$$I_3 = 11ma^2/12 \qquad \boldsymbol{v}_3 = (0{\cdot}408248,\ 0{\cdot}408248,\ -0{\cdot}816496)^{\mathrm{T}}.$$

(The maximum absolute residual was 3·8147E−6, the maximum inner product 4·4226E−7.) The last two principal moments of inertia are the same or *degenerate*. Thus any linear combination of $\boldsymbol{v}_2$ and $\boldsymbol{v}_3$ will give a new vector

$$\boldsymbol{v}_2' = \alpha\boldsymbol{v}_2 + (1-\alpha^2)^{1/2}\boldsymbol{v}_3$$

which forms a new set of principal axes with $\boldsymbol{v}_1$ and

$$\boldsymbol{v}_3' = (1-\alpha^2)^{1/2}\boldsymbol{v}_2 - \alpha\boldsymbol{v}_3$$

which is orthogonal to $\boldsymbol{v}_2'$. Indeed algorithm 14 on the same system found

$$I_1 = 2ma^2/12 \qquad \boldsymbol{v}_1 = (0{\cdot}57735,\ 0{\cdot}57735,\ 0{\cdot}57735)^{\mathrm{T}}$$
$$I_2 = 11ma^2/12 \qquad \boldsymbol{v}_2 = (0{\cdot}732793,\ -0{\cdot}678262,\ -5{\cdot}45312\mathrm{E}-2)^{\mathrm{T}}$$
$$I_3 = 11ma^2/12 \qquad \boldsymbol{v}_3 = (0{\cdot}360111,\ 0{\cdot}454562,\ -0{\cdot}814674)^{\mathrm{T}}$$

with a maximum absolute residual of 7·62939E−6 and a maximum inner product 2·38419E−7.

10.3. THE JACOBI ALGORITHM FOR THE EIGENSOLUTIONS OF A REAL SYMMETRIC MATRIX

Equation (10.1) immediately leads to

$$\mathbf{V}^{\mathrm{T}}\mathbf{A}\mathbf{V}=\mathbf{E} \tag{10.31}$$

(using $\mathbf{V}$ in place of $\mathbf{X}$). The fact that a real symmetric matrix can be *diagonalised* by its eigenvectors gives rise to a number of approaches to the algebraic eigenvalue problem for such matrices. One of the earliest of these was suggested by Jacobi (1846). This proposes the formation of the sequence of matrices

$$\begin{aligned}\mathbf{A}^{(0)}&=\mathbf{A}\\ \mathbf{A}^{(k+1)}&=(\mathbf{V}^{(k)})^{\mathrm{T}}\mathbf{A}^{(k)}\mathbf{V}^{(k)}\end{aligned} \tag{10.32}$$

where the $\mathbf{V}^{(k)}$ are the plane rotations introduced in §3.3. The limit of the sequence is a diagonal matrix under some conditions on the angles of rotation. Each rotation is chosen to set one off-diagonal element of the matrix $\mathbf{A}^{(k)}$ to zero. In general an element made zero by one rotation will be made non-zero by another so that a series of *sweeps* through the off-diagonal elements are needed to reduce the matrix to diagonal form. Note that the rotations in equation (10.32) preserve symmetry, so that there are $n(n-1)/2$ rotations in one sweep if $\mathbf{A}$ is of order n.

Consider now the effect of a single rotation, equation (3.11), in the ij plane. Then for $m\neq i, j$

$$A_{mi}^{(k+1)}=A_{mi}^{(k)}\cos\phi+A_{mj}^{(k)}\sin\phi=A_{im}^{(k+1)} \tag{10.33}$$

$$A_{mj}^{(k+1)}=-A_{mi}^{(k)}\sin\phi+A_{mj}^{(k)}\cos\phi=A_{jm}^{(k+1)} \tag{10.34}$$

while

$$A_{ii}^{(k+1)}=A_{ii}^{(k)}\cos^2\phi+2A_{ij}^{(k)}\cos\phi\sin\phi+A_{jj}^{(k)}\sin^2\phi \tag{10.35}$$

$$A_{jj}^{(k+1)}=A_{ii}^{(k)}\cos^2\phi-2A_{ij}^{(k)}\cos\phi\sin\phi+A_{jj}^{(k)}\sin^2\phi \tag{10.36}$$

$$A_{ij}^{(k+1)}=A_{ji}^{(k+1)}=-(A_{ii}^{(k)}-A_{jj}^{(k)})\cos\phi\sin\phi+A_{ij}^{(k)}(\cos^2\phi-\sin^2\phi). \tag{10.37}$$

By allowing

$$p=A_{ij}^{(k)} \tag{10.38}$$

and

$$q=A_{ii}^{(k)}-A_{jj}^{(k)} \tag{10.39}$$

the angle calculation defined by equations (3.22)–(3.27) will cause $A_{ij}^{(k+1)}$ to be zero. By letting

$$Z^{(k)}=\sum_{i=1}^{n-1}\sum_{j=i+1}^{n}(A_{ij}^{(k)})^2 \tag{10.40}$$

be the measure of the non-diagonal character of $\mathbf{A}^{(k)}$ in a similar fashion to the non-orthogonality measure (3.17), it is straightforward (if a little tedious) to show that

$$Z^{(k+1)} = Z^{(k)} - (A_{ij}^{(k)})^2 \qquad (10.41)$$

so that each rotation causes $\mathbf{A}^{(k+1)}$ to be 'more diagonal' than $\mathbf{A}^{(k)}$.

Specification of the order in which the off-diagonal elements are to be made zero defines a particular variant of the Jacobi algorithm. One of the simplest is to take the elements in row-wise fashion: (1, 2), (1, 3), . . . , (1, n), (2, 3), (2, 4), . . . , (2, n), . . . , ($n-1$, n). Such *cyclic Jacobi algorithms* have not been proved to converge in general, except in the case where the angle of rotation ϕ is constrained so that

$$-\pi/4 \leqslant \phi \leqslant \pi/4. \qquad (10.42)$$

Similarly to the orthogonalisation algorithm of §3.3, the difficulty lies in demonstrating that the ordering of the diagonal elements is stable. For ϕ restricted as in (10.42), Forsythe and Henrici (1960) have carried out the complicated proof, and most authors (Wilkinson 1965, Rutishauser 1966, Schwarz *et al* 1973, Ralston 1965) discuss algorithms using angle calculations which satisfy the restriction (10.42). In fact, among the variety of texts available on numerical linear algebra, the author has noted only one (Fröberg 1965) which uses the calculation based on equations (10.38), (10.39) and (3.22)–(3.27). The advantage of the calculation given here is that the eigenvalues, which are produced as the diagonal elements of the matrix that is the limit of the sequence $\mathbf{A}^{(k)}$, are ordered from most positive to most negative. Most applications which require eigensolutions also require them ordered in some way and the ordering that the current process yields is the one I have almost invariably been required to produce. No extra program code is therefore needed to order the eigenvalues and eigenvectors and there may be some savings in execution time as well.

We have already noted in chapter 3 the research of Booker (1985) relating to the convergence of the cyclic Jacobi method with the ordering angle calculation.

Jacobi (1846) did not use a cyclic pattern, but chose to zero the largest off-diagonal element in the current matrix. This process has generally been considered inappropriate for automatic computation due to the time required for the search before each plane rotation. However, for comparison purposes I modified a BASIC precursor of algorithm 14. The changes made were only as many as were needed to obtain the zeroing of the largest off-diagonal element in the present matrix, and no attempt was made to remove computations present in the algorithm which were used to provide convergence test information. The processor time required for the order-5 Moler matrix (appendix 1) on a Data General NOVA operating in six hexadecimal digit arithmetic was 15·7 seconds for algorithm 14 while the Jacobi method required 13·7 seconds. Indeed the latter required only 30 rotations while algorithm 14 took 5 sweeps (up to 50 rotations). The comparison of timings on this one example may be misleading, especially as the system uses an interpreter, that is, programs are executed by translating the BASIC at run time instead of compiling it first. (This has an analogy in the translation of a speech either in total or simultaneously.) However, for matrices with only a few large off-diagonal elements, the relatively simple changes to algorithm 14 to perform the search are probably worthwhile. In a larger set of timings run on both a Data General ECLIPSE and an IBM/370 model 168 in six hexadecimal digit arithmetic, the 'original Jacobi' method was on average the slowest of five methods tested. (Note that the convergence test at STEP 17 below cannot be used.)

10.4. ORGANISATION OF THE JACOBI ALGORITHM

To reduce program code length, the following procedure performs left and right multiplications on the matrix $\mathbf{A}^{(k)}$ separately, rather than use the formulae (10.33)–(10.37). This may cause some slight loss in accuracy of the computed results (compared, for instance, to the procedure *jacobi* discussed in §10.5).

Convergence testing is a relatively complicated matter. In the following algorithm, convergence is assessed by counting the number of rotations skipped during one sweep of the matrix because the rotation angle is too small to have any effect on the current matrix. Rotations needed to order the diagonal elements of the matrix (hence the eigenvalues) are always performed. The test that the sine of the rotation angle is smaller than the machine precision is used to decide (at STEP 10) if the rotation is to be carried out when the diagonal elements are in order. Unfortunately, if two diagonal elements are close in value, then the rotation angle may not be small even if the off-diagonal element to be set to zero is quite small, so that it is unnecessary to perform the rotation. Thus at STEP 7, the algorithm tests to see if the off-diagonal element has a magnitude which when multiplied by 100 is incapable of adjusting the diagonal elements, in which case the rotation is skipped. A safety check is provided on the number of sweeps which are carried out since it does not seem possible to guarantee the convergence of the algorithm to the satisfaction of the above tests. Even if the algorithm terminates after 30 sweeps (my arbitrary choice for the safety check limit) the approximations to the eigenvalues and eigenvectors may still be good, and it is recommended that they be checked by computing residuals and other tests.

Algorithm 14. A Jacobi algorithm for eigensolutions of a real symmetric matrix

```
Procedure evJacobi(n: integer; {order of matrices}
                        var A,V : rmatrix; {matrix and eigenvector array}
                        initev: boolean); {flag to initialize eigenvector
                        array to a unit matrix if TRUE}
{alg14.pas ==
        a variant of the Jacobi algorithm for computing eigensolutions of a
        real symmetric matrix
        n is the order of the eigenproblem
        A is the real symmetric matrix in full
        V will be rotated by the Jacobi transformations.
        initev is TRUE if we wish this procedure to initialize
            V to a unit matrix before commencing; otherwise,
            V is assumed to be initialized outside the procedure,
            e.g. for computing the eigenvectors of a generalized
            eigenproblem as in alg15.pas.
                        Copyright 1988 J.C.Nash
```

Algorithm 14. A Jacobi algorithm for eigensolutions of a real symmetric matrix (cont.)

```
}
{STEP 0 -- via the calling sequence of the procedure, we supply the matrix
and its dimensions to the program.}
var
    count, i, j, k, limit, skipped : integer;
    c, p, q, s, t : real;
    ch : char;
    oki, okj, rotn : boolean;
begin
    writeln('alg14.pas -- eigensolutions of a real symmetric');
    writeln(' matrix via a Jacobi method');
    if initev then {Do we initialize the eigenvectors to columns of
                    the identity?}
    begin
        for i := 1 to n do
        begin
            for j := 1 to n do V[i,j] := 0.0;
            V[i,i] := 1.0; {to set V to a unit matrix -- rotated to become
                    the eigenvectors}
        end; {loop on i;}
    end; {initialize eigenvectors}
    count := 0;
    limit := 30; {an arbitrary choice following lead of Eberlein}
    skipped := 0; {so far no rotations have been skipped. We need to set
            skipped here because the while loop below tests this variable.}
    {main loop}
    while (count<=limit) and (skipped<((n*(n-1)) div 2) ) do
    {This is a safety check to avoid indefinite execution of the algorithm.
    The figure used for limit here is arbitrary. If the program terminates
    by exceeding the sweep limit, the eigenvalues and eigenvectors computed
    may still be good to the limitations of the machine in use, though
    residuals should be calculated and other tests made. Such tests are
    always useful, though they are time- and space-consuming.}
    begin
        count := count+1; {to count sweeps -- STEP 1}
        write('sweep ',count,' ');
        skipped := 0; {to count rotations skipped during the sweep.}
        for i := 1 to (n-1) do {STEP 2}
        begin {STEP 3}
            for j := (i+1) to n do {STEP 4}
            begin
                rotn := true; {to indicate that we carry out a rotation unless
                        calculations show it unnecessary}
                p := 0.5*(A[i,j]+A[j,i]); {An average of the off-diagonal elements
                    is used because the right and left multiplications by the
                    rotation matrices are non-commutative in most cases due to
                    differences between floating-point and exact arithmetic.}
                q := A[i,i]-A[j,j]; {Note: this may suffer from digit cancellation
                    when nearly equal eigenvalues exist. This cancellation is not
                    normally a problem, but may cause the algorithm to perform more
                    work than necessary when the off-diagonal elements are very
                    small.}
```

Algorithm 14. A Jacobi algorithm for eigensolutions of a real symmetric matrix (cont.)

```
          t := sqrt(4.0*p*p+q*q);
          if t=0.0 then {STEP 5}
          begin {STEP 11 -- If t is zero, no rotation is needed.}
            rotn := false; {to indicate no rotation is to be performed.}
          end
          else
          begin {t>0.0}
            if q>=0.0 then {STEP 6}
            begin {rotation for eigenvalue approximations already in order}
              {STEP 7 -- test for small rotation}
              oki := (abs(A[i,i])=abs(A[i,i])+100.0*abs(p));
              okj := (abs(A[j,j])=abs(A[j,j])+100.0*abs(p));
              if oki and okj then rotn := false
              else rotn := true;
              {This test for a small rotation uses an arbitrary factor of
              100 for scaling the off-diagonal elements. It is chosen to
              ensure "small but not very small" rotations are performed.}
              if rotn then
              begin {STEP 8}
              c := sqrt((t+q)/(2.0*t)); s := p/(t*c);
              end;
            end {if q>=0.0}
            else
            begin {q<0.0 -- always rotate to bring eigenvalues into order}
              rotn := true; {STEP 9}
              s := sqrt((t-q)/(2.0*t));
              if p<0.0 then s := -s;
              c := p/(t*s);
            end; {STEP 10}
            if 1.0=(1.0+abs(s)) then rotn := false; {test for small angle}
          end; {if t=0.0}
          if rotn then {STEP 11 -- rotate if necessary}
          begin {STEP 12}
            for k := 1 to n do
            begin
              q := A[i,k]; A[i,k] := c*q+s*A[j,k]; A[j,k] := -s*q+c*A[j,k];
            end; {left multiplication of matrix A}
            {STEP 13}
            for k := 1 to n do
            begin {right multiplication of A and V}
              q := A[k,i]; A[k,i] := c*q+s*A[k,j]; A[k,j] := -s*q+c*A[k,j];
              {STEP 14 -- can be omitted if eigenvectors not needed}
              q := V[k,i]; V[k,i] := c*q+s*V[k,j]; V[k,j] := -s*q+c*V[k,j];
            end; {loop on k for right multiplication of matrices A and V}
          end {rotation carried out}
          else
          {STEP 11 -- count skipped rotations}
              skipped := skipped+1; {to count the skipped rotations}
        end; {loop on j} {STEP 15}
      end; {loop on i. This is also the end of the sweep. -- STEP 16}
      writeln(' ',skipped,' / ',n*(n-1) div 2,' rotations skipped');
    end; {while -- main loop}
end; {alg14.pas == evJacobi -- STEP 17}
```

Example 10.2. Application of the Jacobi algorithm in celestial mechanics

It is appropriate to illustrate the use of algorithm 14 by Jacobi's (1846) own example. This arises in the study of orbital perturbations of the planets to compute corrections to some of the parameters of the solar system. Unfortunately at the time Jacobi was writing his paper, Neptune had not been discovered. Leverrier reported calculations suggesting the existence of this planet on 31 August 1846, to *l'Académie des Sciences* in Paris, and Galle in Berlin confirmed this hypothesis by observation less than three weeks later on 18 September. The derivation of the eigenproblem in this particular case is lengthy and irrelevant to the present illustration, so we will begin with Jacobi's equations V. These give a non-symmetric matrix $\tilde{\mathbf{A}}$ which can be symmetrised by a diagonal transformation resulting in Jacobi's equations VIII, where the off-diagonal elements are expressed in terms of their common logarithms to which 10 has been added. I decided to work with the non-symmetric form and symmetrised it by means of

$$A_{ij} = A_{ji} = (\tilde{A}_{ij}\tilde{A}_{ji})^{1/2}.$$

The output from a Hewlett–Packard 9830 (machine precision = 1E−11) is given below, and includes the logarithmic elements which in every case approximated very closely Jacobi's equations VIII. For comparison, he computed eigenvalues −2·2584562, −3·7151584, −5·2986987, −7·5740431, −17·1524687, −17·8632192 and −22·4267712 after 10 rotations. At this point the largest off-diagonal element (which is marked as being negative) had a logarithm (8·8528628 − 10), which has the approximate antilog −7·124E−2. Jacobi used as a computing system his student Ludwig Seidel, apparently operating in eight-digit decimal arithmetic!

```
ENHJCB JACOBI WITH ORDERING MAR 5 75
ORDER= 7
INPUT JACOBI'S MATRIX
ROW 1     :
-5.509882     1.870086     0.422908     8.81400E-03     0.148711
 3.90800E-03     4.50000E-05
ROW 2     :
 0.287865    -11.811654     5.7119     0.058717     0.728088     0.018788
 2.24000E-04
ROW 3     :
 0.049099     4.308033    -12.970687     0.229326     1.689087     0.04258
 5.04000E-04
ROW 4     :
 6.23500E-03     0.269851     1.397369    -17.596207     5.304038     0.125346
 1.45100E-03
ROW 5     :
 2.23100E-05     7.09480E-04     2.18227E-03     1.12462E-03    -7.489041
 4.815454     0.035319
ROW 6     :
 1.45000E-06     4.52200E-05     1.35880E-04     6.56500E-05     11.893979
-18.58541     0.232241
ROW 7     :
 6.00000E-08     1.94000E-06     5.79000E-06     2.73000E-06     0.313829
 0.835482    -2.325935

SYMMETRIZE A(I,J)=A(J,I):=SQR(A(I,J)*A(J,I))
LOG(S) GIVEN FOR COMPARISON WITH JACOBI'S TABLE VIII
S=A( 1     , 2     )= 0.733711324     LOG10(S)+10= 9.865525222
S=A( 1     , 3     )= 0.144098438     LOG10(S)+10= 9.158659274
```

```
S=A( 1    , 4    )= 7.41318E-03      LOG10(S)+10= 7.870004752
S=A( 1    , 5    )= 1.82147E-03      LOG10(S)+10= 7.260421332
S=A( 1    , 6    )= 7.52768E-05      LOG10(S)+10= 5.876661279
S=A( 1    , 7    )= 1.64317E-06      LOG10(S)+10= 4.215681882
S=A( 2    , 3    )= 4.960549737      LOG10(S)+10= 10.6955298
S=A( 2    , 4    )= 0.125876293      LOG10(S)+10= 9.099943945
S=A( 2    , 5    )= 0.022728042      LOG10(S)+10= 8.356562015
S=A( 2    , 6    )= 9.21734E-04      LOG10(S)+10= 6.964605555
S=A( 2    , 7    )= 2.08461E-05      LOG10(S)+10= 5.319024874
S=A( 3    , 4    )= 0.566085721      LOG10(S)+10= 9.7528822
S=A( 3    , 5    )= 0.060712798      LOG10(S)+10= 8.78328025
S=A( 3    , 6    )= 2.40536E-03      LOG10(S)+10= 7.381180598
S=A( 3    , 7    )= 5.40200E-05      LOG10(S)+10= 5.73255455
S=A( 4    , 5    )= 0.077233589      LOG10(S)+10= 8.887806215
S=A( 4    , 6    )= 2.86862E-03      LOG10(S)+10= 7.457672605
S=A( 4    , 7    )= 6.29383E-05      LOG10(S)+10= 5.798915030
S=A( 5    , 6    )= 7.568018813      LOG10(S)+10= 10.8789822
S=A( 5    , 7    )= 0.105281178      LOG10(S)+10= 9.022350736
S=A( 6    , 7    )= 0.440491969      LOG10(S)+10= 9.643937995

MATRIX
ROW 1    :
-5.509882     0.733711324    0.144098438    7.41318E-03    1.82147E-03
 7.52768E-05    1.64317E-06
ROW 2    :
 0.733711324   -11.811654    4.960549737    0.125876293    0.022728042
 9.21734E-04    2.08461E-05
ROW 3    :
 0.144098438    4.960549737   -12.970687    0.566085721    0.060712798
 2.40536E-03    5.40200E-05
ROW 4    :
 7.41318E-03    0.125876293    0.566085721   -17.596207    0.077233589
 2.86862E-03    6.29383E-05
ROW 5    :
 1.82147E-03    0.022728042    0.060712798    0.077233589   -7.489041
 7.568018813    0.105281178
ROW 6    :
 7.52768E-05    9.21734E-04    2.40536E-03    2.86862E-03    7.568018813
-18.58541     0.440491969
ROW 7    :
 1.64317E-06    2.08461E-05    5.40200E-05    6.29383E-05    0.105281178
 0.440491969   -2.325935

NASH JACOBI ALG. 14 DEC 13/77
 0      ROTATIONS SKIPPED
 0      ROTATIONS SKIPPED
 0      ROTATIONS SKIPPED
 5      ROTATIONS SKIPPED
 19     ROTATIONS SKIPPED
 21     ROTATIONS SKIPPED
CONVERGED

EIGENVALUE 1    =-2.258417596    VECTOR:
 4.90537E-04    1.37576E-03    1.72184E-03    9.85037E-04    0.175902256
 0.107934624    0.978469440
EIGENVALUE 2    =-3.713643588    VECTOR:
 6.13203E-03    0.010411191    0.011861238    5.53618E-03    0.874486723
 0.438925002   -0.205670808
EIGENVALUE 3    =-5.298872615    VECTOR:
-0.954835405   -0.240328992   -0.174066634   -0.010989746    9.16066E-03
 5.12620E-03   -1.07833E-03
EIGENVALUE 4    =-7.574719192    VECTOR:
 0.295304747   -0.704148469   -0.644023478   -0.044915626    0.011341429
 7.55197E-03   -8.51399E-04
EIGENVALUE 5    =-17.15255764    VECTOR:
-0.027700000    0.560202581   -0.584183372   -0.586587905    1.65605E-03
 6.88542E-03   -2.12488E-04
```

```
EIGENVALUE 6     =-17.86329687     VECTOR:
 0.016712951    -0.363977437     0.462076406     -0.808519448     6.62711E-04
 4.72676E-03    -1.36341E-04
EIGENVALUE 7     =-22.42730849     VECTOR:
 3.56838E-05    -3.42339E-04     2.46894E-03      6.41285E-03    -0.451791590
 0.891931555    -0.017179128

MAXIMUM ABSOLUTE RESIDUAL= 9.51794E-10
MAXIMUM ABSOLUTE INNER PRODUCT= 1.11326E-11
```

10.5. A BRIEF COMPARISON OF METHODS FOR THE EIGENPROBLEM OF A REAL SYMMETRIC MATRIX

The programmer who wishes to solve the eigenproblem of a real symmetric matrix must first choose which algorithm he will use. The literature presents a vast array of methods of which Wilkinson (1965) and Wilkinson and Reinsch (1971) give a fairly large sample. The method that is chosen will depend on the size of the matrix, its structure and whether or not eigenvectors are required. Suppose that all eigenvalues and eigenvectors are wanted for a matrix whose order is less than 20 with no particular structure. There are two families of algorithm which are candidates to solve this problem. The first reduces the square matrix to a tridiagonal form of which the eigenproblem is then solved by means of one of several variants of either the **QR** algorithm or bisection. (Bisection will find only eigenvalues; inverse iteration is used to obtain the eigenvectors.) The eigenvectors of the tridiagonal matrix must then be back-transformed to find those of the original matrix. Because this family of methods—tridiagonalisation, solution of reduced form, back-transformation—requires fairly complicated codes from the standpoint of this work, they will not be discussed in detail. For matrices of order greater than 10, however, the Householder tridiagonalisation with the **QL** algorithm as described by Wilkinson and Reinsch (1971) is probably the most efficient method for the solution of the complete eigenproblem of a real symmetric matrix. (Some studies by Dr Maurice Cox of the UK National Physical Laboratory show that Givens' tridiagonalisation, if carefully coded, usually involves less work than that of Householder.) It is probably worthwhile for the user with many eigenproblems to solve of order greater than 10 to implement such a method.

The other family of methods is based on the Jacobi algorithm already discussed. Wilkinson, in Wilkinson and Reinsch (1971, pp 192–3), describes Rutishauser's variant of this, called *jacobi*:

> 'The method of Jacobi is the *most elegant* of those developed for solving the complete eigenproblem. The procedure *jacobi* . . . is an extremely compact procedure and considerable care has been taken to ensure that both eigenvalues and eigenvectors are of the highest precision attainable with the word length that is used.'

The last sentence above implies that while *jacobi* uses very little storage, steps have been taken to optimise the program with respect to precision of the results. This is accomplished in part by saving and accumulating some intermediate results in a working vector. The problem requires, formally, n storage locations for the eigenvalues in addition to two n by n arrays to store the original matrix and the eigenvectors. Thus $2n^2+n$ elements appear to be needed. Rutishauser's program requires $2n^2+2n$ elements (that is, an extra vector of length n) as well as a number of individual variables to handle the extra operations used to guarantee the high precision of the computed eigensolutions. Furthermore, the program does not order the eigenvalues and in a great many applications this is a necessary requirement before the eigensolutions can be used in further calculations. The ordering the author has most frequently required is from most positive eigenvalue to most negative one (or vice versa).

The extra code required to order the eigenvalues and their associated eigenvectors can be avoided by using different formulae to calculate intermediate results. That is to say, a slightly different organisation of the program permits an extra function to be incorporated without increasing the code length. This is illustrated by the second column of table 10.1. It should be emphasised that the program responsible for these results was a simple combination of Rutishauser's algorithm and some of the ideas that have been presented in §3.3 with little attention to how well these meshed to preserve the high-precision qualities Rutishauser has carefully built into his routine. The results of the mongrel program are nonetheless quite precise.

If the precision required is reduced a little further, both the extra vector of length n and about a third to a half of the code can be removed. Here the uncertainty in measuring the reduction in the code is due to various choices such as which DIMENSION-ing, input–output or constant setting operations are included in the count.

It may also seem attractive to save space by using the symmetry of the matrix $\mathbf{A}$ as well as that of the intermediate matrices $\mathbf{A}^{(k)}$. This reduces the workspace by $n(n-1)/2$ elements. Unfortunately, the modification introduces sufficient extra code that it is only useful when the order, n, of $\mathbf{A}$ is greater than approximately 10. However, 10 is roughly the order at which the Jacobi methods become non-competitive with approaches such as those mentioned earlier in this section. Still worse, on a single-precision machine, the changes appear to reduce the precision of the results, though the program used to produce column 4 of table 10.1 was not analysed extensively to discover and rectify sources of precision loss. Note that at order 10, the memory capacity of a small computer may already be used up, especially if the eigenproblem is part of a larger computation.

If the storage requirement is critical, then the methods of Hestenes (1958), Chartres (1962) and Kaiser (1972) as modified by Nash (1975) should be considered. This latter method is outlined in §§10.1 and 10.2, and is one which transforms the original matrix $\mathbf{A}$ into another, $\mathbf{B}$, whose columns are the eigenvectors of $\mathbf{A}$ each multiplied by its corresponding eigenvalue, that is

$$\mathbf{A} \rightarrow \mathbf{B} = \mathbf{VE} \tag{10.43}$$

where $\mathbf{E}$ is the diagonal matrix of eigenvalues. Thus only n^2 storage locations are required and the code is, moreover, very short. Column 5 of table 10.1 shows the precision that one may expect to obtain, which is comparable to that found using simpler forms of the traditional Jacobi method. Note that the residual and inner-product computations for table 10.1 were all computed in single precision.

Chapter 11

THE GENERALISED SYMMETRIC MATRIX EIGENVALUE PROBLEM

Consider now the generalised matrix eigenvalue problem

$$\mathbf{A}x = e\mathbf{B}x \tag{2.63}$$

where $\mathbf{A}$ and $\mathbf{B}$ are symmetric and $\mathbf{B}$ is positive definite. A solution to this problem will be sought by transforming it into a conventional eigenvalue problem. The trivial approach

$$\mathbf{B}^{-1}\mathbf{A}x = ex \tag{11.1}$$

gives an eigenvalue problem of the single matrix $\mathbf{B}^{-1}\mathbf{A}$ which is unfortunately not symmetric. The approximation to $\mathbf{B}^{-1}\mathbf{A}$ generated in a computation may therefore have complex eigenvalues. Furthermore, methods for solving the eigenproblem of non-symmetric matrices require much more work than their symmetric matrix counterparts. Ford and Hall (1974) discuss several transformations that convert (2.63) into a symmetric matrix eigenproblem.

Since $\mathbf{B}$ is positive definite, its eigenvalues can be written as the squares D_{ii}^2, $i = 1, 2, \ldots, n$, so that

$$\mathbf{B} = \mathbf{Z}\mathbf{D}^2\mathbf{Z}^{\mathrm{T}} \tag{11.2}$$

where $\mathbf{Z}$ is the matrix of eigenvectors of $\mathbf{B}$ normalised so that

$$\mathbf{Z}\mathbf{Z}^{\mathrm{T}} = \mathbf{Z}^{\mathrm{T}}\mathbf{Z} = \mathbf{1}_n. \tag{11.3}$$

Then

$$\mathbf{B}^{-1/2} = \mathbf{Z}\mathbf{D}^{-1}\mathbf{Z}^{\mathrm{T}} \tag{11.4}$$

and

$$(\mathbf{B}^{-1/2}\mathbf{A}\mathbf{B}^{-1/2})(\mathbf{B}^{1/2}\mathbf{X}) = \mathbf{B}^{1/2}\mathbf{X}\mathbf{E} \tag{11.5}$$

is equivalent to the complete eigenproblem

$$\mathbf{A}\mathbf{X} = \mathbf{B}\mathbf{X}\mathbf{E} \tag{11.5a}$$

which is simply a matrix form which collects together all solutions to (2.63).

Equation (11.5) can be solved as a conventional symmetric eigenproblem

$$\mathbf{A}_1\mathbf{V} = \mathbf{V}\mathbf{E} \tag{11.6}$$

where

$$\mathbf{A}_1 = \mathbf{B}^{-1/2}\mathbf{A}\mathbf{B}^{-1/2} \tag{11.7a}$$

and

$$\mathbf{V} = \mathbf{B}^{1/2}\mathbf{X}. \tag{11.7b}$$

However, it is possible to use the decomposition (11.2) in a simpler fashion since

$$\mathbf{AX} = \mathbf{ZD}^2\mathbf{Z}^{\mathrm{T}}\mathbf{XE} \tag{11.8}$$

so that

$$(\mathbf{D}^{-1}\mathbf{Z}^{\mathrm{T}}\mathbf{AZD}^{-1})(\mathbf{DZ}^{\mathrm{T}}\mathbf{X}) = (\mathbf{DZ}^{\mathrm{T}}\mathbf{X})\mathbf{E} \tag{11.9}$$

or

$$\mathbf{A}_2\mathbf{Y} = \mathbf{YE} \tag{11.10}$$

where

$$\mathbf{Y} = \mathbf{DZ}^{\mathrm{T}}\mathbf{X} \tag{11.11a}$$

and

$$\mathbf{A}_2 = \mathbf{D}^{-1}\mathbf{Z}^{\mathrm{T}}\mathbf{AZD}^{-1}. \tag{11.11b}$$

Another approach is to apply the Choleski decomposition (algorithm 7) to $\mathbf{B}$ so that

$$\mathbf{AX} = \mathbf{LL}^{\mathrm{T}}\mathbf{XE} \tag{11.12}$$

where $\mathbf{L}$ is lower-triangular. Thus we have

$$(\mathbf{L}^{-1}\mathbf{AL}^{-\mathrm{T}})(\mathbf{L}^{\mathrm{T}}\mathbf{X}) = (\mathbf{L}^{\mathrm{T}}\mathbf{X})\mathbf{E} \tag{11.13}$$

or

$$\mathbf{A}_3\mathbf{W} = \mathbf{WE}. \tag{11.14}$$

Note that $\mathbf{A}_3$ can be formed by solving the sets of equations

$$\mathbf{LG} = \mathbf{A} \tag{11.15}$$

and

$$\mathbf{A}_3\mathbf{L}^{\mathrm{T}} = \mathbf{G} \tag{11.16}$$

or

$$\mathbf{LA}_3^{\mathrm{T}} = \mathbf{G}^{\mathrm{T}} \tag{11.17}$$

so that only the forward-substitutions are needed from the Choleski back-solution algorithm 8. Also, the eigenvector transformation can be accomplished by solving

$$\mathbf{L}^{\mathrm{T}}\mathbf{X} = \mathbf{W} \tag{11.18}$$

requiring only back-substitutions from this same algorithm.

While the variants of the Choleski decomposition method are probably the most efficient way to solve the generalised eigenproblem (2.63) in terms of the number of arithmetic operations required, any program based on such a method must contain two different types of algorithm, one for the decomposition and one to solve the eigenproblem (11.13). The eigenvalue decomposition (11.2), on the other hand, requires only a matrix eigenvalue algorithm such as the Jacobi algorithm 14.

Here the one-sided rotation method of algorithm 13 is less useful since there is no simple analogue of the Gerschgorin bound enabling a shifted matrix

$$\mathbf{A}' = \mathbf{A} + k\mathbf{B} \tag{11.19}$$

to be made positive definite by means of an easily calculated shift k. Furthermore, the vectors in the Jacobi algorithm are computed as the product of individual plane rotations. These post-multiply the current approximation to the vectors, which is usually initialised to the unit matrix $\mathbf{1}_n$. From (11.11a), however, we have

$$\mathbf{X}=\mathbf{Z}\mathbf{D}^{-1}\mathbf{Y} \tag{11.20}$$

where $\mathbf{Y}$ is the matrix of eigenvectors of $\mathbf{A}_2$, and hence a product of plane rotations. Therefore, by setting the initial approximation of the vectors to $\mathbf{Z}\mathbf{D}^{-1}$ when solving the eigenproblem (11.10) of $\mathbf{A}_2$, it is not necessary to solve (11.11a) explicitly, nor to save $\mathbf{Z}$ or $\mathbf{D}$.

The form (2.63) is not the only generalised eigenproblem which may arise. Several are discussed in Wilkinson and Reinsch (1971, pp 303–14). This treatment is based mainly on the Choleski decomposition. Nash (1974) also uses a Choleski-type decomposition for the generalised eigenproblem (2.63) in the case where $\mathbf{A}$ and $\mathbf{B}$ are complex Hermitian matrices. This can be solved by 'doubling-up' the matrices to give the real symmetric eigenproblem

$$\begin{bmatrix}\mathrm{Re}(\mathbf{A}) & -\mathrm{Im}(\mathbf{A})\\ \mathrm{Im}(\mathbf{A}) & \mathrm{Re}(\mathbf{A})\end{bmatrix}\begin{bmatrix}\mathrm{Re}(\boldsymbol{x})\\ \mathrm{Im}(\boldsymbol{x})\end{bmatrix}=e\begin{bmatrix}\mathrm{Re}(\mathbf{B}) & -\mathrm{Im}(\mathbf{B})\\ \mathrm{Im}(\mathbf{B}) & \mathrm{Re}(\mathbf{B})\end{bmatrix}\begin{bmatrix}\mathrm{Re}(\boldsymbol{x})\\ \mathrm{Im}(\boldsymbol{x})\end{bmatrix} \tag{11.21}$$

which therefore requires $12n^2$ matrix elements to take the expanded matrices and resulting eigenvectors. Nash (1974) shows how it may be solved using only $4n^2+4n$ matrix elements.

Algorithm 15. Solution of a generalised matrix eigenvalue problem by two applications of the Jacobi algorithm

```
procedure genevJac( n : integer; {order of problem}
                    var A, B, V : rmatrix); {matrices and eigenvectors}
{alg15.pas ==
      To solve the generalized symmetric eigenvalue problem
      A x = e B x
      where A and B are symmetric and B is positive-definite.
Method: Eigenvalue decomposition of B via Jacobi method to form the
'matrix square root' B-half. The Jacobi method is applied a second time
to the matrix
            C = Bihalf A Bihalf
where Bihalf is the inverse of B-half. The eigenvectors x are the columns
of the matrix
            X = Bihalf V
where V is the matrix whose columns are the eigenvectors of matrix C.
                    Copyright 1988 J.C.Nash
}
var
   i,j,k,m : integer;
   s : real;
   initev : boolean;
begin {STEPS 0 and 1}
   writeln('alg15.pas -- generalized symmetric matrix eigenproblem');
   initev := true; {to indicate eigenvectors must be initialized}
   writeln('Eigensolutions of metric B');
   evJacobi(n, B, V, initev); {eigensolutions of B -- STEP 2}
```

Algorithm 15. Solution of a generalised matrix eigenvalue problem by two applications of the Jacobi algorithm (cont.)

```
  {**** WARNING **** No check is made to see that the sweep limit
  has not been exceeded. Cautious users may wish to pass back the
  number of sweeps and the limit value and perform a test here.}
  {STEP 3 is not needed in this version of the algorithm.}
  {STEP 4 -- transformation of initial eigenvectors and creation of matrix
    C, which we store in array B}
  for i := 1 to n do
  begin
    if B[i,i]<=0.0 then halt; {matrix B is not computationally
                    positive definite.}
    s := 1.0/sqrt(B[i,i]);
    for j := 1 to n do V[j,i] := s * V[j,i]; {to form Bihalf}
  end; {loop on i}
  {STEP 5 -- not needed as matrix A already entered}
  {STEP 6 -- perform the transformation 11.11b}
  for i := 1 to n do
  begin
    for j := i to n do {NOTE: i to n NOT 1 to n}
    begin
      s := 0.0;
      for k := 1 to n do
        for m := 1 to n do
          s := s+V[k,i]*A[k,m]*V[m,j];
      B[i,j] := s; B[j,i] := s;
    end; {loop on j}
  end; {loop on i}
  {STEP 7 -- revised to provide simpler Pascal code}
  initev := false; {Do not initialize eigenvectors, since we have
            provided the initialization in STEP 4.}
  writeln('Eigensolutions of general problem');
  evJacobi( n, B, V, initev); {eigensolutions of generalized problem}
end; {alg15.pas == genevJac}
```

Example 11.1. The generalised symmetric eigenproblem: the anharmonic oscillator

The anharmonic oscillator problem in quantum mechanics provides a system for which the stationary states cannot be found exactly. However, they can be computed quite accurately by the Rayleigh–Ritz method outlined in example 2.5.

The Hamiltonian operator (Newing and Cunningham 1967) for the anharmonic oscillator is written

$$H = -\mathrm{d}^2/\mathrm{d}x^2 + k_2x^2 + k_4x^4. \tag{11.22}$$

The eigenproblem arises by minimising the Rayleigh quotient

$$R = \left(\int_{-\infty}^{\infty} \phi(x)H\phi(x)\mathrm{d}x\right)\left(\int_{-\infty}^{\infty} \phi(x)\phi(x)\mathrm{d}x\right)^{-1} \tag{11.23}$$

where $\phi(x)$ is expressed as some linear combination of basis functions. If these basis functions are orthonormal under the integration over the real line, then a conventional eigenproblem results. However, it is common that these functions are not orthonormal, so that a generalised eigenproblem arises. Because of the nature of the operations these eigenproblems are symmetric.

In order to find suitable functions $f_j(x)$ to expand ϕ as

$$\phi(x) = \sum_{j=1}^{n} c_j f_j(x) \tag{11.24}$$

we note that the oscillator which has coefficients $k_2 = 1$, $k_4 = 0$ in its potential has exact solutions which are polynomials multiplied by

$$\exp(-0{\cdot}5x^2).$$

Therefore, the basis functions which will be used here are

$$f_j(x) = Nx^{j-1}\exp(-\alpha x^2) \tag{11.25}$$

where N is a normalising constant.

The approximation sought will be limited to n terms. Note now that

$$Hf_j(x) = N\exp(-\alpha x^2)[-(j-1)(j-2)x^{j-3} + 2\alpha(2j-1)x^{j-1} + (k_2 - 4\alpha^2)x^{j+1} + k_4 x^{j+3}]. \tag{11.26}$$

The minimisation of the Rayleigh quotient with respect to the coefficients c_j gives the eigenproblem

$$\mathbf{Ac} = e\mathbf{Bc} \tag{11.27}$$

where

$$A_{ij} = \int_{-\infty}^{\infty} f_i(x)Hf_j(x)\mathrm{d}x \tag{11.28}$$

and

$$B_{ij} = \int_{-\infty}^{\infty} f_i(x)f_j(x)\mathrm{d}x. \tag{11.29}$$

These integrals can be decomposed to give expressions which involve only the integrals

$$I_m(\alpha) = N^2 \int_{-\infty}^{\infty} x^m \exp(-2\alpha x^2)\mathrm{d}x$$

$$= \begin{cases} 0 & \text{for } m \text{ odd} \\ \dfrac{1\times 3\times 5\times\cdots\times(m-1)}{2^m \alpha^{m/2}} & \text{for } m \text{ even} \\ 1 & \text{for } m = 0. \end{cases} \tag{11.30}$$

The normalising constant N^2 has been chosen to cancel some awkard constants in the integrals (see, for instance, Pierce and Foster 1956, p 68).

Because of the properties of the integrals (11.30) the eigenvalue problem (11.27) reduces to two smaller ones for the even and the odd functions. If we set a parity indicator w equal to zero for the even case and one for the odd case, we can substitute

$$j-1=2(q-1)+w \tag{11.31a}$$

$$i-1=2(p-1)+w \tag{11.31b}$$

where p and q will be the new indices for the matrices **A** and **B** running from 1 to $n'=n/2$ (assuming n even). Thus the matrix elements are

$$\tilde{A}_{pq}=-(j-1)(j-2)I_s+2\alpha(2j-1)I_{s+2}+(k_2-4\alpha^2)I_{s+4}+k_4I_{s+6} \tag{11.32}$$

and

$$\tilde{B}_{pq}=I_{s+2} \tag{11.33}$$

where

$$s=i+j-4=2(p+q-3+w)$$

and j is given by (11.31a). The tilde is used to indicate the re-numeration of **A** and **B**.

The integrals (11.30) are easily computed recursively.

STEP	DESCRIPTION
0	Enter s, α. *Note* s *is even.*
1	Let v = 1.
2	If s ≤ 0, stop. I_s *is in* v. *For* s < 0 *this is always multiplied by* 0.
3	For k = 1 to s/2. Let v = v * (2 * k − 1) * 0·25/α. End loop on k.
4	End *integral.* I_s *is returned in* v.

As an example, consider the exactly solvable problem using $n'=2$, and $\alpha=0{\cdot}5$ for $w=0$ (even parity). Then the eigenproblem has

$$\tilde{\mathbf{A}}=\begin{pmatrix}1 & 0{\cdot}5\\ 0{\cdot}5 & 2{\cdot}75\end{pmatrix} \qquad \tilde{\mathbf{B}}=\begin{pmatrix}1 & 0{\cdot}5\\ 0{\cdot}5 & 0{\cdot}75\end{pmatrix}$$

with solutions

$$e=1 \qquad \mathbf{c}=(1,0)^{\mathrm{T}}$$

and

$$e=5 \qquad \mathbf{c}=2^{-1/2}(-1,2)^{\mathrm{T}}.$$

The same oscillator ($\alpha=0{\cdot}5$) with $w=1$ and $n'=10$ should also have exact solutions. However, the matrix elements range from 0·5 to 3·2E+17 and the solutions are almost all poor approximations when found by algorithm 15.

Likewise, while the problem defined by $n'=5$, $w=0$, $\alpha=2$, $k_2=0$, $k_4=1$ is solved quite easily to give the smallest eigenvalue $e_1=1{\cdot}06051$ with eigenvector

$$\mathbf{c}=(0{\cdot}747087, 1{\cdot}07358, 0{\cdot}866449, 0{\cdot}086206, 0{\cdot}195257)^{\mathrm{T}}$$

the similar problem with $n'=10$ proves to have a **B** matrix which is computationally singular (step 4 of algorithm 15). Inverse iteration saves the day giving, for $n'=5$, $e=1{\cdot}0651$ and for $n'=10$, $e=1{\cdot}06027$ with eigenvectors having small residuals. These results were found using an inverse iteration program based on

Gauss elimination run on a Data General NOVA in 23-bit binary arithmetic.

The particular basis functions (11.25) are, as it turns out, rather a poor set to choose since no matter what exponential parameter α is chosen, the matrix **B** as given by (11.29) suffers large differences in scale in its elements. To preserve the symmetry of the problem, a diagonal scaling could be applied, but a better idea is to use basic functions which are already orthogonal under the inner product defined by (11.29). These are the Hermite functions (Newing and Cunningham 1967). As the details of the solution are not germane to the discussion, they will not be given.

BASIC program code to compute the matrix elements for this problem has been given in Nash (1984a, §18-4, in particular the code on p 242). The reader is warned that there are numerous misprints in the referenced work, which was published about the time of the failure of its publisher. However, the BASIC code, reproduced photographically, is believed to be correct.

Chapter 12

OPTIMISATION AND NONLINEAR EQUATIONS

12.1. FORMAL PROBLEMS IN UNCONSTRAINED OPTIMISATION AND NONLINEAR EQUATIONS

The material which follows in the next few chapters deals with finding the minima of functions or the roots of equations. The functions or equations will in general be nonlinear in the parameters, that is following Kowalik and Osborne (1968), the problems they generate will not be solvable by means of linear equations (though, as we shall see, iterative methods based on linear subproblems are very important).

A special case of the minimisation problem is the nonlinear least-squares problem which, because of its practical importance, will be presented first. This can be stated: given M nonlinear functions of n parameters

$$f_i(b_1, b_2, \ldots, b_n) \qquad i=1, 2, \ldots, M \tag{12.1}$$

minimise the sum of squares

$$S(b_1, b_2, \ldots, b_n)=\sum_{i=1}^{M} [f_i(b_1, b_2, \ldots, b_n)]^2. \tag{12.2}$$

It is convenient to collect the n parameters into a vector $\boldsymbol{b}$; likewise the functions can be collected as the vector of M elements $\boldsymbol{f}$. The nonlinear least-squares problem commonly, but not exclusively, arises in the fitting of equations to data by the adjustment of parameters. The data, in the form of K variables (where $K=0$ when there are no data), may be thought of as occupying a matrix $\mathbf{Y}$ of which the jth column $\mathbf{y}_j$ gives the value of variable j at each of the M data points. The terms *parameters, variables* and *data points* as used here should be noted, since they lead naturally to the least-squares problem with

$$\boldsymbol{f}(\boldsymbol{b}, \mathbf{Y})=\mathbf{g}(\boldsymbol{b}, \mathbf{y}_1, \mathbf{y}_2, \ldots, \mathbf{y}_{K-1})-\mathbf{y}_K \tag{12.3}$$

in which it is hoped to fit the function(s) $\mathbf{g}$ to the variable $\mathbf{y}_K$ by adjusting the parameters $\boldsymbol{b}$. This will reduce the size of the functions $\boldsymbol{f}$, and hence reduce the sum of squares S. Since the functions $\boldsymbol{f}$ in equation (12.3) are formed as differences, they will be termed *residuals*. Note that the sign of these is the opposite of that usually chosen. The sum of squares is the same of course. My preference for the form used here stems from the fact that the partial derivatives of $\boldsymbol{f}$ with respect to the parameters $\boldsymbol{b}$ are identical to those of $\mathbf{g}$ with respect to the same parameters and the possibility of the sign error in their computation is avoided.

By using the shorthand of vector notation, the nonlinear least-squares problem is written: minimise

$$S(\boldsymbol{b}) = \boldsymbol{f}^{\mathrm{T}}\boldsymbol{f} = \boldsymbol{f}^{\mathrm{T}}(\boldsymbol{b}, \mathbf{Y})\boldsymbol{f}(\boldsymbol{b}, \mathbf{Y}) \tag{12.4}$$

with respect to the parameters $\boldsymbol{b}$. Once again, K is the number of variables, M is the number of data points and n is the number of parameters.

Every nonlinear least-squares problem as defined above is an unconstrained minimisation problem, though the converse is not true. In later sections methods will be presented with aim to minimise $S(\boldsymbol{b})$ where S is any function of the parameters. Some ways of handling constraints will also be mentioned. Unfortunately, the mathematical programming problem, in which a minimum is sought for a function subject to many constraints, will only be touched upon briefly since very little progress has been made to date in developing algorithms with minimal storage requirements.

There is also a close relationship between the nonlinear least-squares problem and the problem of finding solutions of systems of nonlinear equations. A system of nonlinear equations

$$\boldsymbol{f}(\boldsymbol{b}, \mathbf{Y}) = \mathbf{0} \tag{12.5}$$

having $n = M$ (number of parameters equal to the number of equations) can be approached as the nonlinear least-squares problem: minimise

$$S = \boldsymbol{f}^{\mathrm{T}}\boldsymbol{f} \tag{12.6}$$

with respect to $\boldsymbol{b}$. For M greater than n, solutions can be sought in the least-squares sense; from this viewpoint the problems are then indistinguishable. The minimum in (12.6) should be found with $S = 0$ if the system of equations has a solution. Conversely, the derivatives

$$\partial S(\boldsymbol{b})/\partial b_j \qquad j = 1, 2, \ldots, n \tag{12.7}$$

for an unconstrained minimisation problem, and in particular a least-squares problem, should be zero at the minimum of the function $S(\boldsymbol{b})$, so that these problems may be solved by a method for nonlinear equations, though local maxima and saddle points of the function will also have zero derivatives and are acceptable solutions of the nonlinear equations. In fact, very little research has been done on the general minimisation or nonlinear-equation problem where either all solutions or extrema are required or a global minimum is to be found.

The minimisation problem when $n = 1$ is of particular interest as a subproblem in some of the methods to be discussed. Because it has only one parameter it is usually termed the linear search problem. The comparable nonlinear-equation problem is usually called root-finding. For the case that $f(b)$ is a polynomial of degree $(K-1)$, that is

$$f(b) = \sum_{j=1}^{K} y_{1j} b^{j-1} \tag{12.8}$$

the problem has a particularly large literature (see, for instance, Jenkins and Traub 1975).

Example 12.1. Function minimisation—optimal operation of a public lottery

Perry and Soland (1975) discuss the problem of deciding values for the main variables under the control of the organisers of a public lottery. These are p, the price per ticket; v, the value of the first prize; w, the total value of all other prizes; and t, the time interval between draws. If N is the number of tickets sold, the expected cost of a single draw is

$$K_1+K_2N$$

that is, a fixed amount plus some cost per ticket sold. In addition, a fixed cost per time K_3 is assumed. The number of tickets sold is assumed to obey a Cobb–Douglas-type production function

$$N=Ft^{\alpha}v^{\beta}w^{\gamma}p^{-\delta}$$

where F is a scale factor (to avoid confusion the notation has been changed from that of Perry and Soland). There are a number of assumptions that have not been stated in this summary, but from the information given it is fairly easy to see that each draw will generate a revenue

$$R=Np-(v+w+K_1+K_2N+K_3t).$$

Thus the revenue per unit time is

$$R/t=-S=[(p-K_2)N-(v+w+K_1)]/t-K_3.$$

Therefore, maximum revenue per unit time is found by minimising $S(\boldsymbol{b})$ where

$$\boldsymbol{b}=\begin{pmatrix} t \\ p \\ v \\ w \end{pmatrix}.$$

Example 12.2. Nonlinear least squares

The data of table 12.1 are thought to approximate the logistic growth function (Oliver 1964)

$$g(x)=b_1/[1+b_2\exp(xb_3)] \tag{12.9}$$

for each point $x=i$. Thus the residuals for this problem are

$$f_i=b_1/[1+b_2\exp(ib_3)]-Y_{i1}. \tag{12.10}$$

Example 12.3. An illustration of a system of simultaneous nonlinear equations

In the econometric study of the behaviour of a market for a given commodity, the following relationships are observed:

$$\text{quantity produced}=q=Kp^{\alpha}$$

where p is the price of the commodity and K and α are constants, and

$$\text{quantity consumed or demanded}=q=Zp^{-\beta}$$

TABLE 12.1. Nonlinear least-squares data of example 12.2.

i	Y_{i1}
1	5·308
2	7·24
3	9·638
4	12·866
5	17·069
6	23·192
7	31·443
8	38·558
9	50·156
10	62·948
11	75·995
12	91·972

where Z and β are constants. In this simple example, the equations reduce to

$$Kp^{\alpha} = Zp^{-\beta}$$

or

$$p = (Z/K)^{1/(\alpha+\beta)}$$

so that

$$q = K(Z/K)^{\alpha/(\alpha+\beta)}.$$

However, in general, the system will involve more than one commodity and will not offer a simple analytic solution.

Example 12.4. Root-finding

In the economic analysis of capital projects, a measure of return on investment that is commonly used is the *internal rate of return r*. This is the rate of interest applicable over the life of the project which causes the net present value of the project at the time of the first investment to be zero. Let y_{1i} be the net revenue of the project, that is, revenue or income minus loss or investment, in the ith time period. This has a present value at the first time period of

$$y_{1i}/(1+0{\cdot}01r)^{i-1}$$

where r is the interest rate in per cent per period. Thus the total present value at the beginning of the first time period is

$$P(r) = \sum_{i=1}^{K} y_{1i}/(1+0{\cdot}01r)^{i-1}$$

where K is the number of time periods in the life of the project. By setting

$$b = 1/(1+0{\cdot}01r)$$

this problem is identified as a polynomial root-finding problem (12.8).

Example 12.5. Minimum of a function of one variable

Suppose we wish to buy some relatively expensive item, say a car, a house or a new computer. The present era being afflicted by inflation at some rate r, we will pay a price

$$P(1+r)^t$$

at some time t after the present. We can save at s dollars (pounds) per unit time, and when we buy we can borrow money at an overall cost of $(F-1)$ dollars per dollar borrowed, that is, we must pay back F dollars for every dollar borrowed. F can be evaluated given the interest rate R and number of periods N of a loan as

$$F = NR(1+R)^N/[(1+R)^N - 1].$$

Then, to minimise the total cost of our purchase, we must minimise

$$\begin{aligned} S(t) &= ts + [P(1+r)^t - ts]F \\ &= ts(1-F) + FP(1+r)^t. \end{aligned}$$

This has an analytic solution

$$t = \ln\{(F-1)s/[FP\ln(1+r)]\}/\ln(1+r).$$

However, it is easy to construct examples for which analytic solutions are harder to obtain, for instance by changing inflation rate r with time.

12.2. DIFFICULTIES ENCOUNTERED IN THE SOLUTION OF OPTIMISATION AND NONLINEAR-EQUATION PROBLEMS

It is usually relatively easy to state an optimisation or nonlinear-equation problem. It may even be straightforward, if tedious, to find one solution. However, to find the solution of interest may be very nearly impossible.

In unconstrained minimisation problems the principal difficulty arises due to *local* minima. If the *global* minimum is sought, these local minima will tend to attract algorithms to themselves much as sand bunkers attract golf balls on the course. That is to say, there may be no reason why a particular local minimum will be found; equally there is no reason why it will not. The nonlinear-equation problem which arises by setting the derivatives of the function to zero will have solutions at each of these local minima. Local maxima and saddle points will also cause these equations to be satisfied.

Unfortunately, very little research has been done on how to get the desired solution. One of the very few studies of this problem is that of Brown and Gearhart (1971) who discuss deflation techniques to calculate several solutions of simultaneous nonlinear equations. These methods seek to change the equations being solved so that solutions already found are not solutions of the modified equations. The interested reader will also find an excellent discussion of the difficulties attendant on solving nonlinear equations and minimisation problems in Acton (1970, chaps 2 and 14). The traditional advice given to users wishing to avoid unwanted solutions is always to provide starting values for the parameters which are close to the expected answer. As Brown and Gearhart (1971) have pointed out, however, starting points can be close to the desired solution without guaranteeing convergence to that solution. They found that certain problems in combination with certain methods have what they termed *magnetic zeros* to which the method in use converged almost regardless of the starting parameters employed. However, I did not discover this 'magnetism' when attempting to solve the cubic–parabola problem of Brown and Gearhart using a version of algorithm 23. In cases where one root appears to be magnetic, the only course of action once several deflation methods have been tried is to reformulate the problem so the desired solution dominates. This may be asking the impossible!

Another approach to 'global' minimisation is to use a pseudo-random-number generator to generate points in the domain of the function (see Bremmerman (1970) for discussion of such a procedure including a FORTRAN program). Such methods are primarily heuristic and are designed to sample the surface defined by the function. They are probably more efficient than an n-dimensional grid search, especially if used to generate starting points for more sophisticated minimisation algorithms. However, they cannot be presumed to be reliable, and there is a lack of elegance in the need for the shot-gun quality of the pseudo-random-number generator. It is my opinion that wherever possible the properties of the function should be examined to gain insight into the nature of a global minimum, and whatever information is available about the problem should be used to increase the chance that the desired solution is found. Good starting values can greatly reduce the cost of finding a solution and greatly enhance the likelihood that the desired solution will be found.

Chapter 13

ONE-DIMENSIONAL PROBLEMS

13.1. INTRODUCTION

One-dimensional problems are important less in their own right than as a part of larger problems. 'Minimisation' along a line is a part of both the conjugate gradients and variable metric methods for solution of general function minimisation problems, though in this book the search for a minimum will only proceed until a satisfactory new point has been found. Alternatively a linear search is useful when only one parameter is varied in a complicated function, for instance when trying to discover the behaviour of some model of a system to changes in one of the controls. Roots of functions of one variable are less commonly needed as a part of larger algorithms. They arise in attempts to minimise functions by setting derivatives to zero. This means maxima and saddle points are also found, so I do not recommend this approach in normal circumstances. Roots of polynomials are another problem which I normally avoid, as some clients have a nasty habit of trying to solve eigenproblems by means of the characteristic equation. The polynomial root-finding problem is very often inherently unstable in that very small changes in the polynomial coefficients give rise to large changes in the roots. Furthermore, this situation is easily worsened by ill chosen solution methods. The only genuine polynomial root-finding problem I have encountered in practice is the internal rate of return (example 12.4). However, accountants and economists have very good ideas about where they would like the root to be found, so I have not tried to develop general methods for finding all the roots of a polynomial, for instance by methods such as those discussed by Jenkins and Traub (1975). Some experiments I have performed with S G Nash (unpublished) on the use of matrix eigenvalue algorithms applied to the companion matrices of polynomials were not very encouraging as to accuracy or speed, even though we had expected such methods to be slow.

13.2. THE LINEAR SEARCH PROBLEM

The linear search problem can be stated:

$$\text{minimise } S(b) \text{ with respect to } b. \tag{13.1}$$

However, it is not usual to leave the domain of b unrestricted. In all cases considered here, b is real and will often be confined to some interval $[u, v]$.

If $S(b)$ is differentiable, this problem can be approached by applying a root-finding algorithm to

$$S'(b) = \mathrm{d}S(b)/\mathrm{d}b. \tag{13.2}$$

Since local maxima also zero the derivative of the function $S(b)$, such solutions will have to be checked either by ensuring the second derivative $S''(b)$ is positive or equivalently by examining the values of the function at points near the supposed minimum.

When the derivative of S is not available or is expensive to compute, a method for minimisation along a line is needed which depends only on function values. Obviously, any method which evaluates the function at a finite number of points is not guaranteed to detect deep, narrow wells in the function. Therefore, some assumption must be made about the function on the interval $[u, v]$. Here it will be assumed that the function $S(b)$ is *unimodal* in $[u, v]$, that is, that there is only one stationary value (either a maximum or a minimum) in the interval. In the case where the stationary value is a maximum, the minimum of the function will be either at u or at v.

Given that $S(b)$ is unimodal in $[u, v]$, a grid or equal-interval search could be used to decrease the size of the interval. For instance, the function values could be computed for each of the points

$$b_j = u + jh \qquad j = 0, 1, 2, \ldots, n \tag{13.3}$$

where

$$h = (v - u)/n. \tag{13.4}$$

If the smallest of these values is $S(b_k)$, then the minimum lies in the interval $[b_{k-1}, b_{k+1}]$. Equation (13.3) can be used to compute the endpoints of this interval. If $S(b_0)$ or $S(b_n)$ is the smallest value, then the interval is $[u, b_1]$ or $[b_{n-1}, v]$, though for simplicity this could be left out of a program. The search can now be repeated, noting, of course, that the function values at the endpoints of the interval have already been computed.

Algorithm 16. Grid search along a line

```
procedure gridsrch( var lbound, ubound : real; {the lower and
                        upper bounds to the interval to be tested}
                        nint : integer; {the number of grid intervals}
                        var fmin: real; {the lowest function value found}
                        var minarg: integer; {the grid point in the set
                        0,1,...,nint at which the minimum function value
                        was found}
                        var changarg: integer {the grid point in the set
                        1,2,...,nint which is nearest the upper bound
                        ubound such that there has been a sign change
                        in the function values f(lbound+(changarg-1)*h)
                        and f(lbound+changarg*h) where h is the step
                        == (ubound - lbound)/nint } );
{alg16.pas == one-dimensional grid search over function values
  This version halts execution if the function is not computable at all
  the grid points. Note that it is not equivalent to the version in the
  first edition of Compact Numerical Methods.
                        Copyright 1988 J.C.Nash
}
```

Algorithm 16. Grid search along a line (cont.)

```
var
    j : integer;
    h, p, t : real;
    notcomp : boolean;
begin
    writeln('alg16.pas -- one-dimensional grid search');
    {STEP 0 via the procedure call}
    writeln('In gridsrch lbound=',lbound,' ubound=',ubound);
    notcomp:=false; {function must be called with notcomp false or
                root will be displayed -- STEP 1}
    t:=fn1d(lbound, notcomp); {compute function at lower bound and set
                pointer k to lowest function value found so far i.e. the 0'th}
    writeln(' lb f(',lbound,')=',t);
    if notcomp then halt;
    fmin:=t; {to save the function value}
    minarg:=0; {so far this is the lowest value found}
    changarg:=0; {as a safety setting of this value}
    h:=(ubound-lbound)/nint;
    for j:=1 to nint do {STEP 2}
    {Note: we increase the number of steps so that the upper bound ubound is
        now a function argument. Warning: because the argument is now
        calculated, we may get a value slightly different from ubound in
        forming (lbound+nint*h). }
    begin
        p:=fn1d(lbound+j*h, notcomp); {STEP 3}
        write(' f(',lbound+j*h,')=',p);
        if notcomp then halt;
        if p<fmin then {STEP 4}
        begin {STEP 5}
            fmin:=p; minarg:=j;
        end; {if p<fmin}
        if p*t<=0 then
        begin
            writeln(' *** sign change ***');
            changarg:=j; {to save change point argument}
        end
        else
        begin
            writeln; {no action since sign change}
        end;
        t:=p; {to save latest function value} {STEP 6}
    end; {loop on j}
    writeln('Minimum so far is f(',lbound+minarg*h,')=',fmin);
    if changarg>0 then
    begin
        writeln('Sign change observed last in interval ');
        writeln(' [',lbound+(changarg-1)*h,',',lbound+changarg*h,']');
    end
    else
        writeln('Apparently no sign change in [',lbound,',',ubound,']');
end; {alg16.pas == gridsrch}
```

A grid or equal-interval search is not a particularly efficient method for finding a minimum. However, it is very simple to program and provides a set of points for plotting the function along the line. Even a crude plot, such as that produced using a line printer or teletypewriter, can be used to gain a visual appreciation of the function studied. It is in this fashion that a grid search can be used most effectively, not to find the minimum of a function but to carry out a preliminary investigation of it. Furthermore, if the grid search is used in this way, there is no necessity that the function be unimodal.

Consider now the case where instead of the interval $[u, v]$, only a starting position u and a step h are given. If $S(u+h)<S(u)$, the step will be termed a success, otherwise it will be called a failure. This is the basis of Rosenbrock's (1960) algorithm for minimising a function of several variables. Here, however, the goal will be to find points at which the function value is reduced. A procedure which does this can be repeated until a minimum is found.

Suppose $S(u+h)<S(u)$, then an obvious tactic is to replace u by $(u+h)$ and try again, perhaps increasing the step somewhat. If, however, $S(u+h)\geqslant S(u)$, then either the step is too large or it is in the wrong direction. In this case a possibility is to reduce the size of h *and* change its sign. (If only the size is incorrect, two operations of this sort will find a lower function value.)

The above paragraph defines the success–failure algorithm except for the factors to be used to increase or decrease the size of h. Notice that so far neither the grid search nor the success–failure algorithm take into account the function values apart from the fact that one function value is larger than another. How much larger it is, which indicates how rapidly the function is changing with respect to h, is not considered. Using the function values can, however, lead to more efficient minimisation procedures. The idea is to use the function values and their associated points (as well as the derivatives of the function at these points if available) to generate some simple function, call it $I(b)$, *which interpolates the function* $S(b)$ between the points at which S has been computed. The most popular choices for $I(b)$ are polynomials, in particular of degree 2 (parabolic approximation) or degree 3 (cubic approximation). Consider the case where $S(b)$ is known at three points b_0, b_1 and b_2. Then using a parabola to interpolate the function requires

$$S(b_j)=A+Bb_j+Cb_j^2 \qquad \text{for } j=0, 1, 2 \tag{13.5}$$

which provides three linear equations for the three unknowns A, B and C. Once these are found, it is simple to set the derivative of the interpolant to zero

$$\mathrm{d}I(b)/\mathrm{d}b=2Cb+B=0 \tag{13.6}$$

to find the value of b which minimises the interpolating polynomial $I(b)$. This presumes that the parabola has a minimum at the stationary point, and upon this presumption hang many *ifs, ands* and *buts,* and no doubt miles of computer listing of useless results. Indeed, the largest part of many procedures for minimising a function along a line by minimising an interpolant $I(b)$ is concerned with ensuring that the interpolant has the proper shape to guarantee a minimum. Furthermore $I(b)$ cannot be used to *extrapolate* very far from the region in which it has been generated without risking, at the very least, unnecessary function evaluations which, after all, one is trying to avoid by using more of the information about the function $S(b)$ than the rather conservative direct search methods above.

At this juncture we can consider combining the success–failure algorithm with inverse interpolation. One choice, made in algorithm 22, is to take the initial point in the success–failure procedure as b_0 Then 'successes' lower the function value. Ultimately, however, a 'failure' will follow a 'success' unless $u = b_0$ is minimal, even if the procedure begins with a string of 'failures'. Call the position at the last 'failure' b_2, and the position at the 'success' which preceded it b_1. (Note that b_1 has to be the lowest point found so far.) Then $S(b_1)$ is less than either $S(b_0)$ or $S(b_2)$ and the three points define a V-shape which ensures that the interpolating parabola (13.5) will not be used to extrapolate.

Alternatively, as in algorithm 17 below, any of the following combinations of events lead to a V-shaped triple of points with b_1 the lowest of the three:

(i) initial point, success, failure (b_0, b_1, b_2)
(ii) success, success, failure (b_0, b_1, b_2)
(iii) initial point, failure, failure (b_1, b_0, b_2).

Consider then the three points defined by b_0, b_1, b_2 where b_1 occupies the middle position and where, using the notation

$$S_j = S(b_j) \tag{13.7}$$

for brevity,

$$S_1 \leqslant S_0 \tag{13.8}$$

and

$$S_1 \leqslant S_2. \tag{13.9}$$

Excluding the exceptional cases that the function is flat or otherwise perverse, so that at least one of the conditions (13.8) or (13.9) is an inequality, the interpolating parabola will have its minimum between b_0 and b_2. Note now that we can measure all distances from b_1, so that equations (13.5) can be rewritten

$$\begin{aligned} S_0 &= A + Bx_0 + Cx_0^2 \\ S_1 &= A \\ S_2 &= A + Bx_2 + Cx_2^2 \end{aligned} \tag{13.10}$$

where

$$x_j = b_j - b_1 \qquad \text{for } j = 0, 1, 2. \tag{13.11}$$

Equations (13.10) can then be solved by elimination to give

$$B = \frac{(S_0 - S_1)x_2^2 - (S_2 - S_1)x_0^2}{x_0 x_2^2 - x_2 x_0^2} \tag{13.12}$$

and

$$C = \frac{(S_0 - S_1)x_2 - (S_2 - S_1)x_0}{x_0^2 x_2 - x_2^2 x_0}. \tag{13.13}$$

(Note that the denominators differ only in their signs.) Hence the minimum of the parabola is found at

$$x = \frac{-B}{2C} = 0{\cdot}5\,\frac{(S_0 - S_1)x_2^2 - (S_2 - S_1)x_0^2}{(S_0 - S_1)x_2 - (S_2 - S_1)x_0}. \tag{13.14}$$

The success–failure algorithm always leaves the step length equal to x_2. The length x_0 can be recovered if the steps from some initial point to the previous two evaluation points are saved. One of these points will be b_1; the other is taken as b_0. The expression on the right-hand side of equation (13.14) can be evaluated in a number of ways. In the algorithm below, both numerator and denominator have been multiplied by -1.

To find the minimum of a function of one parameter, several cycles of success–failure and parabolic inverse interpolation are usually needed. Note that algorithm 17 recognises that some functions are not computable at certain points b. (This feature has been left out of the program FMIN given by Forsythe *et al* (1977), and caused some failures of that program to minimise fairly simple functions in tests run by B Henderson and the author, though this comment reflects differences in design philosophy rather than weaknesses in FMIN.) Algorithm 17 continues to try to reduce the value of the computed function until $(b+h)$ is not different from b in the machine arithmetic. This avoids the requirement for machine-dependent tolerances, but may cause the algorithm to execute indefinitely in environments where arithmetic is performed in extended-precision accumulators if a storage of $(b+h)$ is not forced to shorten the number of digits carried.

In tests which I have run with B Henderson, algorithm 17 has always been more efficient in terms of both time and number of function evaluations than a linear search procedure based on that in algorithm 22. The reasons for retaining the simpler approach in algorithm 22 were as follows.

(i) A true minimisation along the line requires repeated cycles of success–failure/inverse interpolation. In algorithm 22 only one such cycle is used as part of a larger conjugate gradients minimisation of a function of several parameters. Therefore, it is important that the inverse interpolation not be performed until at least some progress has been made in reducing the function value, and the procedure used insists that at least one 'success' be observed before interpolation is attempted.

(ii) While one-dimensional trials and preliminary tests of algorithm 17-like cycles in conjugate gradients minimisation of a function of several parameters showed some efficiency gains were possible with this method, it was not possible to carry out the extensive set of comparisons presented in chapter 18 for the function minimisation algorithms due to the demise of the Data General NOVA; the replacement ECLIPSE uses a different arithmetic and operating system. In view of the reasonable performance of algorithm 22, I decided to keep it in the collection of algorithms. On the basis of our experiences with the problem of minimising a function of one parameter, however, algorithm 17 has been chosen for linear search problems. A FORTRAN version of this algorithm performed competitively with the program FMIN due to Brent as given in Forsythe *et al* (1977) when several tests were timed on an IBM 370/168.

The choice of the step adjustment factors A1 and A2 to enlarge the step length or to reduce it and change its sign can be important in that poor choices will obviously cause the success–failure process to be inefficient. Systematic optimisation of these two parameters over the class of one-dimensional functions which may have to be minimised is not feasible, and one is left with the rather unsatisfactory situation of having to make a judgement from experience. Dixon (1972) mentions the choices (2, −0·25) and (3, −0·5). In the present application, however, where the success–failure steps are followed by inverse interpolation, I have found the set (1·5, −0·25) to be slightly more efficient, though this may merely reflect problems I have been required to solve.

Algorithm 17. Minimisation of a function of one variable

```
procedure min1d(var bb : real; {initial value of argument of function
                    to be minimised, and resulting minimum position}
                    var st: real; {initial and final step-length}
                    var ifn : integer; {function evaluation counter}
                    var fnminval : real {minimum function value on return});
{alg17.pas ==
      One-dimensional minimisation of a function using success-failure
      search and parabolic inverse interpolation
                    Copyright 1988 J.C.Nash
}
{No check is made that abs(st)>0.0. Algorithm will still converge.}
var
   a1, a2, fii, s0, s1, s2, tt0, tt1, tt2, x0, x1, x2, xii : real;
   notcomp, tripleok: boolean;
begin
   writeln('alg17.pas -- One dimensional function minimisation');
   {STEP 0 -- partly in procedure call}
   ifn := 0; {to initialize function evaluation count}
   a1 := 1.5; {to set the growth parameter of the success-failure search}
   a2 := -0.25; {to set the shrink parameter of the success-failure search}
   x1 := bb; {to start, we must have a current 'best' argument}
   notcomp := false; {is set TRUE when function cannot be computed. We set it
                    here otherwise minimum or root of fn1d may be displayed}
   s0 := fn1d(x1,notcomp); ifn := ifn+1; {Compute the function at x1.}
   if notcomp then
   begin
      writeln('*** FAILURE *** Function cannot be computed at initial point');
      halt;
   end;
   repeat {Main minimisation loop}
      x0 := x1; {to save value of argument}
      bb := x0;
      x1 := x0+st; {to set second value of argument}
      s1 := fn1d(x1,notcomp); if notcomp then s1 := big; ifn := ifn+1;
      {Note mechanism for handling non-computability of the function.}
      tripleok := false; {As yet, we do not have a triple of points in a V.}
```

Algorithm 17. Minimisation of a function of one variable (cont.)

```
if s1<s0 then
begin {Here we can proceed to try to find s2 now in same direction.}
    repeat {success-failure search loop}
        st := st*a1; {increase the stepsize after a success}
        x2 := x1+st; {get next point in the series}
        s2 := fn1d(x2,notcomp); if notcomp then s2 := big; ifn := ifn+1;
        if s2<s1 then
        begin {another success}
            s0 := s1; s1 := s2; {In order to continue search,}
            x0 := x1; x1 := x2; {we copy over the points.}
            write('Success1 ');
        end
        else {'failure' after 'success' ==> V-shaped triple of points}
        begin
            tripleok := true; {to record existence of the triple}
            write('Failure1');
        end;
    until tripleok; {End of the success-failure search for}
end {s1<s0 on first pair of evaluations}
else
begin {s1>=s0 on first pair of evaluations in this major cycle, so we
                must look for a third point in the reverse search direction.}
    st := a2*st; {to reverse direction of search and reduce the step size}
    tt2 := s0; s0 := s1; s1 := tt2; {to swap function values}
    tt2 := x0; x0 := x1; x1 := tt2; {to swap arguments}
    repeat
        x2 := x1+st; {get the potential third point}
        s2 := fn1d(x2,notcomp); if notcomp then s2 := big; ifn := ifn+1;
        if s2<s1 then
        begin {success in reducing function -- keep going}
            s0 := s1; s1 := s2; x0 := x1; x1 := x2; {reorder points}
            st := st*a1; {increase the stepsize maintaining direction}
            write('Success2 ');
        end
        else
        begin {two failures in a row ensures a triple in a V}
            tripleok := true; write('Failure2');
        end;
    until tripleok; {End of success-failure search}
end; {if s1<s0 for first pair of test points}
{Now have a V of points (x0,s0), (x1,s1), (x2,s2).}
writeln; writeln('Triple (',x0,',',s0,')');
writeln(' (',x1,',',s1,')'); writeln(' (',x2,',',s2,')');
tt0 := x0-x1; {to get deviation from best point found. Note that st
                holds value of x2-x1.}
tt1 := (s0-s1)*st; tt2 := (s2-s1)*tt0; {temporary accumulators}
if tt1<>tt2 then {avoid zero divide in parabolic inverse interpolation}
begin
    st := 0.5*(tt2*tt0-tt1*st)/(tt2-tt1); {calculate the step}
    xii := x1+st;
    writeln('Paramin step and argument :',st,' ',xii);
    if (reltest+xii)<>(reltest+x1) then
```

Algorithm 17. Minimisation of a function of one variable (cont.)

```
          begin {evaluate function if argument has been changed}
            fii := fn1d(xii,notcomp); ifn := ifn+1;
            if notcomp then fii := big;
            if fii<s1 then
            begin
              s1 := fii; x1 := xii; {save new & better function, argument}
              writeln('New min f(',x1,')=',s1);
            end;
          end; {evaluate function for parabolic inverse interpolation}
        end; {if not zerodivide situation}
        writeln(ifn,' evalns f(',x1,')=',s1);
        s0 := s1; {to save function value in case of new iteration}
      until (bb=x1); {Iterate until minimum does not change. We could
                      use reltest in this termination condition.}
      writeln('Apparent minimum is f(',bb,')=',s1);
      writeln(' after ',ifn,' function evaluations');
      fnminval := s1; {store value for return to calling program}
end;{alg17.pas == min1d}
```

The driver program DR1617.PAS *on the software diskette allows grid search to be used to localise a minimum, with its precise location found using the one-dimensional minimiser above.*

Example 13.1. Grid and linear search

The expenditure minimisation example 12.5 provides a problem on which to test algorithms 16 and 17. Recall that

$$S(t) = ts(1-F) + FP(1+r)^t$$

where t is the time until purchase (in months), P is the purchase price of the item we propose to buy (in \$), F is the payback ratio on the loan we will have to obtain, s is the amount we can save each month, and r is the inflation rate per month until the purchase date.

Typical figures might be $P = \$10\,000$, $F = 2{\cdot}25$ (try 1·5% per month for 5 years), $s = \$200$ per month, and $r = 0{\cdot}00949$ (equivalent to 12% per annum). Thus the function to be minimised is

$$S(t) = -250t + 22500(1{\cdot}00949)^t$$

which has the analytic solution (see example 12·5) $t^* = 17{\cdot}1973522154$ as computed on a Tektronix 4051.

NOVA	ECLIPSE
$F(0) = 22500$	$F(0) = 22500$
$F(10) = 22228{\cdot}7$	$F(10) = 22228{\cdot}5$
$F(20) = 22178{\cdot}1$	$F(20) = 22177{\cdot}7$
$F(30) = 22370{\cdot}2$	$F(30) = 22369{\cdot}4$
$F(40) = 22829$	$F(40) = 22827{\cdot}8$
$F(50) = 23580{\cdot}8$	$F(50) = 23579{\cdot}2$

For both sets, the endpoints are $u = 0$, $v = 50$ and number of points is $n = 5$.

Simple grid search was applied to this function on Data General NOVA and ECLIPSE computers operating in 23-bit binary and six-digit hexadecimal arithmetic, respectively. The table at the bottom of the previous page gives the results of this exercise. Note the difference in the computed function values!

An extended grid search (on the ECLIPSE) uses 26 function evaluations to localise the minimum to within a tolerance of 0·1.

```
 NEW
*ENTER"SGRID"
*RUN
SGRID NOV 23 77
 3  5  1978    16  44  31
 ENTER SEARCH INTERVAL ENDPOINTS
 AND TOLERANCE OF ANSWER'S PRECISION
 ? 10 ? 30 ? 1
 ENTER THE NUMBER OF GRID DIVISIONS
 ? 5
F( 14 )= 22180.5
F( 18 )= 22169.2
F( 22 )= 22195.9
F( 26 )= 22262.1
THE MINIMUM LIES IN THE INTERVAL [ 14 , 22 ]
F( 15.6 )= 22171.6
F( 17.2 )= 22168.6
F( 18.8 )= 22171.6
F( 20.4 )= 22180.7
THE MINIMUM LIES IN THE INTERVAL [ 15.6 , 18.8 ]
F( 16.24 )= 22169.7
F( 16.88 )= 22168.7
F( 17.52 )= 22168.7
F( 18.16 )= 22169.7
THE MINIMUM LIES IN THE INTERVAL [ 16.24 , 17.52 ]
F( 16.496 )= 22169.2
F( 16.752 )= 22168.8
F( 17.008 )= 22168.7
F( 17.264 )= 22168.6
THE MINIMUM LIES IN THE INTERVAL [ 17.008 , 17.52 ]
   18  FUNCTION EVALUATIONS
NEW TOLERANCE ? .1
F( 17.1104 )= 22168.6
F( 17.2128 )= 22168.6
F( 17.3152 )= 22168.6
F( 17.4176 )= 22168.6
THE MINIMUM LIES IN THE INTERVAL [ 17.1104 , 17.3152 ]
F( 17.1513 )= 22168.6
F( 17.1923 )= 22168.6
F( 17.2332 )= 22168.6
F( 17.2742 )= 22168.6
THE MINIMUM LIES IN THE INTERVAL [ 17.1923 , 17.2742 ]
   26  FUNCTION EVALUATIONS
NEW TOLERANCE ? -1

STOP AT 0420
*
```

Algorithm 17 requires a starting point and a step length. The ECLIPSE gives

```
*
*RUN
NEWMIN JULY 7 77
STARTING VALUE= ? 10  STEP ? 5
```

```
F( 10 )= 22228.5
F( 15 )= 22174.2
SUCCESS
F( 22.5 )= 22202.2
PARAMIN STEP= 2.15087
F( 17.1509 )= 22168.6
NEW K4=-1.78772
F( 15.3631 )= 22172.6
FAILURE
F( 17.5978 )= 22168.8
PARAMIN STEP= 7.44882E-02
F( 17.2253 )= 22168.6
NEW K4=-.018622
F( 17.2067 )= 22168.6
SUCCESS
F( 17.1788 )= 22168.6
PARAMIN STEP=-4.65551E-03
F( 17.2021 )= 22168.6
PARAMIN FAILS
NEW K4= 4.65551E-03
F( 17.2114 )= 22168.6
FAILURE
F( 17.2055 )= 22168.6
PARAMIN FAILS
NEW K4= 0
MIN AT   17.2067    =    22168.6
 12  FN EVALS

STOP AT 0060
*
```

The effect of step length choice is possibly important. Therefore, consider the following applications of algorithm 17 using a starting value of $t=10$.

Step length	Minimum at	Function evaluations
1	17·2264	13
5	17·2067	12
10	17·2314	10
20	17·1774	11

The differences in the minima are due to the flatness of this particular function, which may cause difficulties in deciding when the minimum has been located. By way of comparison, a linear search based on the success–failure/inverse interpolation sequence in algorithm 22 found the following minima starting from $t=10$.

Step length	Minimum at	Function evaluations
1	17·2063	23
5	17·2207	23
10	17·2388	21
20	17·2531	24

A cubic inverse interpolation algorithm requiring both the function and derivative to be computed, but which employs a convergence test based solely on the change in the parameter, used considerably more effort to locate a minimum from $t=10$.

Step length	Minimum at	Function and derivative evaluations
1	17·2083	38+38
5	17·2082	23+23
10	17·2082	36+36
20	17·2081	38+38

Most of the work in this algorithm is done near the minimum, since the region of the minimum is located very quickly.

If we can be confident of the accuracy of the derivative calculation, then a root-finder is a very effective way to locate a minimum of this function. However, we should check that the point found is not a maximum or saddle. Algorithm 18 gives

```
*
*RUN200
ROOTFINDER
U= ? 10    V= ? 30

BISECTION EVERY ? 5
TOLERANCE ? 0
F( 10 )=-16.4537       F( 30 )= 32.0994
FP   ITN 1    U= 10      V= 30     F( 16.7776 )=-1.01735
FP   ITN 2    U= 16.7776      V= 30     F( 17.1838 )=-.0582123
FP   ITN 3    U= 17.1838      V= 30     F( 17.207 )=-.00361633
FP   ITN 4    U= 17.207     V= 30     F( 17.2084 )=-2.28882E-04
FP   ITN 5    U= 17.2084      V= 30     F( 17.2085 )=-3.05176E-05
BI   ITN 6    U= 17.2085      V= 30     F( 23.6042 )= 15.5647
FP   CONVERGED
ROOT: F( 17.2085 )=-3.05176E-05

STOP AT 0340
*
```

Unless otherwise stated all of the above results were obtained using a Data General ECLIPSE operating in six hexadecimal digit arithmetic.

It may appear that this treatment is biased against using derivative information. For instance, the cubic inverse interpolation uses a convergence test which does not take it into account at all. The reason for this is that in a single-precision environment (with a short word length) it is difficult to evaluate the projection of a gradient along a line since inner-product calculations cannot be made in extended precision. However, if the linear search is part of a larger algorithm to minimise a function for several parameters, derivative information must usually be computed in this way. The function values may still be well determined, but inaccuracy in the derivative has, in my experience, appeared to upset the performance of either the inverse interpolation or the convergence test.

13.3. REAL ROOTS OF FUNCTIONS OF ONE VARIABLE

The linear search problem may, as mentioned, be approached by finding the roots of the derivative of the function to be minimised. The problem of finding one or more values b for which $f(b)$ is zero arises directly in a variety of disciplines. Formally the problem may be stated:

find the (real) value or values b such that $f(b)=0$.

However, in practice there is almost always more information available about the roots other than that they are real. For instance, the roots may be restricted to an interval $[u, v]$ and the number of roots in this interval may be known. In such cases, a grid search may be used to reduce the length of the interval in which the roots are to be sought. Choosing some integer n to define a step

$$h=(v-u)/(n+1) \tag{13.15}$$

gives a grid upon which function values

$$f(u+jh) \qquad j=0, 1, 2, \ldots, (n+1) \tag{13.16}$$

can be computed. If

$$f(u+jh) * f(u+(j+1)h) \leqslant 0 \tag{13.17}$$

then the interval $[u+jh, u+(j+1)h]$ contains at least one root and the search can be repeated on this smaller interval if necessary. Roots which occur with

$$f(b)=0 \qquad f'(b)=0 \tag{13.18}$$

simultaneously are more difficult to detect, since the function may not cross the b axis and a sign change will not be observed. Grid searches are notoriously expensive in computer time if used indiscriminately. As in the previous section, they are a useful tool for discovering some of the properties of a function by 'taking a look', particularly in conjunction with some plotting device. As such, the grid parameter, n, should not be large; $n<10$ is probably a reasonable bound. Unfortunately, it is necessary to caution that if the inequality (13.17) is not satisfied, there may still be an even number of roots in $[u+jh, u+(j+1)h]$.

Suppose now that a single root is sought in the interval $[u, v]$ which has been specified so that

$$f(u) * f(v)<0. \tag{13.19}$$

Thus the function has at least one root in the interval if the function is continuous. One possible way to find the root is to *bisect* the interval, evaluating the function at

$$b=(u+v)/2. \tag{13.20}$$

If

$$f(u) * f(b)<0 \tag{13.21}$$

then the root lies in $[u, b]$; otherwise in $[b, v]$. This *bisection* can be repeated as many times as desired. After t bisections the interval length will be

$$2^{-t}(u-v) \tag{13.22}$$

so that the root can always be located by this method in a fixed number of steps. Since the function values are only examined for their signs, however, an unnecessarily large number of function evaluations may be required. By reversing the process of interpolation—that is, estimating function values between tabulated values by means of an interpolating function fitted to the tabular points—one may expect to find the root with fewer evaluations of the function. Many interpolants exist, but the simplest, a straight line, is perhaps the easiest and most reliable to use. This inverse linear interpolation seeks the zero of the line

$$y-f(u)=[f(v)-f(u)](b-u)/(v-u) \tag{13.23}$$

which by substitution of $y=0$ gives

$$b=[uf(v)-vf(u)]/[f(v)-f(u)]. \tag{13.24}$$

Once $f(b)$ has been evaluated, the interval can be reduced by testing for condition (13.21) as in the bisection process.

All the above discussion is very much a matter of common sense. However, in computer arithmetic the formulae may not give the expected results. For instance, consider the function depicted in figure 13.1(a). Suppose that the magnitude of $f(v)$ is very much greater than that of $f(u)$. Then formula (13.24) will return a value b very close to u. In fact, to the precision used, it may happen that b and u are identical and an iteration based on (13.24) may never converge. For this reason it is suggested that this procedure, known as the method of *False Position*, be combined with bisection to ensure convergence. In other words, after every few iterations using the False Position formula (13.24), a step is made using the bisection formula (13.20). In all of this discussion the practical problems of evaluating the function correctly have been discretely ignored. The algorithms will, of course, only work correctly if the function is accurately computed. Acton (1970, chap 1) discusses some of the difficulties involved in computing functions.

A final possibility which can be troublesome in practice is that either of the formulae (13.20) or (13.24) may result in a value for b *outside* the interval $[u, v]$ when applied in finite-precision arithmetic, a clearly impossible situation when the calculations are exact. This has already been discussed in §1.2 (p 6). Appropriate tests must be included to detect this, which can only occur when u and v are close to each other, that is, when the iteration is nearing convergence. The author has had several unfortunate experiences with algorithms which, lacking the appropriate tests, continued to iterate for hundreds of steps.

Some readers may wonder where the famous Newton's algorithm has disappeared in this discussion. In a manner similar to False Position, Newton's method seeks a root at the zero of the line

$$y-f(u)=f'(u)(b-u) \tag{13.25}$$

where the point $(u, f(u))$ has been used with $f'(u)$, the derivative at that point, to generate the straight line. Thus

$$b = u - f(u)/f'(u) \tag{13.26}$$

which gives the zero of the line, is suggested as the next point approximating the root of f and defines an iteration if b replaces u. The iteration converges very rapidly except in a variety of delightful cases which have occupied many authors (see, for instance, Acton 1970, chap 2, Henrici 1964, chap 4) and countless careless programmers for many hours. The difficulty occurs principally when $f'(u)$ becomes small. The bisection/False Position combination requires no derivatives and is thoroughly reliable if carefully implemented. The only situation which may upset it is that in which there is a discontinuity in the function in the interval $[u, v]$, when the algorithm may converge to the discontinuity if $f(b_-)f(b_+)<0$, where b_- and b_+ refer to values of b as approached from below and above.

It should be noted that the False Position formula (13.24) is an approximation to Newton's formula (13.26) by approximating

$$f'(u) = [f(v) - f(u)]/(v - u). \tag{13.27}$$

The root-finding algorithm based on (13.24) with any two points u, v instead of a pair which straddle at least one root is called the *secant algorithm.*

Algorithm 18. Root-finding by bisection and False Position

```
procedure root1d(var lbound, ubound: real; {range in which
                    root is to be found -- refined by procedure}
                    var ifn: integer; {to count function evaluations}
                    tol : real; {the width of the final interval
                    [lbound, ubound] within which a root is to be
                    located. Zero is an acceptable value.}
                    var noroot: boolean {to indicate that interval
                    on entry may not contain a root since both
                    function values have the same sign});
{alg18.pas == a root of a function of one variable
                    Copyright 1988 J.C.Nash
}
var
   nbis: integer;
   b, fb, flow, fup : real;
   notcomp: boolean;
begin
   writeln('alg18.pas -- root of a function of one variable');
   {STEP 0 -- partly in the procedure call}
   notcomp := false; {to set flag or else the 'known' root will be displayed
                    by the function routine}
   ifn := 2; {to initialize the function evaluation count}
   nbis := 5; {ensure a bisection every 5 function evaluations}
   fup := fn1d(ubound,notcomp);
   if notcomp then halt;
   flow := fn1d(lbound,notcomp);
   if notcomp then halt; {safety check}
   writeln('f(',lbound:8:5,')=',flow,' f(',ubound:8:5,')=',fup);
```

Algorithm 18. Root-finding by bisection and False Position (cont.)

```
    if fup*flow>0 then noroot := true else noroot := false; {STEP 1}
    while (not noroot) and ((ubound-lbound)>tol) do
    begin {main body of procedure to find root. Note that the test on
                noroot only occurs once.}
      {STEP 9b -- now placed here in Pascal version}
      if (nbis * ((ifn - 2) div nbis) = (ifn - 2)) then
      begin {STEP 10 -- order changed}
        write('Bisect ');
        b := lbound + 0.5*(ubound - lbound) {bisection of interval}
      end
      else
      begin {STEP 2}
        write('False P ');
        b := (lbound*fup-ubound*flow)/(fup-flow);{to compute false position b}
      end;
      {STEP 3 -- begin convergence tests}
      if b<=lbound then
      begin
        b := lbound; {to bring argument within interval again}
        ubound := lbound; {to force convergence, since the function
          argument b cannot be outside the interval [lbound, ubound]
          in exact arithmetic by either false position or bisection}
      end;
      if b>=ubound then {STEP 4}
      begin
        b := ubound; lbound := ubound; {as in STEP 3}
      end;
      ifn := ifn+1; {to count function evaluations} {STEP 5}
      fb := fn1d(b, notcomp);
      if notcomp then halt; {safety check}
      write(ifn,' evalns: f(',b:16,')=',fb:10);
      writeln(' width interval= ',(ubound-lbound):10);
      if (ubound-lbound)>tol then
      begin {update of interval}
        if fb*flow<0.0 then {STEP 6}
        begin {STEP 7}
          fup := fb; ubound := b; {since root is in [lbound, b]}
        end
        else {we could check the equal to zero case -- root found,
                but it will be picked up at STEPs 2, 3, or 4}
        begin
          flow := fb; lbound := b; {since root is in [b, ubound]}
        end; {else}
      end; {update of interval}
    end;{while loop}
    writeln('Converged to f(',b,')=',fb);
    writeln(' Final interval width =',ubound-lbound);
  end; {alg18.pas == root1d}
```

The algorithm above is able to halt if a point is encountered where the function is not computable. In later minimisation codes we will be able to continue the search for a minimum by presuming such points are not *local minima. However, in the present context of one-dimensional root-*

finding, we prefer to require the user to provide an interval in which at least one root exists and upon which the function is defined. The driver program DR1618.PAS *on the software diskette is intended to allow users to approximately localise roots of functions using grid search, followed by a call to algorithm 18 to refine the position of a suspected root.*

Example 13.2. A test of root-finding algorithms

In order to test the algorithm above it is useful to construct a problem which can be adapted to cause either the bisection or the False Position methods some relative difficulty. Therefore, consider finding the root of

$$f(b) = z * [\tanh(y) + w] \tag{13.28}$$

where

$$y = s * (b - t) \tag{13.29}$$

which has a root at

$$b^* = t + \ln[(1-w)/(1+w)]/(2s). \tag{13.30}$$

Note that z is used to provide a scale for the function, s scales the abscissa while t translates the root to right or left. The position of the root is determined by w to within the scaling s. Suppose now that s is large, for example $s = 100$. Then the function $f(b)$ will change very rapidly with b near the root, but otherwise will be approximated very well by

$$f(b) \simeq \begin{cases} z * (w-1) & \text{for } b \ll b^* \\ z * (w+1) & \text{for } b \gg b^*. \end{cases} \tag{13.31}$$

In fact using the grid search procedure (algorithm 16) we find the values in table 13.1 given $t = 0{\cdot}5$, $z = 100$, $s = 100$ and $w = 0{\cdot}99$. (The results in the table have been computed on a Data General NOVA having 23-bit binary arithmetic.)

The consequences of such behaviour can be quite serious for the False Position algorithm. This is because the linear approximation used is not valid, and a typical step using $u = 0$, $v = 1$ gives

$$b = 1{\cdot}00001/200 = 5{\cdot}00005\text{E}-3.$$

Since the root is near 0·473533, the progress is painfully slow and the method requires 143 iterations to converge. Bisection, on the other hand, converges in 24 iterations (nbis = 1 in the algorithm above). For nbis = 2, 25 iterations are required, while for nbis = 5, which is the suggested value, 41 iterations are needed. This may indicate that bisection should be a permanent strategy. However, the function (13.28) can be smoothed considerably by setting $w = 0{\cdot}2$ and $s = 1$, for which the root is found near 0·297268. In this case the number of iterations needed is again 24 for nbis = 1 (it is a function only of the number of bits in the machine arithmetic), 6 for nbis = 5 and also 6 if nbis is set to a large number so no bisections are performed. Figure 13.1 shows plots of the two functions obtained on a Hewlett–Packard 9830 calculator.

TABLE 13.1. Values found in example 13.2.

b	$f(b)$
0	−1·00001
0·1	−1·00001
0·2	−1·00001
0·3	−1·00001
0·4	−1·00001
0·41	−1·00001
0·42	−0·999987
0·43	−0·999844
0·44	−0·998783
0·45	−0·990939
0·46	−0·932944
0·47	−0·505471
0·48	2·5972
0·49	22·8404
0·5	98·9994
0·6	199
0·7	199
0·8	199
0·9	199
1·0	199

Example 13.3. Actuarial calculations

The computation of the premium for a given insurance benefit or the level of benefit for a given premium are also root-finding problems. To avoid over-simplifying a real-world situation and thereby presenting a possibly misleading image of what is a very difficult problem, consider the situation faced by some enterprising burglars who wish to protect their income in case of arrest. In their foresight, the criminals establish a cooperative fund into which each pays a premium p every period that the scheme operates. If a burglar is arrested he is paid a fixed benefit b. For simplicity, let the number of members of the scheme be fixed at m. This can of course be altered to reflect the arrests and/or admission of new members. The fund, in addition to moneys mp received as premiums in each period, may borrow money at a rate r_b to meet its obligations and may also earn money at rate r_e. The fund is started at some level f_0. The scheme, to operate effectively, should attain some target f_T after T periods of operation. However, in order to do this, it must set the premium p to offset the benefits $n_i b$ in each period i, where n_i is the number of arrests in this period. If historical data are available, then these could be used to *simulate* the behaviour of the scheme. However, historical data may reflect particular events whose timing may profoundly influence the profitability of the scheme. That is to say, the equation

$$F(p, \boldsymbol{n}) - f_T = 0 \tag{13.32}$$

may require a very much higher premium to be satisfied if all the arrests occur

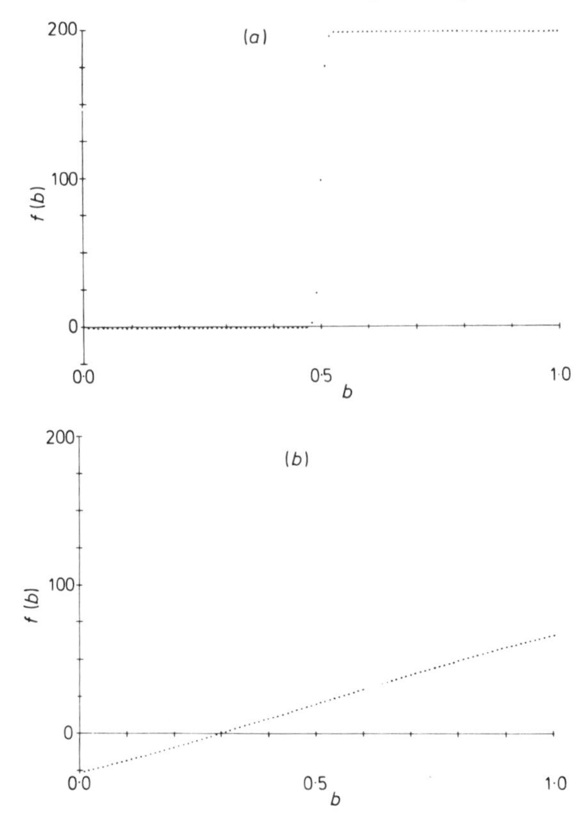

FIGURE 13.1. Function (13.28) for (*a*) $t=0{\cdot}5$, $z=100$, $s=100$, $w=0{\cdot}99$, and (*b*) $t=0{\cdot}5$, $z=100$, $s=1$, $w=0{\cdot}2$.

early in the simulation period than if they occur at the end. Therefore, it is likely that any sensible simulation will use root-finding to solve (13.32) for p for a variety of sets of arrest figures $\boldsymbol{n}$. In particular, a pseudo-random-number generator can be used to provide such sets of numbers chosen from some distribution or other. The function is then computed via one of the two recurrence relations

$$f_{i+1}(p)=f_i(p)(1+r_e)+mp(1+0{\cdot}5r_e)-n_ib \qquad \text{for } f_i(p)\geqslant 0 \qquad (13.33)$$

or

$$f_{i+1}(p)=f_i(p)(1+r_b)+mp(1+0{\cdot}5r_e)-n_ib \qquad \text{for } f_i(p)<0. \qquad (13.34)$$

Note that our shrewd criminals invest their premium money to increase the fund. The rate $0{\cdot}5r_e$ is used to take account of the continuous collection of premium payments over a period.

To give a specific example consider the following parameters: benefit $b=1200$, membership $m=2000$, interest rates $r_e=0{\cdot}08$ and $r_b=0{\cdot}15$, initial fund $f_0=0$ and after 10 periods $f_{10}=0$ (a non-profit scheme!). The root-finding algorithm is then applied using $u=0$, $v=2000$. Three sets of arrest figures were used to simulate the operation of the scheme. The results are given in table 13.2. The arrests are drawn from a uniform distribution on (0,400).

TABLE 13.2. Simulated operation of an income insurance program.

Premium =	92·90		86·07		109·92	
Period	n_i	$f_i(p)$	n_i	$f_i(p)$	n_i	$f_i(p)$
1	0	193237·50	17	158630·94	188	3029·50
2	2	399533·94	232	71952·31	315	146098·69
3	279	289934·12	317	−123660·62	194	−172183·94
4	124	357566·31	67	−43578·75	313	−344982·00
5	374	130609·06	74	40115·38	35	−210099·75
6	356	−92904·75	55	156355·50	7	−21385·19
7	101	−34802·94	152	165494·81	127	51636·50
8	281	−183985·87	304	−7034·69	387	−180003·12
9	23	−49546·25	113	35341·00	55	−44374·06
10	117	−0·69	181	−0·81	148	−0·69
Total	1657		1512		1769	
Function evaluations to find root		10		14		11
(Total benefits)/(number of premiums paid)		99·42		90·72		106·14

The last entry in each column is an approximation based on no interest paid or earned in the fund management. Thus

$$\text{approximate premium} = \text{total arrests} * b/(n * T)$$
$$= \text{total arrests} * 0{\cdot}06.$$

These examples were run in FORTRAN on an IBM 370/168.

Chapter 14

DIRECT SEARCH METHODS

14.1. THE NELDER–MEAD SIMPLEX SEARCH FOR THE MINIMUM OF A FUNCTION OF SEVERAL PARAMETERS

The first method to be examined for minimising a function of n parameters is a search procedure based on heuristic ideas. Its strengths are that it requires no derivatives to be computed, so it can cope with functions which are not easily written as analytic expressions (for instance, the result of simulations), and that it always increases the information available concerning the function by reporting its value at a number of points. Its weakness is primarily that it does not use this information very effectively, so may take an unnecessarily large number of function evaluations to locate a solution. Furthermore, the method requires $(n+1)$ by $(n+2)$ storage locations to hold intermediate results in the algorithm, so is not well adapted to problems in a large number of parameters. However, for more than about five parameters the procedure appears to become inefficient. Justification of the choice of this algorithm is postponed until §14.4.

The original paper of Nelder and Mead (1965) outlines quite clearly and succinctly their method, which is based on that of Spendley *et al* (1962). The method will be described here to point out some particular modifications needed to improve its reliability on small machines, particularly those with arithmetic having few digits in the mantissa.

A *simplex* is the structure formed by $(n+1)$ points, not in the same plane, in an n-dimensional space. The essence of the algorithm is as follows: the function is evaluated at each point (vertex) of the simplex and the vertex having the highest function value is replaced by a new point with a lower function value. This is done in such a way that 'the simplex adapts itself to the local landscape, and contracts on to the final minimum.' There are four main operations which are made on the simplex: reflection, expansion, reduction and contraction. In order to operate on the simplex, it is necessary to order the points so that the highest is $\boldsymbol{b}_H$, the next-to-highest $\boldsymbol{b}_N$, and the lowest $\boldsymbol{b}_L$. Thus the associated function values obey

$$S(\boldsymbol{b}_H) \geqslant S(\boldsymbol{b}_N) \geqslant S(\boldsymbol{b}_i) \geqslant S(\boldsymbol{b}_L) \tag{14.1}$$

for all $i \neq$ H, N or L. Figure 14.1 illustrates the situation.

The centroid of all the points other than $\boldsymbol{b}_H$ is defined by

$$\boldsymbol{b}_C = \left(\sum_{\substack{j=1 \\ j \neq H}}^{n+1} \boldsymbol{b}_j \right) \Big/ n. \tag{14.2}$$

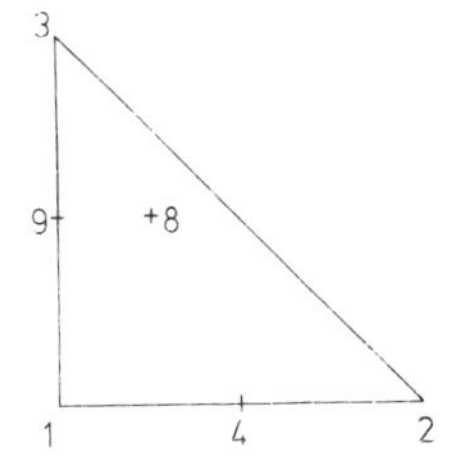

FIGURE 14.1. Points generated by the Nelder–Mead simplex algorithm in two dimensions. Point 1, $\boldsymbol{b}_{\mathrm{L}}$, the lowest vertex in the simplex; point 2, $\boldsymbol{b}_{\mathrm{N}}$, the next-to-highest vertex; and point 3, $\boldsymbol{b}_{\mathrm{H}}$, the highest vertex. Point 4, $\boldsymbol{b}_{\mathrm{C}}$, the centroid of all points except $\boldsymbol{b}_{\mathrm{H}}$, that is, of $\boldsymbol{b}_{\mathrm{N}}$ and $\boldsymbol{b}_{\mathrm{L}}$. Also one of the points generated by a general contraction of the simplex towards $\boldsymbol{b}_{\mathrm{L}}$. Point 5, $\boldsymbol{b}_{\mathrm{R}}$, the reflection of $\boldsymbol{b}_{\mathrm{H}}$ through $\boldsymbol{b}_{\mathrm{C}}$; point 6, $\boldsymbol{b}_{\mathrm{E}}$, the result of extension of the line ($\boldsymbol{b}_{\mathrm{C}}$, $\boldsymbol{b}_{\mathrm{R}}$); point 7, the result of reduction of the line ($\boldsymbol{b}_{\mathrm{C}}$, $\boldsymbol{b}_{\mathrm{R}}$) which occurs when $\boldsymbol{b}_{\mathrm{R}}$ is lower than $\boldsymbol{b}_{\mathrm{H}}$ but higher than $\boldsymbol{b}_{\mathrm{N}}$; and point 8, the result of reduction of the line ($\boldsymbol{b}_{\mathrm{C}}$, $\boldsymbol{b}_{\mathrm{H}}$). Point 9, one of the points of the simplex generated by a general contraction of the simplex made up of vertices 1, 2 and 3 towards $\boldsymbol{b}_{\mathrm{L}}$.

The operation of *reflection* then reflects $\boldsymbol{b}_{\mathrm{H}}$ through $\boldsymbol{b}_{\mathrm{C}}$ using a reflection factor α, that is

$$\begin{aligned}\boldsymbol{b}_{\mathrm{R}} &= \boldsymbol{b}_{\mathrm{C}} + \alpha(\boldsymbol{b}_{\mathrm{C}} - \boldsymbol{b}_{\mathrm{H}}) \\ &= (1+\alpha)\boldsymbol{b}_{\mathrm{C}} - \alpha\boldsymbol{b}_{\mathrm{H}}. \qquad (14.3)\end{aligned}$$

If $S(\boldsymbol{b}_{\mathrm{R}})$ is less than $S(\boldsymbol{b}_{\mathrm{L}})$ a new lowest point has been found, and the simplex can be expanded by extending the line $(\boldsymbol{b}_{\mathrm{R}} - \boldsymbol{b}_{\mathrm{C}})$ to give the point

$$\begin{aligned}\boldsymbol{b}_{\mathrm{E}} &= \boldsymbol{b}_{\mathrm{R}} + (\gamma - 1)(\boldsymbol{b}_{\mathrm{R}} - \boldsymbol{b}_{\mathrm{C}}) \\ &= \gamma\boldsymbol{b}_{\mathrm{R}} + (1-\gamma)\boldsymbol{b}_{\mathrm{C}} \\ &= (1+\alpha\gamma)\boldsymbol{b}_{\mathrm{C}} - \alpha\gamma\boldsymbol{b}_{\mathrm{H}} \qquad (14.4)\end{aligned}$$

where γ, the expansion factor, is greater than unity or else (14.4) represents a contraction. If $S(\boldsymbol{b}_{\mathrm{E}}) < S(\boldsymbol{b}_{\mathrm{R}})$ then $\boldsymbol{b}_{\mathrm{H}}$ is replaced by $\boldsymbol{b}_{\mathrm{E}}$ and the procedure repeated by finding a new highest point and a new centroid of n points $\boldsymbol{b}_{\mathrm{C}}$. Otherwise $\boldsymbol{b}_{\mathrm{R}}$ is the new lowest point and it replaces $\boldsymbol{b}_{\mathrm{H}}$.

In the case where $\boldsymbol{b}_R$ is not a new lowest point, but is less than $\boldsymbol{b}_N$, the next-to-highest point, that is

$$S(\boldsymbol{b}_L) \leqslant S(\boldsymbol{b}_R) < S(\boldsymbol{b}_N) \tag{14.5}$$

$\boldsymbol{b}_H$ is replaced by $\boldsymbol{b}_R$ and the procedure repeated. In the remaining situation, we have $S(\boldsymbol{b}_R)$ at least as great as $S(\boldsymbol{b}_N)$ and should *reduce* the simplex.

There are two possibilities. (*a*) If

$$S(\boldsymbol{b}_N) \leqslant S(\boldsymbol{b}_R) < S(\boldsymbol{b}_H) \tag{14.6}$$

then the reduction is made by replacing $\boldsymbol{b}_H$ by $\boldsymbol{b}_R$ and finding a new vertex between $\boldsymbol{b}_C$ and $\boldsymbol{b}_R$ (now $\boldsymbol{b}_H$). This is a reduction on the side of the reflection ('low' side). (*b*) If

$$S(\boldsymbol{b}_R) > S(\boldsymbol{b}_H) \tag{14.7}$$

the reduction is made by finding a new vertex between $\boldsymbol{b}_C$ and $\boldsymbol{b}_H$ ('high' side).

In either of the above cases the reduction is controlled by a factor β between 0 and 1. Since case (*a*) above replaces $\boldsymbol{b}_H$ by $\boldsymbol{b}_R$, the same formula applies for the new point $\boldsymbol{b}_S$ ('S' denotes that the simplex is smaller) in both cases. $\boldsymbol{b}_H$ is used to denote both $\boldsymbol{b}_R$ and $\boldsymbol{b}_H$ since in case (*a*) $\boldsymbol{b}_R$ has become the new highest point in the simplex

$$\begin{aligned} \boldsymbol{b}_S &= \boldsymbol{b}_C + \beta(\boldsymbol{b}_H - \boldsymbol{b}_C) \\ &= \beta\boldsymbol{b}_H + (1-\beta)\boldsymbol{b}_C. \end{aligned} \tag{14.8}$$

The new point $\boldsymbol{b}_S$ then replaces the current $\boldsymbol{b}_H$, which in case (*a*) is, in fact, $\boldsymbol{b}_R$, unless

$$S(\boldsymbol{b}_S) > \min(S(\boldsymbol{b}_H), S(\boldsymbol{b}_R)). \tag{14.9}$$

The replacement of $\boldsymbol{b}_H$ by $\boldsymbol{b}_R$ in case (*a*) will, in an implementation, mean that this minimum has already been saved with its associated point. When (14.9) is satisfied a reduction has given a point higher than $S(\boldsymbol{b}_N)$, so a general contraction of the simplex about the lowest point so far, $\boldsymbol{b}_L$, is suggested. That is

$$\begin{aligned} \boldsymbol{b}_i' &= \boldsymbol{b}_L + \beta'(\boldsymbol{b}_i - \boldsymbol{b}_L) \\ &= \beta'\boldsymbol{b}_i + (1-\beta')\boldsymbol{b}_L \end{aligned} \tag{14.10}$$

for all $i \neq$ L. In exact arithmetic, (14.10) is acceptable for all points, and the author has in some implementations omitted the test for $i =$ L. Some caution is warranted, however, since some machines can form a mean of two numbers which is *not* between those two numbers. Hence, the point $\boldsymbol{b}_L$ may be altered in the operations of formula (14.10).

Different contraction factors β and β' may be used in (14.8) and (14.10). In practice these, as well as α and γ can be chosen to try to improve the rate of convergence of this procedure either for a specific class or for a wide range of problems. Following Nelder and Mead (1965), I have found the strategy

$$\alpha = 1 \qquad \gamma = 2 \qquad \beta' = \beta = 0{\cdot}5 \tag{14.11}$$

to be effective. It should be noted, however, that the choice of these values is based on limited testing. In fact, every aspect of this procedure has been evolved heuristically based on a largely intuitive conception of the minimisation problem. As such there will always be functions which cannot be minimised by this method because they do not conform to the idealisation. Despite this, the algorithm is surprisingly robust and, if permitted to continue long enough, almost always finds the minimum.

The thorniest question concerning minimisation algorithms must, therefore, be addressed: when has the minimum been found? Nelder and Mead suggest using the 'standard error' of the function values

$$\text{test} = \left[\left(\sum_{j=1}^{n+1} [S(\boldsymbol{b}_j) - \bar{S}]^2\right) \Big/ n\right]^{1/2} \tag{14.12}$$

where

$$\bar{S} = \sum_{j=1}^{n+1} S(\boldsymbol{b}_j)/(n+1). \tag{14.13}$$

The procedure is taken to have converged when the test value falls below some preassigned tolerance. In the statistical applications which interested Nelder and Mead, this approach is reasonable. However, the author has found this criterion to cause premature termination of the procedure on problems with fairly flat areas on the function surface. In a statistical context one might wish to stop if such a region were encountered, but presuming the minimum is sought, it seems logical to use the simpler test for equality between $S(\boldsymbol{b}_L)$ and $S(\boldsymbol{b}_H)$, that is, a test for equal height of all points in the simplex.

An additional concern on machines with low-precision arithmetic is that it is possible for a general contraction (14.10) *not* to reduce the simplex size. Therefore, it is advisable to compute some measure of the simplex size during the contraction to ensure a decrease in the simplex size, as there is no point in continuing if the contraction has not been effective. A very simple measure is the sum

$$\text{size} = \sum_{j=1}^{n+1} \|\boldsymbol{b}_j - \boldsymbol{b}_L\| \tag{14.14}$$

where

$$\|\boldsymbol{b}_j - \boldsymbol{b}_L\| = \sum_{i=1}^{n} |b_{ij} - b_{iL}|. \tag{14.15}$$

Finally, it is still possible to converge at a point which is not the minimum. If, for instance, the $(n+1)$ points of the simplex are all in one plane (which is a line in two dimensions), the simplex can only move in $(n-1)$ directions in the n-dimensional space and may not be able to proceed towards the minimum. O'Neill (1971), in a FORTRAN implementation of the Nelder–Mead ideas, tests the function value at either side of the supposed minimum along each of the parameter axes. If any function value is found lower than the current supposed minimum, then the procedure is restarted.

The author has found the axial search to be useful in several cases in avoiding false convergence. For instance, in a set of 74 tests, six failures of the procedure were observed. This figure would have been 11 failures without the restart facility.

14.2. POSSIBLE MODIFICATIONS OF THE NELDER–MEAD ALGORITHM

Besides choices for α, β, β' and γ other than (14.11) there are many minor variations on the basic theme of Nelder and Mead. The author has examined several of these, mainly using the Rosenbrock (1960) test function of two parameters

$$S(\boldsymbol{b}) = 100(b_2 - b_1^2)^2 + (1 - b_1)^2 \tag{14.16}$$

starting at the point (−1·2, 1).

(i) The function value $S(\boldsymbol{b}_C)$ can be computed at each iteration. If $S(\boldsymbol{b}_C) < S(\boldsymbol{b}_L)$, $\boldsymbol{b}_L$ is replaced by $\boldsymbol{b}_C$. The rest of the procedure is unaffected by this change, which is effectively a contraction of the simplex. If there are more than two parameters, the computation of $\boldsymbol{b}_C$ can be repeated. In cases where the minimum lies within the current simplex, this modification is likely to permit rapid progress towards the minimum. Since, however, the simplex moves by means of reflection and expansion, the extra function evaluation is often unnecessary, and in tests run by the author the cost of this evaluation outweighed the benefit.
(ii) In the case that $S(\boldsymbol{b}_R) < S(\boldsymbol{b}_L)$ the simplex is normally expanded by extension along the line $(\boldsymbol{b}_R - \boldsymbol{b}_C)$. If $\boldsymbol{b}_R$ is replaced by $\boldsymbol{b}_E$, the formulae contained in the first two lines of equation (14.4) permit the expansion to be repeated. This modification suffers the same disadvantages as the previous one; the advantages of the repeated extension are not great enough—in fact do not occur often enough—to offset the cost of additional function evaluations.
(iii) Instead of movement of the simplex by reflection of $\boldsymbol{b}_H$ through $\boldsymbol{b}_C$, one could consider extensions along the line $(\boldsymbol{b}_L - \boldsymbol{b}_C)$, that is, from the 'low' vertex of the simplex. Simple drawings of the two-dimensional case show that this tends to stretch the simplex so that the points become coplanar, forcing restarts. Indeed, a test of this idea produced precisely this behaviour.
(iv) For some sets of parameters $\boldsymbol{b}$, the function may not be computable, or a constraint may be violated (if constraints are included in the problem). In such cases, a very large value may be returned for the function to prevent motion in the direction of forbidden points. Box (1965) has enlarged on this idea in his Complex Method which uses more than $(n+1)$ points in an attempt to prevent all the points collapsing onto the constraint.
(v) The portion of the algorithm for which modifications remain to be suggested is the starting (and restarting) of the procedure. Until now, little mention has been made of the manner in which the original simplex should be generated. Nelder and Mead (1965) performed a variety of tests using initial simplexes generated by equal step lengths along the parameter axes and various 'arrangements of the initial simplex.' The exact meaning of this is not specified. They found the rate of convergence to be influenced by the step length chosen to generate an initial simplex. O'Neill (1971) in his FORTRAN implementation permits the step along each parameter axis to be specified separately, which permits differences in the scale of the parameters to be accommodated by the program. On restarting, these steps are reduced by a factor of 1000. General rules on how step lengths should

be chosen are unfortunately difficult to state. Quite obviously any starting step should appreciably alter the function. In many ways this is an $(n+1)$-fold repetition of the necessity of good initial estimates for the parameters as in §12.2.

More recently other workers have tried to improve upon the Nelder–Mead strategies, for example Craig *et al* (1980). A parallel computer version reported by Virginia Torczon seems to hold promise for the solution of problems in relatively large numbers of parameters. Here we have been content to stay close to the original Nelder–Mead procedure, though we have simplified the method for ranking the vertices of the polytope, in particular the selection of the point $\boldsymbol{b}_N$.

Algorithm 19. A Nelder–Mead minimisation procedure

```
procedure nmmin(n: integer; {the number of parameters in the
                    function to be minimised}
                    var Bvec,X: rvector; {the parameter values on
                    input (Bvec) and output (X) from minmeth}
                    var Fmin: real; {'minimum' function value}
                    Workdata: probdata; {user defined data area}
                    var fail: boolean; {true if method has failed}
                    var intol: real); {user-initialized convergence
                    tolerance; zero on entry if it is not set yet.}
{alg19.pas == Nelder Mead minimisation of a function of n parameters.
        Original method due to J. Nelder and R. Mead, Computer Journal,
          vol 7, 1965 pp. 308-313.
        Modification as per Nash J and Walker-Smith M, Nonlinear Parameter
        Estimation: an Integrated System in BASIC, Marcel Dekker: New York,
        1987.
        Modifications are principally
        - in the computation of the "next to highest" vertex of the current
        polytope,
        - in the verification that the shrink operation truly reduces the size
        of the polytope, and
        - in form of calculation of some of the search points.
        We further recommend an axial search to verify convergence. This can
        be called outside the present code. If placed in-line, the code can
        be restarted at STEP3.
        If space is at a premium, vector X is not needed except to return
        final values of parameters.
                    Copyright 1988 J.C.Nash
}
const
    Pcol = 27; {Maxparm + 2 == maximum number of columns in polytope}
    Prow = 26; {Maxparm + 1 == maximum number of rows in polytope}
    alpha = 1.0; {reflection factor}
    beta = 0.5; {contraction and reduction factor}
    gamma = 2.0; {extension factor}
var
    action : string[15]; {description of action attempted on polytope. The
                    program does not inform the user of the success of
                    the attempt. However, the modifications to do
                    this are straightforward.}
```

Algorithm 19. A Nelder–Mead minimisation procedure (cont.)

```
    C : integer; {pointer column in workspace P which stores the
                        centroid of the polytope. C is set to n+2) here.}
    calcvert : boolean; {true if vertices to be calculated, as at start
                        or after a shrink operation}
    convtol : real; {a convergence tolerance based on function
                        value differences}
    f : real; {temporary function value}
    funcount : integer; {count of function evaluations}
    H : integer; {pointer to highest vertex in polytope}
    i,j : integer; {working integers}
    L : integer; {pointers to lowest vertex in polytope}
    notcomp : boolean; {non-computability flag}
    n1 : integer; {n+1}
    oldsize : real; {former size measure of polytope}
    P : array[1..Prow,1..Pcol] of real; {polytope workspace}
    shrinkfail: boolean; {true if shrink has not reduced polytope size}
    size : real; { a size measure for the polytope }
    step : real; {stepsize used to build polytope}
    temp : real; {a temporary variable}
    trystep : real; {a trial stepsize}
    tstr : string[5]; {storage for function counter in string form
                        to control display width}
    VH,VL,VN : real; {function values of 'highest','lowest' and
                        'next' vertices}
    VR : real; {function value at Reflection}
begin
    writeln('Nash Algorithm 19 version 2 1988-03-17');
    writeln(' Nelder Mead polytope direct search function minimiser');
    fail := false; {the method has not yet failed!}
    f := fminfn(n,Bvec,Workdata,notcomp); {initial fn calculation -- STEP 1}
    if notcomp then
    begin
        writeln('**** Function cannot be evaluated at initial parameters ****');
        fail := true; {method has failed -- cannot get started}
    end
    else
    begin {proceed with minimisation}
        writeln('Function value for initial parameters = ',f);
        if intol<0.0 then intol := Calceps;
        funcount := 1; {function count initialized to 1}
        convtol := intol*(abs(f)+intol); {ensures small value relative to
                        function value -- note that we can restart the procedure if
                        this is too big due to a large initial function value.}
        writeln(' Scaled convergence tolerance is ',convtol);
        n1 := n+1; C := n+2; P[n1,1] := f; {STEP 2}
        for i := 1 to n do P[i,1] := Bvec[i];
        {This saves the initial point as vertex 1 of the polytope.}
        L := 1; {We indicate that it is the 'lowest' vertex at the moment, so
                        that its funtion value is not recomputed later in STEP 10}
        size := 0.0; {STEP 3}
        {STEP 4: build the initial polytope using a fixed step size}
        step := 0.0;
```

Algorithm 19. A Nelder–Mead minimisation procedure (cont.)

```
        for i := 1 to n do if 0.1*abs(Bvec[i])>step then step := 0.1*abs(Bvec[i]);
        writeln('Stepsize computed as ',step);
        for j := 2 to n1 do {main loop to build polytope} {STEP 5}
        begin {STEP 6}
            action := 'BUILD ';
            for i := 1 to n do P[i,j] := Bvec[i]; {set the parameters}
            {alternative strategy -- variable step size -- in the build phase}
            { step := 0.1*abs(Bvec[j-1])+0.001; }
            {Note the scaling and avoidance of zero stepsize.}
            trystep := step; {trial step -- STEP 7}
            while P[j-1,j]=Bvec[j-1] do
            begin
                P[j-1,j] := Bvec[j-1]+trystep; trystep := trystep*10;
            end; {while}
            size := size+trystep; {to compute a size measure for polytope -- STEP 8}
        end; {loop on j for parameters}
        oldsize := size; {to save the size measure -- STEP 9}
        calcvert := true; {must calculate vertices when polytope is new}
        shrinkfail := false; {initialize shrink failure flag so we don't have
                        false convergence}
        repeat {main loop for Nelder-Mead operations -- STEP 10}
            if calcvert then
            begin
                for j := 1 to n1 do {compute the function at each vertex}
                begin
                    if j<>L then {We already have function value for L(owest) vertex.}
                    begin
                        for i := 1 to n do Bvec[i] := P[i,j]; {get the parameter values}
                        f := fminfn(n,Bvec,Workdata,notcomp); {function calculation}
                        if notcomp then f := big; funcount := funcount+1; P[n1,j] := f;
                    end; {if j<>L clause}
                end; {loop on j to compute polytope vertices}
                calcvert := false; {remember to reset flag so we don't calculate
                        vertices every cycle of algorithm}
            end; {calculation of vertices}
            {STEP 11: find the highest and lowest vertices in current polytope}
            VL := P[n1,L]; {supposedly lowest value}
            VH := VL; {highest value must hopefully be higher}
            H := L; {pointer to highest vertex initialized to L}
            {Now perform the search}
            for j := 1 to n1 do
            begin
                if j<>L then
                begin
                    f := P[n1,j]; {function value at vertex j}
                    if f<VL then
                    begin
                        L := j; VL := f; {save new 'low'}
                    end;
                    if f>VH then
                    begin
```

Algorithm 19. A Nelder–Mead minimisation procedure (cont.)

```
                H := j; VH := f; {save new 'high'}
            end;
        end; {if j<>L}
    end; {search for highest and lowest}
    {STEP 12: test and display current polytope information}
    if VH>VL+convtol then
    begin {major cycle of the method}
        str(funcount:5,tstr); {this is purely for orderly display of results
                in aligned columns}
        writeln(action,tstr,' ',VH,' ',VL);
        VN := beta*VL+(1.0-beta)*VH;
        {interpolate to get "next to highest" function value -- there are
            many options here, we have chosen a fairly conservative one.}
        for i := 1 to n do {compute centroid of all but point H -- STEP 13}
        begin
            temp := -P[i,H]; {leave out point H by subtraction}
            for j := 1 to n1 do temp := temp+P[i,j];
            P[i,C] := temp/n; {centroid parameter i}
        end; {loop on i for centroid}
        for i := 1 to n do {compute reflection in Bvec[] -- STEP 14}
                Bvec[i] := (1.0+alpha)*P[i,C]-alpha*P[i,H];
        f := fminfn(n,Bvec,Workdata,notcomp); {function value at refln point}
        if notcomp then f := big; {When function is not computable, a very
                large value is assigned.}
        funcount := funcount+1; {increment function count}
        action := 'REFLECTION '; {label the action taken}
        VR := f; {STEP 15: test if extension should be tried}
        if VR<VL then
        begin {STEP 16: try extension}
            P[n1,C] := f; {save the function value at reflection point}
            for i := 1 to n do
            begin
                f := gamma*Bvec[i]+(1-gamma)*P[i,C];
                P[i,C] := Bvec[i]; {save the reflection point in case we need it}
                Bvec[i] := f;
            end; {loop on i for extension point}
            f := fminfn(n,Bvec,Workdata,notcomp); {function calculation}
            if notcomp then f := big; funcount := funcount+1;
            if f<VR then {STEP 17: test extension}
            begin {STEP 18: save extension point, replacing H}
                for i := 1 to n do P[i,H] := Bvec[i];
                P[n1,H] := f; {and its function value}
                action := 'EXTENSION '; {change the action label}
            end {replace H}
            else
            begin {STEP 19: save reflection point}
                for i := 1 to n do P[i,H] := P[i,C];
                P[n1,H] := VR; {save reflection function value}
            end {save reflection point in H; note action is still reflection}
        end {try extension}
        else {reflection point not lower than current lowest point}
        begin {reduction and shrink -- STEP 20}
```

Algorithm 19. A Nelder–Mead minimisation procedure (cont.)

```
action := 'HI-REDUCTION '; { default to hi-side reduction}
if VR<VH then {save reflection -- then try reduction on lo-side
    if function value not also < VN}
begin {STEP 21: replace H with reflection point}
    for i := 1 to n do P[i,H] := Bvec[i];
    P[n1,H] := VR; {and save its function value}
    action := 'LO-REDUCTION ';{re-label action taken}
end; {R replaces H so reduction on lo-side}
{STEP 22: carry out the reduction step}
for i := 1 to n do Bvec[i] := (1-beta)*P[i,H]+beta*P[i,C];
f := fminfn(n,Bvec,Workdata,notcomp); {function calculation}
if notcomp then f := big; funcount := funcount+1;
{STEP 23: test reduction point}
if f<P[n1,H] then {replace H -- may be old R in this case,
    so we do not use VH in this comparison}
begin {STEP 24: save new point}
    for i := 1 to n do P[i,H] := Bvec[i];
    P[n1,H] := f; {and its function value, which may now not be
    the highest in polytope}
end {replace H}
else {not a new point during reduction}
{STEP 25: test for failure of all tactics so far to reduce the
    function value. Note that this cannot be an 'else' statement
    from the 'if VR<VH' since we have used that statement in STEP
    21 to save the reflection point as a prelude to lo-reduction
    tactic, a step which is omitted when we try the hi-reduction,
    which has failed to reduce the function value.}
if VR>=VH then {hi-reduction has failed to find a point lower
    than H, and reflection point was also higher}
begin {STEP 26: shrink polytope toward point L}
    action := 'SHRINK ';
    calcvert := true; {must recalculate the vertices after this}
    size := 0.0;
    for j := 1 to n1 do
    begin
    if j<>L then {ignore the low vertex}
    for i := 1 to n do
    begin
    P[i,j] := beta*(P[i,j]-P[i,L])+P[i,L]; {note the form of
    expression used to avoid rounding errors}
    size := size+abs(P[i,j]-P[i,L]);
    end; {loop on i and if j<>L}
    end; {loop on j}
    if size<oldsize then {STEP 27 -- test if shrink reduced size}
    begin {the new polytope is 'smaller', so we can proceed}
    shrinkfail := false; {restart after shrink}
    oldsize := size;
    end
    else {shrink failed -- polytope has not shrunk}
    begin {STEP 28 -- exit on failure}
    writeln('Polytope size measure not decreased in shrink');
    shrinkfail := true;
```

Algorithm 19. A Nelder–Mead minimisation procedure (cont.)

```
                    end;{ else shrink failed}
                  end; {if VR>=VH -- shrink}
               end; {reduction}
            end; {if VH>VL+...}
            {STEP 29 -- end of major cycle of method}
         until ((VH<=VL+convtol) or shrinkfail );
         {STEP 30: if progress made, or polytope shrunk successfully, try
                    another major cycle from STEP 10}
      end; {begin minimisation}
      {STEP 31}{save best parameters and function value found}
      writeln('Exiting from Alg19.pas Nelder Mead polytope minimiser');
      writeln(' ',funcount,' function evaluations used');
      Fmin := P[n1,L]; {save best value found}
      for i := 1 to n do X[i] := P[i,L];
      if shrinkfail then fail := true;
      {STEP 32}{exit}
   end; {alg19.pas == nmmin}
```

14.3. AN AXIAL SEARCH PROCEDURE

The following procedure is a device designed primarily to generate information concerning the shape of the surface $S(\boldsymbol{b})$ near the minimum, which will be labelled $\boldsymbol{b}^*$. In some instances, where a minimisation algorithm has converged prematurely, the axial search may reveal a point having a lower function value than that at the supposed minimum. The step taken along each axis is somewhat arbitrary. O'Neill (1971), for instance, in applying the axial search after a Nelder–Mead program, uses $\pm 0{\cdot}001$ times the initial steps used to generate the simplex. However, these latter increments must be supplied by the program user. I prefer to adjust the parameter b_i by an increment

$$s = e(|b_i| + e) \tag{14.17}$$

where e is some small number such as the square root of the machine precision. Section 18.2 gives a discussion of this choice. Its principal advantage is that the increment always adjusts the parameter. Alternatively, I have employed the assignment

$$b_i := b_i(1 \pm 0{\cdot}001) \tag{14.18}$$

unless these fail to change the parameter, in which case I use

$$b_i := \pm 0{\cdot}001. \tag{14.19}$$

The latter axial search was used in the set of tests used to compare algorithms for function minimisation which were included in the first edition and reported in §18.4 of both editions. What follows below reflects our current usage, including some measures of the curvature and symmetry of the functional surface near the presumed minimum.

Algorithm 20. Axial search

```
procedure axissrch(n: integer; {the number of parameters in the
                         function to be minimised}
                         var Bvec: rvector; {the parameter values on
                         input (Bvec) and output (X) from minmeth}
                         var Fmin: real; {'minimum' function value}
                         var lowerfn: boolean; {set true if lower value
                         is found during the axial search}
                         Workdata: probdata); {user defined data area}
{alg20.pas == axial search verification of function minimum
                         for a function of n parameters.
        Note: in this version, function evaluations are not counted.
                         Copyright 1988 J.C.Nash
}
var
    cradius, eps, f, fplus, step, temp, tilt : real;
    i : integer;
    notcomp : boolean;
begin
    writeln('alg20.pas -- axial search');
    eps := calceps; {machine precision}
    eps := sqrt(eps); {take its square root for a step scaling}
    writeln(' Axis':6,' Stepsize ':14,'function + ':14,
              'function - ':14,' rad. of curv.':14,' tilt');
    lowerfn := false; {initially no lower function value than fmin exists
                         for the function at hand}
    for i := 1 to n do {STEP 1}
    begin
        if (not lowerfn) then
        begin {STEP 2}
            temp := Bvec[i]; {to save the parameter value}
            step := eps*(abs(temp)+eps); {the change in the parameter -- STEP 3}
            Bvec[i] := temp+step; {STEP 4}
            f := fminfn(n, Bvec, Workdata, notcomp); {function calculation}
            if notcomp then f := big; {substitution of a large value for
                         non-computable function}
            write(i:5,' ',step:12,' ',f:12,' ');
        end; {step forwards}
        if f<fmin then lowerfn := true; {STEP 5}
        if (not lowerfn) then
        begin
            fplus := f; {to save the function value after forward step}
            Bvec[i] := temp-step; {STEP 6}
            f := fminfn(n,Bvec,Workdata,notcomp); {function calculation}
            if notcomp then f := big; {substitution of a large value for
                         non-computable function}
            write(f:12,' ');
        end; {step backwards}
        if f<fmin then lowerfn := true; {STEP 7}
        if (not lowerfn) then {STEP 8}
        begin
            Bvec[i] := temp; {to restore parameter value}
            {compute tilt and radius of curvature}
```

Algorithm 20. Axial search (cont.)

```
            temp := 0.5*(fplus-f)/step; {first order parabola coefficient}
            fplus := 0.5*(fplus+f-2.0*fmin)/(step*step);
            {2nd order parabolic coefficient - 0th order is fmin}
            if fplus<>0.0 then {avoid zero divide}
            begin
              cradius := 1.0+temp*temp;
              cradius := cradius*sqrt(cradius)/fplus; {radius of curvature}
            end
            else
              cradius := big; {a safety measure}
            tilt := 45.0*arctan(temp)/arctan(1.0); {rem tilt in degrees}
            write(cradius:12,' ',tilt:12);
        end;
        writeln; {to advance printer to next line}
    end; {loop on i -- STEP 9}
  end; {alg20.pas == axissrch -- STEP 10}
```

Example 14.1. Using the Nelder–Mead simplex procedure (algorithm 19)

Consider a (time) series of numbers

$$Q_t \qquad t=1,2,\ldots,m.$$

A transformation of this series to

$$P_t = Q_t - b_1 Q_{t-1} - b_2 Q_{t-2} \qquad t=1,2,\ldots,m$$

will have properties different from those of the original series. In particular, the *autocorrelation coefficient* of order k is defined (Kendall 1973) as

$$r_k = (m-2)\left(\sum_{t=k+3}^{m} [(P_t-u)(P_{t-k}-u)]\right)\left((m-k-2)\sum_{t=3}^{m}(P-u)^2\right)^{-1}.$$

The following output was produced with driver program DR1920, but has been edited for brevity. The final parameters are slightly different from those given in the first edition, where algorithm 19 was run in much lower precision. Use of the final parameters from the first edition (1·1104, −0·387185) as starting parameters for the present code gives an apparent minimum.

```
Minimum function value found =  2.5734305415E-24
At parameters
B[1]=  1.1060491080E+00
B[2]= -3.7996531780E-01
```

```
dr1920.pas -- driver for Nelder-Mead minimisation
1989/01/25   15:21:37
File for input of control data ([cr] for keyboard) ex14-1
File for console image ([cr] = nul) d:test14-1.
Function: Jaffrelot Minimisation of First Order ACF
  25.02000  25.13000  25.16000  23.70000  22.09000  23.39000  26.96000
```

```
 27.56000  28.95000  27.32000  29.38000  27.81000  26.78000  28.75000
 32.16000  30.10000  29.02000  26.76000  29.18000  26.59000  26.79000
 26.95000  28.25000  27.29000  27.82000  31.07000  36.59000  38.37000
 40.68000  36.01000  34.34000  33.58000  32.46000  31.61000  30.26000
 28.57000  28.23000  28.69000  33.60000  33.63000  33.61000  34.19000
 37.13000  37.81000  38.34000  33.40000  30.85000  26.99000  25.63000
 23.74000  26.21000  27.58000  33.32000  34.91000  39.95000  41.97000
 49.36000  48.98000  63.49000  57.74000  50.78000  42.25000  53.00000
Enter starting parameters
starting point (  1.0000000000E+00,  1.0000000000E+00)
Nash Algorithm 19 version 2 1988-03-17
  Nelder Mead polytope direct search function minimiser
Function value for initial parameters =   6.2908504924E-01
  Scaled convergence tolerance is   1.1442990385E-12
Stepsize computed as   1.0000000000E-01
BUILD             3   6.5569689140E-01   6.2908504924E-01
EXTENSION         5   6.5006359306E-01   5.9698860501E-01
EXTENSION         7   6.2908504924E-01   5.1102081991E-01
EXTENSION         9   5.9698860501E-01   2.7249463178E-01
EXTENSION        11   5.1102081991E-01   1.8450323330E-02
LO-REDUCTION     13   2.7249463178E-01   1.8450323330E-02
LO-REDUCTION     15   4.0644968776E-02   1.8450323330E-02
REFLECTION       17   3.2560286564E-02   5.8957995328E-03
HI-REDUCTION     19   1.8450323330E-02   3.5399834634E-04
...
LO-REDUCTION     67   3.0354380079E-11   8.8619333373E-12
HI-REDUCTION     69   1.8902369403E-11   4.9132841645E-13
LO-REDUCTION     71   8.8619333373E-12   1.0921611719E-13
HI-REDUCTION     73   1.7534478538E-12   1.0921611719E-13
Exiting from Alg19.pas Nelder Mead polytope minimiser
    75 function evaluations used

 Minimum function value found =  1.0921611719E-13
 At parameters
 B[1]=  1.1204166839E+00
 B[2]= -4.0364239252E-01
alg20.pas -- axial search
  Axis    Stepsize    function +    function -    rad. of curv.   tilt
    1 1.511270E-06  1.216470E-11  6.654846E-12  2.455700E-01  1.044457E-04
    2 5.446594E-07  1.265285E-12  4.698094E-14
Lower function value found
Nash Algorithm 19 version 2 1988-03-17
  Nelder Mead polytope direct search function minimiser
Function value for initial parameters =   4.6980937735E-14
  Scaled convergence tolerance is   3.3941802781E-24
Stepsize computed as   1.1205378270E-01
```

```
BUILD              3   4.1402114150E-02   4.6980937735E-14
LO-REDUCTION       5   1.2559406706E-02   4.6980937735E-14
HI-REDUCTION       7   3.4988133663E-03   4.6980937735E-14
HI-REDUCTION       9   7.8255935023E-04   4.6980937735E-14
...
SHRINK            59   3.1448995130E-14   1.0373099578E-16
SHRINK            63   2.4400978639E-14   1.0373099578E-16
HI-REDUCTION      65   1.7010223449E-14   1.0373099578E-16
...
HI-REDUCTION     117   6.0920713485E-24   3.9407472806E-25
Exiting from Alg19.pas Nelder Mead polytope minimiser
   119 function evaluations used

 Minimum function value found =  1.7118624554E-25
 At parameters
 B[1]=  1.1213869326E+00
 B[2]= -4.0522273834E-01
alg20.pas -- axial search
  Axis    Stepsize      function +     function -     rad. of curv.   tilt
    1 1.512415E-06  9.226758E-12  9.226726E-12  2.479099E-01  6.159003E-10
    2 5.465253E-07  4.546206E-13  4.546053E-13  6.570202E-01  8.031723E-10
```

14.4. OTHER DIRECT SEARCH METHODS

The Nelder–Mead procedure presented here is by no means the only direct search procedure, nor can it be taken to be the most economical of space or time. Dixon (1972, chap 5) discusses a number of direct search methods. Some of these perform various linear searches then form the resultant or sum of these over, say, n directions. A new set of directions is produced with the first being the resultant and the others orthogonal to this direction and to each other. This is the basis of the method of Rosenbrock and that of Davies, Swann and Campey. These both require storage of at least n vectors of n elements, equivalent to algorithm 19. The major differences in the two methods mentioned occur in the linear search and in the orthogonalisation procedure, which must be made resilient to the occurrence of directions in which no progress can be made, since these will have length zero and will cause a loss of dimensionality.

A method which is very simple to code and which uses only two vectors of working storage is that of Hooke and Jeeves (1961). This is the only method I have tested using the 77 test functions in Nash (1976). At that time, as reported in the first edition, the performance of this method in BASIC on a Data General NOVA computer was not satisfactory. However, for the types of functions encountered in many teaching situations it seems quite reliable, and Eason and Fenton (1973) showed a preference for this method. Furthermore, it is explicable to students whose major interest is not mathematics, such as those in economics or commerce, who nevertheless may need to minimise functions. I have used the Hooke and Jeeves method in a number of forecasting courses, and published it as a step-and-description version with a BASIC code in Nash (1982). A more advanced version appears in Nash and Walker-Smith (1987). I caution that it may prove very slow to find a minimum, and that it is possible to devise quite simple functions (Nash and Walker-Smith 1989) which will defeat its heuristic search. Algorithm 27 below presents a Pascal implementation.

Algorithm 27. Hooke and Jeeves minimiser

```
procedure hjmin(n: integer; {the number of parameters in the
                        function to be minimised}
                var B,X: rvector; {the parameter values on
                        input (B) and output (X) from minmeth}
                var Fmin: real; {'minimum' function value}
                        Workdata: probdata; {user defined data area}
                var fail: boolean; {true if method has failed}
                        intol: real); {user-initialized convergence
                        tolerance}
{alg27.pas == Hooke and Jeeves pattern search function minimisation
        From Interface Age, March 1982 page 34ff.
                Copyright 1988 J.C.Nash
}
var
    i, j: integer; {loop counters}
    stepsize: real; {current step size}
    fold: real; {function value at 'old' base point}
    fval: real; {current function value}
    notcomp: boolean; {set true if function not computable}
    temp: real; {temporary storage value}
    samepoint: boolean; {true if two points are identical}
    ifn: integer; {to count the number of function evaluations}
begin
    if intol<0.0 then intol := calceps; {set convergence tolerance if necessary}
    ifn := 1; {to initialize the count of function evaluations}
    fail := false; {Algorithm has not yet failed.}
    {STEP HJ1: n already entered, but we need an initial stepsize. Note the
    use of the stepredn constant, though possibly 0.1 is faster for
    convergence. Following mechanism used to set stepsize initial value.}
    stepsize := 0.0;
    for i := 1 to n do
        if stepsize < stepredn*abs(B[i]) then stepsize := stepredn*abs(B[i]);
    if stepsize=0.0 then stepsize := stepredn; {for safety}
    {STEP HJ2: Copy parameters into vector X}
    for i := 1 to n do X[i] := B[i];
    {STEP HJ3 not needed. In original code, parameters are entered into X
    and copied to B}
    fval := fminfn(n, B,Workdata,notcomp); {STEP HJ4}
    if notcomp then
    begin
        writeln('*** FAILURE *** Function not computable at initial point');
        fail := true;
```

Algorithm 27. Hooke and Jeeves minimiser (cont.)

```
    end {safety stop for non-computable function}
    else {initial function computed -- proceed}
    begin {main portion of routine}
      writeln('Initial function value =',fval);
      for i := 1 to n do
      begin
        write(B[i]:10:5,' ');
        if (7 * (i div 7) = i) and (i<n) then writeln;
      end;
      writeln;
      fold := fval; Fmin := fval; { to save function value at 'old' base which
                  is also current best function value}
      while stepsize>intol do {STEP HJ9 is now here}
      begin {STEP HJ5 == Axial Search}
        {write('A');} {Indicator output}
        for i := 1 to n do {STEP AS1}
        begin {STEP AS2}
          temp := B[i]; B[i] := temp+stepsize; {save parameter, step 'forward'}
          fval := fminfn(n, B,Workdata,notcomp); ifn := ifn+1; {STEP AS3}
          if notcomp then fval := big; {to allow for non-computable function}
          if fval<Fmin then
            Fmin := fval {STEP AS4}
          else
          begin {STEP AS5}
            B[i] := temp-stepsize; {to step 'backward' if forward step
              unsuccessful in reducing function value}
            fval := fminfn(n, B,Workdata,notcomp); ifn := ifn+1; {STEP AS6}
            if notcomp then fval := big; {for non-computable function}
            if fval<Fmin then {STEP AS7}
              Fmin := fval {STEP AS9 -- re-ordering of original algorithm}
            else {STEP AS8}
              B[i] := temp; {to reset the parameter value to original value}
          end; {else fval>=Fmin}
        end;{loop on i over parameters and end of Axial Search}
        if Fmin<fold then {STEP HJ6}
        begin {Pattern Move} {STEP PM1}
          {write('P');} {Indicator output}
          for i := 1 to n do {loop over parameters}
          begin {STEP PM2}
            temp := 2.0*B[i]-X[i]; {compute new base point component}
            X[i] := B[i]; B[i] := temp; {save current point and new base point}
          end; {loop on i -- STEP PM3}
          fold := Fmin; {to save new base point value}
        end {of Pattern Move -- try a new Axial Search}
        else
        begin
          samepoint := true; {initially assume points are the same}
          i := 1;{set loop counter start}
          repeat
            if B[i]<>X[i] then samepoint := false;
            i := i+1;
          until (not samepoint) or (i>n); {test for equality of points}
```

Algorithm 27. Hooke and Jeeves minimiser (cont.)

```
                if samepoint then
                begin {STEP HJ8}
                    stepsize := stepsize*stepredn; {to reduce stepsize. The reduction
                        factor (0.2) should be chosen to reduce the stepsize
                        reasonably rapidly when the initial point is close to
                        the solution, but not so small that progress cannot
                        still be made towards a solution point which is not
                        nearby.}
                    {writeln;} {Needed if indicator output ON}
                    write('stepsize now ',stepsize:10,' Best fn value=',Fmin);
                    writeln(' after ',ifn);
                    for i := 1 to n do
                    begin
                        write(B[i]:10:5,' ');
                        if (7 * (i div 7) = i) and (i<n) thenwriteln;
                    end;
                    writeln;
                end
                else {not the same point -- return to old basepoint}
                begin {STEP HJ7}
                    for i := 1 to n do B[i] := X[i];
                    {writeln;} {Needed if indicator output ON}
                    writeln('Return to old base point');
                end;{reset base point}
            end; {if Fmin<fold}
        end; {while stepsize>intol} {STEP HJ10}
        writeln('Converged to Fmin=',Fmin,' after ',ifn,' evaluations');
        {Fmin has lowest function value, X[] has parameters.}
    end; {if notcomp on first function evaluation}
end; {alg27.pas == hjmin}
```

In the next two chapters, considerable attention is paid to calculating sets of mutually conjugate directions. Direct search techniques parallelling these developments have been attempted. The most successful is probably that of M J D Powell. This is discussed at length by Brent (1973). These algorithms are more elaborate than any in this monograph, though it is possible they could be simplified. Alternatively, numerical approximation of derivatives (§18.2) can be used in algorithms 21 or 22.

Out of this range of methods I have chosen to present the Nelder–Mead procedure and Hooke and Jeeves method because I feel they are the most easily understood. There is no calculus or linear algebra to explain and for two-dimensional problems the progress of algorithm 19 can be clearly visualised.

The software diskette includes driver programs DR1920.PAS to allow the Nelder–Mead minimiser to be used in conjunction with algorithm 20. Driver DR27.PAS will allow the Hooke and Jeeves minimiser to be used, but this driver does not invoke algorithm 20 since a major part of the minimiser is an axial search.

Chapter 15

DESCENT TO A MINIMUM I: VARIABLE METRIC ALGORITHMS

15.1. DESCENT METHODS FOR MINIMISATION

The Nelder–Mead algorithm is a direct search procedure in that it involves only function values. In the next few sections, methods will be considered which make use of the gradient of the function $S(\boldsymbol{b})$ which will be called $\mathbf{g}$:

$$g_j = \partial S(\boldsymbol{b})/\partial b_j \qquad \text{for } j=1,2,\ldots,n \tag{15.1}$$

evaluated at the point $\boldsymbol{b}$. Descent methods all use the basic iterative step

$$\boldsymbol{b}' = \boldsymbol{b} - k\mathbf{B}\mathbf{g} \tag{15.2}$$

where $\mathbf{B}$ is a matrix defining a transformation of the gradient and k is a step length. The simplest such algorithm, the method of *steepest descents*, was proposed by Cauchy (1848) for the solution of systems of nonlinear equations. This uses

$$\mathbf{B} = \mathbf{1}_n \tag{15.3}$$

and any step length k which reduces the function so that

$$S(\boldsymbol{b}') < S(\boldsymbol{b}). \tag{15.4}$$

The principal difficulty with steepest descents is its tendency to *hemstitch*, that is, to criss-cross a valley on the function $S(\boldsymbol{b})$ instead of following the floor of the valley to a minimum. Kowalik and Osborne (1968, pp 34–9) discuss some of the reasons for this weakness, which is primarily that the search directions generated are not linearly independent. Thus a number of methods have been developed which aim to transform the gradient $\mathbf{g}$ so that the search directions generated in (15.2) are linearly independent or, equivalently, are conjugate to each other with respect to some positive definite matrix $\mathbf{A}$. In other words, if $\boldsymbol{x}_i$ and $\boldsymbol{x}_j$ are search directions, $\boldsymbol{x}_i$ and $\boldsymbol{x}_j$ are conjugate with respect to the positive definite matrix $\mathbf{A}$ if

$$\boldsymbol{x}_i^{\mathrm{T}}\mathbf{A}\boldsymbol{x}_j = 0. \tag{15.5}$$

The *conjugate gradients* algorithm 22 generates such a set of search directions implicitly, avoiding the storage requirements of either a transformation matrix $\mathbf{B}$ or the previous search directions. The *variable metric* algorithm 21 uses a transformation matrix which is adjusted at each step to generate appropriate search directions. There is, however, another way to think of this process. Consider the set of nonlinear equations formed by the gradient at a minimum

$$\mathbf{g}(\boldsymbol{b}') = \mathbf{0}. \tag{15.6}$$

As in the one-dimensional root-finding problem, it is possible to seek such solutions via a linear approximation from the current point $\boldsymbol{b}$ as in equation (13.25), that is

$$\mathbf{g}(\boldsymbol{b}') = \mathbf{g}(\boldsymbol{b}) + \mathbf{H}(\boldsymbol{b})(\boldsymbol{b}' - \boldsymbol{b}) \tag{15.7}$$

where $\mathbf{H}(\boldsymbol{b})$ is the *Hessian* matrix

$$H_{ij} = \partial g_i(\boldsymbol{b})/\partial b_j = \partial^2 S(\boldsymbol{b})/\partial b_i \partial b_j \tag{15.8}$$

of second derivatives of the function to be minimised or first derivatives of the nonlinear equations. For convex functions, $\mathbf{H}$ is at least non-negative definite. For the current purpose, suppose it is positive definite so that its inverse exists. Using the inverse together with equations (15.6) implies

$$\boldsymbol{b}' = \boldsymbol{b} - \mathbf{H}^{-1}(\boldsymbol{b})\mathbf{g}(\boldsymbol{b}) \tag{15.9}$$

which is Newton's method in n parameters. This is equivalent to equation (15.2) with

$$\mathbf{B} = \mathbf{H}^{-1} \qquad k = 1. \tag{15.10}$$

The step parameter k is rarely fixed, however, and usually some form of linear search is used (§13.2). While Newton's method may be useful for solving nonlinear-equation systems if the n-dimensional equivalents of the one-dimensional difficulties of the algorithm are taken care of, for minimisation problems it requires that second derivatives be computed.

This means that n^2 second derivative evaluations, n first derivative evaluations and a matrix inverse are needed even before the linear search is attempted. Furthermore the chances of human error in writing the subprograms to compute these derivatives is very high—the author has found that most of the 'failures' of his algorithms have been due to such errors when only first derivatives were required. For these reasons, Newton's method does not recommend itself for most problems.

Suppose now that a method could be found to approximate $\mathbf{H}^{-1}$ directly from the first derivative information available at each step of the iteration defined by (15.2). This would save a great deal of work in computing both the derivative matrix $\mathbf{H}$ and its inverse. This is precisely the role of the matrix $\mathbf{B}$ in generating the conjugate search directions of the variable metric family of algorithms, and has led to their being known also as *quasi-Newton* methods.

15.2. VARIABLE METRIC ALGORITHMS

Variable metric algorithms, also called quasi-Newton or matrix iteration algorithms, have proved to be the most effective class of general-purpose methods for solving unconstrained minimisation problems. Their development is continuing so rapidly, however, that the vast array of possibilities open to a programmer wishing to implement such a method is daunting. Here an attempt will be made to outline the underlying structure on which such methods are based. The next section will describe one set of choices of strategy.

All the variable metric methods seek to minimise the function $S(\boldsymbol{b})$ of n

parameters by means of a sequence of steps

$$\boldsymbol{b}' = \boldsymbol{b} - k\mathbf{B}\mathbf{g} \tag{15.2}$$

where $\mathbf{g}$ is the gradient of S. The definition of an algorithm of this type consists in specifying (a) how the matrix $\mathbf{B}$ is computed, and (b) how k is chosen. The second of these choices is the linear search problem of §13.2. In practice the algorithm suggested there is unnecessarily rigorous and a simpler search will be used. The rest of this section will be devoted to the task of deciding appropriate conditions on the matrix $\mathbf{B}$.

Firstly, when it works, Newton's method generally converges very rapidly. Thus it would seem desirable that $\mathbf{B}$ tend in some way towards the inverse Hessian matrix $\mathbf{H}^{-1}$. However, the computational requirement in Newton's method for second partial derivatives and for the solution of linear-equation systems at each iteration must be avoided.

Secondly, $\mathbf{B}$ should be positive definite, not only to avoid the possibility that the algorithm will 'get stuck' because $\mathbf{Bg}$ lacks components necessary to reduce the function because they are in the null space of $\mathbf{B}$ but also to permit the algorithm to exhibit quadratic termination, that is, to minimise a quadratic form in at most n steps (15.2). A quadratic form

$$S(\boldsymbol{b}) = \tfrac{1}{2}\boldsymbol{b}^{\mathrm{T}}\mathbf{H}\boldsymbol{b} - \boldsymbol{c}^{\mathrm{T}}\boldsymbol{b} + (\text{any scalar}) \tag{15.11}$$

has a unique minimum if $\mathbf{H}$ is positive definite. Note that $\mathbf{H}$ can always be made symmetric, and will be presumed so here. The merits of quadratic termination and positive definiteness are still being debated (see, for instance, Broyden 1972, p 94). Also, a function with a saddle point or ridge has a Hessian which is not always positive definite. Nevertheless, the discussion here will be confined to methods which generate positive definite iteration matrices.

If n linearly independent directions $\boldsymbol{t}_j$, $j = 1, 2, \ldots, n$, exist for which

$$\mathbf{BH}\boldsymbol{t}_j = \boldsymbol{t}_j \tag{15.12}$$

then

$$\mathbf{BH} = \mathbf{1}_n \qquad \text{or} \qquad \mathbf{B} = \mathbf{H}^{-1}. \tag{15.13}$$

The search directions $\boldsymbol{t}_j$ are sought conjugate to $\mathbf{H}$, leading in an implicit fashion to the expansion

$$\mathbf{H}^{-1} = \sum_{j=1}^{n} (\boldsymbol{t}_j\boldsymbol{t}_j^{\mathrm{T}})/(\boldsymbol{t}_j^{\mathrm{T}}\mathbf{H}\boldsymbol{t}_j). \tag{15.14}$$

Since $\mathbf{B}$ is to be developed as a sequence of matrices $\mathbf{B}^{(m)}$, the condition (15.12) will be stated

$$\mathbf{B}^{(m)}\mathbf{H}\boldsymbol{t}_j = \boldsymbol{t}_j. \tag{15.15}$$

Thus we have

$$\mathbf{B}^{(n+1)} = \mathbf{H}^{-1}. \tag{15.16}$$

For a quadratic form, the change in the gradient at $\boldsymbol{b}_j$, that is

$$\boldsymbol{g}_j = \mathbf{H}\boldsymbol{b}_j - \boldsymbol{c} \tag{15.17}$$

is given by

$$\mathbf{y}_j = \mathbf{g}_{j+1} - \mathbf{g}_j = \mathbf{H}(\boldsymbol{b}_{j+1} - \boldsymbol{b}_j)$$
$$= k_j \mathbf{H} \boldsymbol{t}_j \tag{15.18}$$

since the elements of $\mathbf{H}$ are constant in this case. From this it follows that (15.15) becomes

$$\mathbf{B}^{(m)} \mathbf{y}_j = k_j \boldsymbol{t}_j. \tag{15.19}$$

Assuming (15.15) is correct for $j < m$, a new step

$$\boldsymbol{t}_m = \mathbf{B}^{(m)} \mathbf{g}_m \tag{15.20}$$

is required to be conjugate to all previous directions $\boldsymbol{t}_j$, i.e.

$$\boldsymbol{t}_j^{\mathrm{T}} \mathbf{H} \boldsymbol{t}_m = 0 \qquad \text{for } j < m. \tag{15.21}$$

But from (15.18), (15.19) and (15.20) we obtain

$$\boldsymbol{t}_j^{\mathrm{T}} \mathbf{H} \boldsymbol{t}_m = \mathbf{y}_j^{\mathrm{T}} \boldsymbol{t}_m / k_j = \mathbf{y}_j^{\mathrm{T}} \mathbf{B}^{(m)} \mathbf{g}_m / k_j = \boldsymbol{t}_j^{\mathrm{T}} \mathbf{g}_m. \tag{15.22}$$

Suppose now that the linear searches at each step have been performed accurately. Then the directions

$$\boldsymbol{t}_1, \boldsymbol{t}_2, \ldots, \boldsymbol{t}_{m-1} \tag{15.23}$$

define a hyperplane on which the quadratic form has been minimised. Thus $\mathbf{g}_m$ is orthogonal to $\boldsymbol{t}_j$, $j < m$, and (15.21) is satisfied, so that the new direction $\boldsymbol{t}_m$ is conjugate to the previous ones.

In order that $\mathbf{B}^{(m+1)}$ satisfies the above theory so that the process can continue, the update $\mathbf{C}$ in

$$\mathbf{B}^{(m+1)} = \mathbf{B}^{(m)} + \mathbf{C} \tag{15.24}$$

must satisfy (15.19), that is

$$\mathbf{B}^{(m+1)} \mathbf{y}_m = k_m \boldsymbol{t}_m \tag{15.25}$$

or

$$\mathbf{C} \mathbf{y}_m = k_m \boldsymbol{t}_m - \mathbf{B}^{(m)} \mathbf{y}_m \tag{15.26}$$

and

$$\mathbf{B}^{(m+1)} \mathbf{y}_j = k_j \boldsymbol{t}_j \qquad \text{for } j < m \tag{15.27}$$

or

$$\mathbf{C} \mathbf{y}_j = k_j \boldsymbol{t}_j - \mathbf{B}^{(m)} \mathbf{y}_j$$
$$= k_j \boldsymbol{t}_j - k_j \boldsymbol{t}_j = 0. \tag{15.28}$$

In establishing (15.26) and (15.28), equation (15.19) has been applied for $\mathbf{B}^{(m)}$. There are thus m conditions on the order-n matrix $\mathbf{C}$. This degree of choice in $\mathbf{C}$ has in part been responsible for the large literature on variable metric methods.

The essence of the variable metric methods, i.e. that information regarding the Hessian has been drawn from first derivative computations only is somewhat hidden in the above development. Of course, differences could have been used to generate an approximate Hessian from $(n+1)$ vectors of first derivatives, but a

Newton method based on such a matrix would still require the solution of a linear system of equations at each step. The variable metric methods generate an approximate inverse Hessian as they proceed, requiring only one evaluation of the first derivatives per step and, moreover, reduce the function value at each of these steps.

15.3. A CHOICE OF STRATEGIES

In order to specify a particular variable metric algorithm it is now necessary to choose a matrix-updating formula and a linear search procedure. One set of choices resulting in a compact algorithm is that of Fletcher (1970). Here this will be simplified somewhat and some specific details made clear. First, Fletcher attempts to avoid an inverse interpolation in performing the linear search. He suggests an 'acceptable point' search procedure which takes the first point in some generated sequence which satisfies some acceptance criterion. Suppose that the step taken is

$$\boldsymbol{t} = \boldsymbol{b}' - \boldsymbol{b} = -k\mathbf{B}\mathbf{g} \tag{15.29}$$

with $k = 1$ initially. The decrease in the function

$$\Delta S = S(\boldsymbol{b}') - S(\boldsymbol{b}) < 0 \tag{15.30}$$

will be approximated for small steps $\boldsymbol{t}$ by the first term in the Taylor series for S along $\boldsymbol{t}$ from $\boldsymbol{b}$, that is, by $\boldsymbol{t}^{\mathrm{T}}\mathbf{g}$. This is negative when $\boldsymbol{t}$ is a 'downhill' direction. It is not desirable that ΔS be very different in magnitude from $\boldsymbol{t}^{\mathrm{T}}\mathbf{g}$ since this would imply that the local approximation of the function by a quadratic form is grossly in error. By choosing $k = 1, w, w^2, \ldots,$ for $0 < w < 1$ successively, it is always possible to produce a $\boldsymbol{t}$ such that

$$0 < \text{tolerance} < \Delta S / \boldsymbol{t}^{\mathrm{T}}\mathbf{g} \qquad \text{for tolerance} \ll 1 \tag{15.31}$$

unless the minimum has been found. This presumes

$$\boldsymbol{t}^{\mathrm{T}}\mathbf{g} < 0 \tag{15.32}$$

to ensure a 'downhill' direction. In any practical program a test of the condition (15.32) is advisable before proceeding with any search along $\boldsymbol{t}$. Fletcher recommends the values $w = 0{\cdot}1$, tolerance $= 0{\cdot}0001$. However, if the point is not acceptable, he uses a cubic inverse interpolation to reduce k, with the proviso that $0{\cdot}1k$ be the smallest value permitted to generate the next step in the search. The author retains tolerance $= 0{\cdot}0001$, but uses $w = 0{\cdot}2$ with no interpolation procedure. For a study of acceptable point–interpolation procedures in minimisation algorithms see Sargent and Sebastian (1972).

An updating formula which has received several favourable reports is that of Broyden (1970a, b), Fletcher (1970) and Shanno (1970). This employs the update

$$\mathbf{C} = d_2\boldsymbol{t}\boldsymbol{t}^{\mathrm{T}} - [\boldsymbol{t}(\mathbf{B}\mathbf{y})^{\mathrm{T}} + (\mathbf{B}\mathbf{y})\boldsymbol{t}^{\mathrm{T}}]/d_1 \tag{15.33}$$

where $\mathbf{y}$ is the gradient difference

$$\mathbf{y} = \mathbf{g}(\boldsymbol{b}') - \mathbf{g}(\boldsymbol{b}) \tag{15.34}$$

and the coefficients d_1 and d_2 are given by

$$d_1 = \mathbf{t}^T\mathbf{y} \tag{15.35}$$

and

$$d_2 = (1 + \mathbf{y}^T\mathbf{B}\mathbf{y}/d_1)/d_1. \tag{15.36}$$

There are several ways in which the update can be computed and added into **B**. In practice these may give significantly different convergence patterns due to the manner in which digit cancellation may occur. However, the author has not been able to make any definite conclusion as to which of the few ways he has tried is superior overall. The detailed description in the next section uses a simple form for convenience of implementation. The properties of the Broyden–Fletcher–Shanno update will not be discussed further in this work.

In order to start the algorithm, some initial matrix **B** must be supplied. If it were easy to compute the Hessian **H** for the starting parameters $\mathbf{b}$, $\mathbf{H}^{-1}$ would be the matrix of choice. For general application, however, $\mathbf{B} = \mathbf{1}_n$ (15.37) is a simpler choice and has the advantage that it generates the steepest descent direction in equation (15.29). I have found it useful on a machine having short mantissa arithmetic to apply the final convergence test on the steepest descent

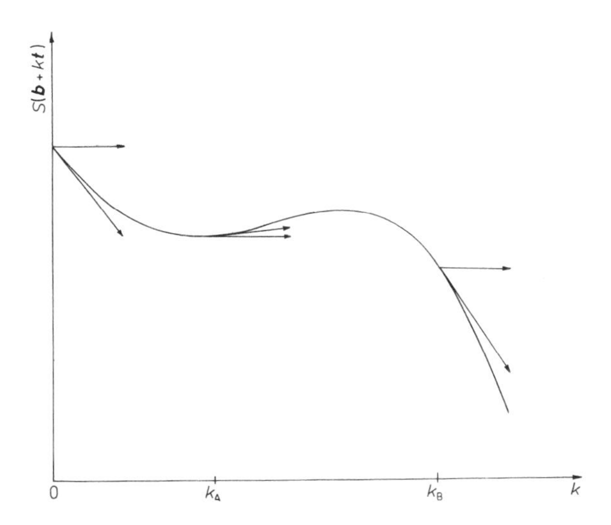

FIGURE 15.1. Illustration of the test at step 14 of algorithm 21. An iteration of the variable metric algorithm 21 consists of a linear search along direction $\mathbf{t}$ using the step parameter k. At a given point along the search line $\mathbf{g}^T\mathbf{t}$ gives the slope of the function with respect to the step length k. If the search finds k_A, then the update can proceed since this slope is increased (made less negative) as we would expect if minimising a quadratic function. If, however, k_B is found, the slope is decreased, and because such behaviour is not consistent with the assumption that the function is approximately represented by a quadratic form, the update cannot be performed and we restart algorithm 21 with a steepest descent step from the point defined by k_B.

direction to ensure that rounding errors in updating **B** and forming t via (15.29) have not accidentally given a direction in which the function S cannot be reduced. Therefore, a restart is suggested in any of the following cases:

(i) $\boldsymbol{t}^{\mathrm{T}}\mathbf{g}\geqslant 0$, that is, the direction of search is 'uphill'.
(ii) $\boldsymbol{b}'=\boldsymbol{b}$, that is, no change is made in the parameters by the linear search along $\boldsymbol{t}$.

If either case (i) or case (ii) occurs during the first step after **B** has been set to the unit matrix, the algorithm is taken to have converged.
(iii) Since the method reduces S along $\boldsymbol{t}$, it is expected that

$$\boldsymbol{t}^{\mathrm{T}}\mathbf{g}(\boldsymbol{b}')$$

will be greater (less negative) than

$$\boldsymbol{t}^{\mathrm{T}}\mathbf{g}(\boldsymbol{b}).$$

Figure 15.1 illustrates this idea. Therefore, $\boldsymbol{t}^{\mathrm{T}}\mathbf{y}=d_1$ should be positive. If it is not there exists the danger that **B** may no longer be positive definite. Thus if $\boldsymbol{t}^{\mathrm{T}}\mathbf{y}\leqslant 0$, the matrix **B** is reset to unity.

This completes the general description of a variable metric algorithm suitable for small computers. I am indebted to Dr R Fletcher of Dundee University for suggesting the basic choices of the Broyden–Fletcher–Shanno updating formula and the acceptable point search. Note particularly that this search does not require the gradient to be evaluated at each trial point.

Algorithm 21. Variable metric minimiser

*The algorithm needs an order-*n *square matrix* B *and five order-*n *vectors* **b**, **x**, **c**, **g** *and* **t**. *Care should be taken when coding to distinguish between* B *and* **b**.

```
procedure vmmin(n: integer; {the number of parameters in the
                        function to be minimised}
                        var Bvec, X: rvector; {the parameter values on
                        input (Bvec) and output (X) from minmeth}
                        var Fmin: real; {'minimum' function value}
                        Workdata: probdata; {user defined data area}
                        var fail: boolean; {true if method has failed}
                        var intol: real); {user-initialized convergence
                        tolerance}
{alg21.pas == modified Fletcher variable metric method,
          Original method due to R. Fletcher, Computer Journal, vol 13,
                        pp. 317-322, 1970
          Unlike Fletcher-Reeves, we do not use a quadratic interpolation,
          since the search is often approximately a Newton step.
                        Copyright 1988 J.C.Nash
}
const
   Maxparm = 25; {maximum allowed number of parameters in the
                        present code. May be changed by the user,
                        along with dependent constants below.}
   stepredn = 0.2; {factor to reduce stepsize in line search}
```

Algorithm 21. Variable metric minimiser (cont.)

```
    acctol = 0.0001; {acceptable point tolerance -- see STEP 11}
    reltest = 10.0; {to check equality of parameters -- see STEP 8}
var
    accpoint : boolean; {to indicate an acceptable point}
    B : array[1..Maxparm, 1..Maxparm] of real;
                    {approximation to inverse Hessian}
    c : rvector; {to store last gradient}
    count : integer; {to check for parameter equality}
    D1, D2 : real; {temporary working storage}
    f : real; {temporary function value}
    funcount : integer; {count of function evaluations}
    g : rvector; {to hold gradient}
    gradcount : integer; {count of gradient evaluations}
    gradproj : real; {gradient projection on search vector}
    i, j : integer; {working integers}
    ilast : integer; {records last step at which B was
                    initialized to a unit matrix}
    notcomp : boolean; {non-computability flag}
    s : real; {inner product accumulator}
    steplength: real; {linear search steplength}
    t : rvector; {to store working vector for line search}
begin
    writeln('alg21.pas -- version 2 1988-03-24');
    writeln(' Variable metric function minimiser');
    fail:=false; {method has yet to fail}
    f:=fminfn(n, Bvec, Workdata, notcomp); {initial fn calculation -- STEP 1}
    if notcomp then
    begin
        writeln('**** Function cannot be evaluated at initial parameters ****');
        fail := true; {method has failed}
    end
    else {proceed with minimisation}
    begin
        Fmin:=f;{save the best value so far}
        funcount:=1; {function count initialized to 1}
        gradcount:=1; {initialize gradient count to 1}
        fmingr(n, Bvec, Workdata, g); {STEP 2}
        ilast:=gradcount; {set count to force initialization of B}
        {STEP 3 -- set B to unit matrix -- top of major cycle}
        repeat {iteration for method -- terminates when no progress can be
                    made, and B is a unit matrix so that search is a steepest
                    descent direction.}
            if ilast=gradcount then
            begin
                for i:=1 to n do
                begin
                    for j:=1 to n do B[i, j]:=0.0; B[i, i]:=1.0;
                end;{loop on i}
            end;{initialize B}
            writeln(gradcount,' ', funcount,' ', Fmin); {STEP 4}
            write('parameters ');
            for i:=1 to n do write(Bvec[i]:10:5,' ');
```

Algorithm 21. Variable metric minimiser (cont.)

```
          writeln;
          for i:=1 to n do
          begin
            X[i]:=Bvec[i];{save best parameters}
            c[i]:=g[i];{save gradient}
          end; {loop on i}
          {STEP 5 -- set t:=-B*g and gradproj=tT*g}
          gradproj:=0.0; {to save tT*g inner product}
          for i:=1 to n do
          begin
            s:=0.0; {to accumulate element of B*g}
            for j:=1 to n do s:=s-B[i, j]*g[j];
            t[i]:=s; gradproj:=gradproj+s*g[i];
          end; {loop on i for STEP 5}
          {STEP 6}{test for descent direction}
          if gradproj<0 then {if gradproj<0 then STEP 7 to perform linear
                      search; other parts of this step follow the 'else' below}
          begin {STEP 7 -- begin linear search}
            steplength:=1.0; {always try full step first}
            {STEP 8 -- step along search direction and test for a change}
            accpoint:=false; {don't have a good point yet}
            repeat {line search loop}
              count:=0; {to count unchanged parameters}
              for i:=1 to n do
              begin
                Bvec[i]:=X[i]+steplength*t[i];
                if (reltest+X[i])=(reltest+Bvec[i]) then count:=count+1;
              end; {loop on i}
              if count<n then {STEP 9 -- main convergence test}
              begin {STEP 10 -- can proceed with linear search}
                f:=fminfn(n, Bvec, Workdata, notcomp); {function calculation}
                funcount:=funcount+1;
                accpoint:=(not notcomp) and
                      (f<=Fmin+gradproj*steplength*acctol);
                {STEP 11 -- a point is acceptable only if function is computable
                (not notcomp) and it satisfies the acceptable point criterion}
                if not accpoint then
                begin
                  steplength:=steplength*stepredn; write('*');
                end;
              end; {compute and test for linear search}
            until (count=n) or accpoint; {end of loop for line search}
            if count<n then
            begin
              Fmin:=f; {save funcion value}
              fmingr(n, Bvec, Workdata, g); {STEP 12}
              gradcount:=gradcount+1;
              D1:=0.0; {STEP 13 -- prepare for matrix update}
              for i:=1 to n do
              begin
                t[i]:=steplength*t[i]; c[i]:=g[i]-c[i]; {to compute vector y}
                D1:=D1+t[i]*c[i]; {to compute inner product t*y}
```

Algorithm 21. Variable metric minimiser (cont.)

```
                end; {loop on i}
                if D1>0 then {STEP 14} {test if update is possible}
                begin {update}
                    D2:=0.0; {STEP 15 -- computation of B*y, yT*B*y}
                    for i:=1 to n do
                    begin
                        s:=0.0;
                        for j:=1 to n do s:=s+B[i, j]*c[j];
                        X[i]:=s; D2:=D2+s*c[i];
                    end; {loop on i}
                    D2:=1.0+D2/D1; {STEP 16 -- complete update}
                    for i:=1 to n do
                    begin
                        for j:=1 to n do
                        begin
                            B[i, j]:=B[i, j]-(t[i]*X[j]+X[i]*t[j]-D2*t[i]*t[j])/D1;
                        end; {loop on j}
                    end; {loop on i -- Update is now complete.}
                end {update}
                else
                begin
                    writeln(' UPDATE NOT POSSIBLE');
                    ilast:=gradcount; {to force a restart with B = 1(n)}
                end;
            end {if count<n}{STEP 17}
            else {count=n, cannot proceed}
            begin
                if ilast<gradcount then
                begin
                    count:=0; {to force a steepest descent try}
                    ilast:=gradcount; {to reset to steepest descent search}
                end; {if ilast}
            end; {count=n}
        end {if gradproj<0 ... do linear search}
        else
        begin
            writeln('UPHILL SEARCH DIRECTION');
            ilast:=gradcount;{to reset B to unit matrix}
            count:=0; {to ensure we try again}
        end;
    until (count=n) and (ilast=gradcount);
  end; {minimisation -- STEP 18}
  {STEP 31 -- save best parameters and function value found}
  writeln('Exiting from alg21.pas variable metric minimiser');
  writeln(' ', funcount,' function evaluations used');
  writeln(' ', gradcount,' gradient evaluations used');
end; {alg21.pas == vmmin}
```

Example 15.1. Illustration of the variable metric algorithm 21

The following output from an IBM 370/168 operating in single precision (six hexadecimal digits) was produced by a FORTRAN version of algorithm 21 as it minimised the Rosenbrock banana-shaped valley function (Nash 1976)

$$S(\boldsymbol{b}) = 100(b_2 - b_1^2)^2 + (b_1 - 1)^2$$

using analytic derivatives. The starting point $b_1 = -1\cdot2$, $b_2 = 1$ was used.

```
# ITNS=   1  # EVALNS=   1  FUNCTION=  0.24199860E+02
# ITNS=   2  # EVALNS=   6  FUNCTION=  0.20226822E+02
# ITNS=   3  # EVALNS=   9  FUNCTION=  0.86069937E+01
# ITNS=   4  # EVALNS=  14  FUNCTION=  0.31230078E+01
# ITNS=   5  # EVALNS=  16  FUNCTION=  0.28306570E+01
# ITNS=   6  # EVALNS=  21  FUNCTION=  0.26346817E+01
# ITNS=   7  # EVALNS=  23  FUNCTION=  0.20069408E+01
# ITNS=   8  # EVALNS=  24  FUNCTION=  0.18900719E+01
# ITNS=   9  # EVALNS=  25  FUNCTION=  0.15198193E+01
# ITNS=  10  # EVALNS=  26  FUNCTION=  0.13677282E+01
# ITNS=  11  # EVALNS=  27  FUNCTION=  0.10138159E+01
# ITNS=  12  # EVALNS=  28  FUNCTION=  0.85555243E+00
# ITNS=  13  # EVALNS=  29  FUNCTION=  0.72980821E+00
# ITNS=  14  # EVALNS=  30  FUNCTION=  0.56827205E+00
# ITNS=  15  # EVALNS=  32  FUNCTION=  0.51492560E+00
# ITNS=  16  # EVALNS=  33  FUNCTION=  0.44735157E+00
# ITNS=  17  # EVALNS=  34  FUNCTION=  0.32320732E+00
# ITNS=  18  # EVALNS=  35  FUNCTION=  0.25737345E+00
# ITNS=  19  # EVALNS=  37  FUNCTION=  0.20997590E+00
# ITNS=  20  # EVALNS=  38  FUNCTION=  0.17693651E+00
# ITNS=  21  # EVALNS=  39  FUNCTION=  0.12203962E+00
# ITNS=  22  # EVALNS=  40  FUNCTION=  0.74170172E-01
# ITNS=  23  # EVALNS=  41  FUNCTION=  0.39149582E-01
# ITNS=  24  # EVALNS=  43  FUNCTION=  0.31218585E-01
# ITNS=  25  # EVALNS=  44  FUNCTION=  0.25947951E-01
# ITNS=  26  # EVALNS=  45  FUNCTION=  0.12625925E-01
# ITNS=  27  # EVALNS=  46  FUNCTION=  0.78500621E-02
# ITNS=  28  # EVALNS=  47  FUNCTION=  0.45955069E-02
# ITNS=  29  # EVALNS=  48  FUNCTION=  0.15429037E-02
# ITNS=  30  # EVALNS=  49  FUNCTION=  0.62955730E-03
# ITNS=  31  # EVALNS=  50  FUNCTION=  0.82553088E-04
# ITNS=  32  # EVALNS=  51  FUNCTION=  0.54429529E-05
# ITNS=  33  # EVALNS=  52  FUNCTION=  0.57958061E-07
# ITNS=  34  # EVALNS=  53  FUNCTION=  0.44057202E-10
# ITNS=  35  # EVALNS=  54  FUNCTION=  0.0
# ITNS=  35  # EVALNS=  54  FUNCTION=  0.0
  B(  1)=   0.10000000E+01
  B(  2)=   0.10000000E+01
# ITNS=  35  # EVALNS=  54  FUNCTION=  0.0
```

Chapter 16

DESCENT TO A MINIMUM II: CONJUGATE GRADIENTS

16.1. CONJUGATE GRADIENTS METHODS

On a small computer, the principal objection to the Nelder–Mead and variable metric methods is their requirement for a working space proportional to n^2, where n is the number of parameters in the function to be minimised. The parameters $\boldsymbol{b}$ and gradient $\mathbf{g}$ require only n elements each, so it is tempting to consider algorithms which make use of this information without the requirement that it be collected in a matrix. In order to derive such a method, consider once again the quadratic form

$$S(\boldsymbol{b})=\tfrac{1}{2}\boldsymbol{b}^{\mathrm{T}}\mathbf{H}\boldsymbol{b}-\boldsymbol{c}^{\mathrm{T}}\boldsymbol{b}+(\text{any scalar}) \tag{15.11}$$

of which the gradient at $\boldsymbol{b}$ is

$$\mathbf{g}=\mathbf{H}\boldsymbol{b}-\boldsymbol{c}. \tag{15.17}$$

Then if the search direction at some iteration j is $\boldsymbol{t}_j$, we have

$$\mathbf{y}_j=\mathbf{g}_{j+1}-\mathbf{g}_j=k_j\mathbf{H}\boldsymbol{t}_j \tag{15.18}$$

where k_j is the step-length parameter.

If any initial step $\boldsymbol{t}_1$ is made subject only to its being 'downhill', that is

$$\boldsymbol{t}_1^{\mathrm{T}}\mathbf{g}_1<0 \tag{16.1}$$

then the construction of search directions $\boldsymbol{t}_i$, $i=1, 2, \ldots, n$, conjugate with respect to the Hessian $\mathbf{H}$, is possible via the Gram–Schmidt process (see Dahlquist and Björck 1974, pp 201–4). That is to say, given an arbitrary new 'downhill' direction $\boldsymbol{q}_i$ at step i, it is possible to construct, by choosing coefficients z_{ij}, a direction

$$\boldsymbol{t}_i=\boldsymbol{q}_i+\sum_{j=1}^{i-1}z_{ij}\boldsymbol{t}_j \tag{16.2}$$

such that

$$\boldsymbol{t}_i^{\mathrm{T}}\mathbf{H}\boldsymbol{t}_j=0 \qquad \text{for } j<i. \tag{16.3}$$

This is achieved by applying $\boldsymbol{t}_j^{\mathrm{T}}\mathbf{H}$ to both sides of equation (16.2), giving

$$z_{ij}=-\boldsymbol{t}_j^{\mathrm{T}}\mathbf{H}\boldsymbol{q}_i/\boldsymbol{t}_j^{\mathrm{T}}\mathbf{H}\boldsymbol{t}_j \tag{16.4}$$

by substitution of the condition (16.3) and the assumed conjugacy of the $\boldsymbol{t}_j$, $j=1, 2, \ldots, (i-1)$. Note that the denominator of (16.4) cannot be zero if $\mathbf{H}$ is positive definite and $\boldsymbol{t}_j$ is not null.

Now if $\boldsymbol{q}_i$ is chosen to be the negative gradient

$$\boldsymbol{q}_i = -\boldsymbol{g}_i \tag{16.5}$$

and $\boldsymbol{t}_j^T\mathbf{H}$ is substituted from (15.18), then we have

$$\begin{aligned} z_{ij} &= \boldsymbol{g}_i^T(\boldsymbol{g}_{j+1} - \boldsymbol{g}_j)/\boldsymbol{t}_j^T(\boldsymbol{g}_{j+1} - \boldsymbol{g}_j) \\ &= \boldsymbol{g}_i^T\boldsymbol{y}_j/\boldsymbol{t}_j^T\boldsymbol{y}_j. \end{aligned} \tag{16.6}$$

Moreover, if accurate line searches have been performed at each of the $(i-1)$ previous steps, then the function S (still the quadratic form (15.11)) has been minimised on a hyperplane spanned by the search directions $\boldsymbol{t}_j$, $j=1, 2, \ldots, (i-1)$, and $\boldsymbol{g}_i$ is orthogonal to each of these directions. Therefore, we have

$$z_{ij} = 0 \qquad \text{for } j<(i-1) \tag{16.7}$$

$$z_{i,i-1} = \boldsymbol{g}_i^T(\boldsymbol{g}_i - \boldsymbol{g}_{i-1})/\boldsymbol{t}_{i-1}^T(\boldsymbol{g}_i - \boldsymbol{g}_{i-1}). \tag{16.8}$$

Alternatively, using

$$\boldsymbol{t}_{i-1} = -\boldsymbol{g}_{i-1} + \sum_{j=1}^{i-2} z_{ij}\boldsymbol{t}_j \tag{16.9}$$

which is a linear combination of $\boldsymbol{g}_j$, $j=1, 2, \ldots, (i-1)$, we obtain

$$z_{i,i-1} = \boldsymbol{g}_i^T(\boldsymbol{g}_i - \boldsymbol{g}_{i-1})/\boldsymbol{g}_{i-1}^T\boldsymbol{g}_{i-1} \tag{16.10}$$

$$= \boldsymbol{g}_i^T\boldsymbol{g}_i/\boldsymbol{g}_{i-1}^T\boldsymbol{g}_{i-1} \tag{16.11}$$

by virtue of the orthogonality mentioned above.

As in the case of variable metric algorithms, the formulae obtained for quadratic forms are applied in somewhat cavalier fashion to the minimisation of general nonlinear functions. The formulae (16.8), (16.10) and (16.11) are now no longer equivalent. For reference, these will be associated with the names: Beale (1972) and Sorenson (1969) for (16.8); Polak and Ribiere (1969) for (16.10); and Fletcher and Reeves (1964) for (16.11). All of these retain the space-saving two-term recurrence which makes the conjugate gradients algorithms so frugal of storage.

In summary, the conjugate gradients algorithm proceeds by setting

$$\boldsymbol{t}_1 = -\boldsymbol{g}(\boldsymbol{b}_1) \tag{16.12}$$

and

$$\boldsymbol{t}_i = z_{i,i-1}\boldsymbol{t}_{i-1} - \boldsymbol{g}_i(\boldsymbol{b}_i) \tag{16.13}$$

with

$$\boldsymbol{b}_{j+1} = \boldsymbol{b}_j + k_j\boldsymbol{t}_j \tag{16.14}$$

where k_j is determined by a linear search for a 'minimum' of $S(\boldsymbol{b}_j + k_j\boldsymbol{t}_j)$ with respect to k_j.

16.2. A PARTICULAR CONJUGATE GRADIENTS ALGORITHM

A program to use the ideas of the previous section requires that a choice be made of (*a*) a recurrence formula to generate the search directions, and (*b*) a linear search.

Since the conjugate gradients methods are derived on the presumption that they minimise a quadratic form in n steps, it is also necessary to suggest a method for continuing the iterations after n steps. Some authors, for instance Polak and Ribiere (1969), continue iterating with the same recurrence formula. However, while the iteration matrix $\mathbf{B}$ in the variable metric algorithms can in a number of situations be shown to tend towards the inverse Hessian $\mathbf{H}^{-1}$ in some sense, there do not seem to be any similar theorems for conjugate gradients algorithms. Fletcher and Reeves (1964) restart their method every $(n+1)$ steps with

$$\boldsymbol{t}_1 = -\boldsymbol{g}_1 \tag{16.15}$$

while Fletcher (1972) does this every n iterations. Powell (1975a, b) has some much more sophisticated procedures for restarting his conjugate gradients method. I have chosen to restart every n steps or whenever the linear search can make no progress along the search direction. If no progress can be made in the first conjugate gradient direction—that of steepest descent—then the algorithm is taken to have converged.

The linear search used in the first edition of this book was that of §13.2. However, this proved to work well in some computing environments but poorly in others. The present code uses a simpler search which first finds an 'acceptable point' by stepsize reduction, using the same ideas as discussed in §15.3. Once an acceptable point has been found, we have sufficient information to fit a parabola to the projection of the function on the search direction. The parabola requires three pieces of information. These are the function value at the end of the last iteration (or the initial point), the projection of the gradient at this point onto the search direction, and the new (lower) function value at the acceptable point. The step length resulting from the quadratic inverse interpolation is used to generate a new trial point for the function. If this proves to give a point lower than the latest acceptable point, it becomes the starting point for the next iteration. Otherwise we use the latest acceptable point, which is the lowest point so far.

A starting step length is needed for the search. In the Newton and variable metric (or quasi-Newton) methods, we can use a unit step length which is ideal for the minimisation of a quadratic function. However, for conjugate gradients, we do not have this theoretical support. The strategy used here is to multiply the best step length found in the line search by some factor to increase the step. Our own usual choice is a factor 1.7. At the start of the conjugate gradients major cycle we set the step length to 1. If the step length exceeds 1 after being increased at the end of each iteration, it is reset to 1.

If the choice of linear search is troublesome, that of a recurrence formula is even more difficult. In some tests by the author on the 23-bit binary NOVA, the Beale–Sorenson formula (16.8) in conjunction with the linear search chosen above required more function and derivative evaluations than either formula (16.10) or formula (16.11). A more complete comparison of the Polak–Ribiere formula (16.10) with that of Fletcher–Reeves (16.11) favoured the former. However, it is worth recording Fletcher's (1972) comment: 'I know of no systematic evidence which indicates how the choice should be resolved in a general-purpose algorithm.' In the current algorithm, the user is given a choice of which approach should be used.

In other sections, conjugate gradients algorithms specially adapted to particular problems will be reported. It is unfortunate that the application to such problems of the general-purpose minimisation algorithm proposed here and detailed in the next section (problems such as, for instance, minimising the Rayleigh quotient to find a matrix eigensolution) may show the method to be extremely slow to converge.

A detail which remains to be dealt with is that the initial step length needs to be established. The value

$$k = 1 \tag{16.16}$$

is probably as good as any in the absence of other information, though Fletcher and Reeves (1964) use an estimate e of the minimum value of the function in the linear approximation

$$k = [e - S(\boldsymbol{b})]/\boldsymbol{g}^{\mathrm{T}}\boldsymbol{t}. \tag{16.17}$$

In any event, it is advisable to check that the step does change the parameters and to increase k until they are altered. After a conjugate gradients iteration, the absolute value of k can be used to provide a step length in the next search.

Algorithm 22. Function minimisation by conjugate gradients

```
procedure cgmin(n: integer; {the number of parameters in the
                    function to be minimised}
                var Bvec, X: rvector; {the parameter values on
                    input (Bvec) and output (X) from minmeth}
                var Fmin: real; {'minimum' function value}
                    Workdata: probdata; {user defined data area}
                var fail: boolean; {true if method has failed}
                var intol: real); {user-initialized convergence
                    tolerance}
{alg22.pas == modified conjugate gradients function
                minimisation method
          Original method due to R. Fletcher & C M Reeves, Computer Journal,
          vol 7, pp. 149-154, 1964
                Copyright 1988 J.C.Nash
}
type
   methodtype= (Fletcher_Reeves, Polak_Ribiere, Beale_Sorenson);
const
   Maxparm = 25; {maximum allowed number of parameters in the
                    present code. May be changed by the user,
                    along with dependent constants below.}
   stepredn = 0.2; {factor to reduce stepsize in line search}
   acctol = 0.0001; {acceptable point tolerance -- see STEP 13}
   reltest = 10.0; {to check equality of parameters -- see STEP 8}
var
   accpoint : boolean; {to indicate an acceptable point}
   c : rvector; {to store last gradient}
   count : integer; {to check for parameter equality}
   cycle : integer; {cycle count of cg process}
   cyclimit : integer; {limit on cycles per major cg sweep}
```

Algorithm 22. Function minimisation by conjugate gradients (cont.)

```
    f : real; {temporary function value}
    funcount : integer; {count of function evaluations}
    g : rvector; {to hold gradient}
    G1, G2 : real; {temporary working storage}
    G3, gradproj : real; {temporary working storage}
    gradcount : integer; {count of gradient evaluations}
    i, j : integer; {working integers}
    method : methodtype;
    newstep : real; {interpolation steplength}
    notcomp : boolean; {non-computability flag}
    oldstep : real; {last stepsize used}
    s : real; {inner product accumulator}
    setstep : real; {to allow for adjustment of best step before
                        next iteration}
    steplength: real; {linear search steplength}
    t : rvector; {to store working vector for line search}
    tol : real; {convergence tolerance}
begin
    writeln('alg22.pas -- Nash Algorithm 22 version 2 1988-03-24');
    writeln(' Conjugate gradients function minimiser');
    {method:=Fletcher_Reeves;} {set here rather than have an input}
    writeln('Steplength saving factor multiplies best steplength found at the');
    writeln(' end of each iteration as a starting value for next search');
    write('Enter a steplength saving factor (sugg. 1.7) -- setstep ');
    readln(infile, setstep);
    if infname<>'con' then writeln(setstep);
    write('Choose method (1=FR, 2=PR, 3=BS) ');
    readln(infile, i); if infname<>'con' then writeln(i);
    case i of
        1: method:=Fletcher_Reeves;
        2: method:=Polak_Ribiere;
        3: method:=Beale_Sorenson;
        else halt;
    end;
    case method of
        Fletcher_Reeves: writeln('Method: Fletcher Reeves');
        Polak_Ribiere: writeln('Method: Polak Ribiere');
        Beale_Sorenson: writeln('Method: Beale Sorenson');
    end;
    fail:=false; {method has not yet failed!}
    cyclimit:=n; {use n steps of cg before re-setting to steepest descent}
    if intol<0.0 then intol:=Calceps; {allows machine to set a value}
    tol:=intol*n*sqrt(intol); {gradient test tolerance}
    {Note: this tolerance should properly be scaled to the problem at hand.
    However, such a scaling presumes knowledge of the function and gradient
    which we do not usually have at this stage of the minimisation.}
    writeln('tolerance used in gradient test=', tol);
    f:=fminfn(n, Bvec, Workdata, notcomp); {initial fn calculation}{STEP 1}
    if notcomp then
    begin
        writeln('**** Function cannot be evaluated at initial parameters ****');
        fail := true; {method has failed}
```

Algorithm 22. Function minimisation by conjugate gradients (cont.)

```
end
else {proceed with minimisation}
begin
    Fmin:=f;{save the best value so far}
    funcount:=1; {function count initialized to 1}
    gradcount:=0; {initialise gradient count}
    repeat {STEP 2: iteration for method -- terminates when no progress
                can be made, and search is a steepest descent.}
        for i:=1 to n do
        begin
            t[i]:=0.0; {to zero step vector}
            c[i]:=0.0; {to zero 'last' gradient}
        end;
        cycle:=0; {STEP 3: main loop of cg process}
        oldstep:=1.0; {initially a full step}
        count:=0;{to ensure this is set < n}
        repeat {until one cg cycle complete}
            cycle:=cycle+1;
            writeln(gradcount, ' ', funcount, ' ', Fmin);
            write('parameters ');
            for i:=1 to n do
            begin
                write(Bvec[i]:10:5, ' ');
                if (7 * (i div 7) = i) and (i<n) then writeln;
            end;
            writeln;
            gradcount:=gradcount+1; {STEP 4: initialize gradient count to 0}
            fmingr(n, Bvec, Workdata, g);
            G1:=0.0; G2:=0.0; {STEP 5}
            for i:=1 to n do
            begin
                X[i]:=Bvec[i];{save best parameters}
                case method of
                    Fletcher_Reeves: begin
                        G1:=G1+sqr(g[i]); G2:=G2+sqr(c[i]);
                    end;
                    Polak_Ribiere : begin
                        G1:=G1+g[i]*(g[i]-c[i]); G2:=G2+sqr(c[i]);
                    end;
                    Beale_Sorenson : begin
                        G1:=G1+g[i]*(g[i]-c[i]); G2:=G2+t[i]*(g[i]-c[i]);
                    end;
                end; {case statement for method selection}
                c[i]:=g[i];{save gradient}
            end; {loop on i}
            if G1>tol then {STEP 6: descent sufficient to proceed}
            begin {STEP 7: generate direction}
                if G2>0.0 then G3:=G1/G2 else G3:=1.0; {ensure G3 defined}
                gradproj:=0.0; {STEP 8}
                for i:=1 to n do
                begin
                    t[i]:=t[i]*G3-g[i]; gradproj:=gradproj+t[i]*g[i];
```

Algorithm 22. Function minimisation by conjugate gradients (cont.)

```
        end;
        steplength:=oldstep; {STEP 9}
        {STEP 10: step along search direction}
        accpoint:=false; {don't have a good point yet}
        repeat {line search}
          count:=0;{to count unchanged parameters}
          for i:=1 to n do
          begin
            Bvec[i]:=X[i]+steplength*t[i];
            if (reltest+X[i])=(reltest+Bvec[i]) then count:=count+1;
          end; {loop on i}
          if count<n then {STEP 11}{main convergence test}
          begin {STEP 12} {can proceed with linear search}
            f:=fminfn(n, Bvec, Workdata, notcomp); {function calculation}
            funcount:=funcount+1;
            accpoint:=(not notcomp) and
                (f<=Fmin+gradproj*steplength*acctol);
            {STEP 13: a point is acceptable only if function is computable
            (not notcomp) and it satisfies the acceptable point criterion}
            if not accpoint then
            begin
              steplength:=steplength*stepredn;  write('*');
            end;
          end; {compute and test for linear search}
        until (count=n) or accpoint; {end of loop for line search}
        if count<n then {STEP 14}
        begin {replacements for STEPS 15 onward}
          newstep:=2*((f-Fmin)-gradproj*steplength);
            {quadratic inverse interpolation}
          if newstep>0 then
          begin {cacl interp}
            newstep:=-gradproj*sqr(steplength)/newstep;
            for i:=1 to n do
            begin
              Bvec[i]:=X[i]+newstep*t[i];
            end; {no check yet on change in parameters}
            Fmin:=f; {save new lowest point}
            f:=fminfn(n, Bvec, Workdata, notcomp);
            funcount:=funcount+1;
            if  f<Fmin then
            begin
              Fmin:=f; write(' i< ');
            end
            else {reset to best Bvec}
            begin
              write(' i> ');
              for i:=1 to n do Bvec[i]:=X[i]+steplength*t[i];
            end;
          end; {interpolation}
        end;{if count < n}
      end; {if G1>tol}
      oldstep:=setstep*steplength; {a heuristic to prepare next iteration}
```

Algorithm 22. Function minimisation by conjugate gradients (cont.)

```
            if oldstep>1.0 then oldstep:=1.0; {with a limitation to prevent too
                  large a step being taken. This strategy follows Nash &
                  Walker-Smith in multiplying the best step by 1.7 and then
                  limiting the resulting step to length = 1 }
        until (count=n) or (G1<=tol) or (cycle=cyclimit);
            {this ends the cg cycle loop}
    until (cycle=1) and ((count=n) or (G1<=tol));
      {this is the convergence condition to end loop at STEP 2}
  end; {begin minimisation}
  writeln('Exiting from Alg22.pas conjugate gradients minimiser');
  writeln(' ', funcount, ' function evaluations used');
  writeln(' ', gradcount, ' gradient evaluations used');
end; {alg22.pas == cgmin}
```

Example 16.1. Conjugate gradients minimisation

In establishing a tentative regional model of hog supply in Canada, Dr J J Jaffrelot wanted to reconcile the sum of supply by region for a given year (an annual figure) with the sum of quarterly supply figures (not reported by regions!). The required sum can, by attaching minus signs, be written

$$T=\sum_{j=1}^{n} w_j b_j$$

where the b_j are the parameters which should give $T=0$ when summed as shown with the weights $\boldsymbol{w}$. Given w_j, $j=1, 2, \ldots, n$, T can easily be made zero since there are $(n-1)$ degrees of freedom in $\boldsymbol{b}$. However, some degree of confidence must be placed in the published figures, which we shall call p_j, $j=1, 2, \ldots, n$. Thus, we wish to limit each b_j so that

$$|b_j-p_j|\leqslant d_j \qquad \text{for } j=1,2,\ldots,n$$

where d_j is some tolerance. Further, we shall try to make $\boldsymbol{b}$ close to $\boldsymbol{p}$ by minimising the function

$$S(b)=100\left(\sum_{j=1}^{n} w_j b_j\right)^2+\sum_{j=1}^{n}(b_j-p_j)^2$$
$$=100(\boldsymbol{w}^{\mathrm{T}}\boldsymbol{b})^2+(\boldsymbol{b}-\boldsymbol{p})^{\mathrm{T}}(\boldsymbol{b}-\boldsymbol{p}).$$

The factor 100 is arbitrary. Note that this is in fact a linear least-squares problem, subject to the constraints above. However, the conjugate gradients method is quite well suited to the particular problem in 23 parameters which was presented, since it can easily incorporate the tolerances d_j by declaring the function to be 'not computable' if the constraints are violated. (In this example they do not in fact appear to come into play.) The output below was produced on a Data General ECLIPSE operating in six hexadecimal digit arithmetic. Variable 1 is used to hold the values $\boldsymbol{p}$, variable 2 to hold the tolerances $\boldsymbol{d}$ and variable 3 to hold the weights $\boldsymbol{w}$. The number of data points is reported to be 24 and a zero has been appended to each of the above-mentioned variables to accommodate the particular way in which the objective function was computed. The starting values for the parameters ***b*** are the obvious choices, that is, the reported values ***p***. The minimisation program and data for this problem used less than 4000 bytes (characters) of memory.

```
*
*NEW
*ENTER"JJJRUN"
*RUN
 12  7  1978   9  15  2
NCG JULY 26 77
CG + SUCCESS FAILURE
DATA FILE NAME ? D16.12
# OF VARIABLES 3
# DATA POINTS 24
# OF PARAMETERS 23
ENTER VARIABLES
VARIABLE 1  - COMMENT -THE PUBLISHED VALUES P
 167.85  .895  167.85  .895 -99.69
 167.85  .895 -74.33  167.85  .895
-4.8 -1.03 -1  3.42 -68.155
-.73 -.12 -20.85  1.2 -2.85
 31.6 -20.66 -8.55  0
VARIABLE 2  - COMMENT -THE TOLERANCES D
 65.2  5.5E-02  65.2  5.5E-02  20
 65.2  5.5E-02  19.9  65.2  5.5E-02
 1.6  .36  .34  1.5  10.188
 .51  .26  9.57  .27  .56
 14.7  3.9  4.8  0
VARIABLE 3  - COMMENT -THE WEIGHTS W
 1  1309.67  1  1388.87  1
 1  1377.69  1  1  1251.02
 15  119.197  215  29.776  15
 806.229  1260  23.62  2761  2075
 29.776  33.4  51.58  0
ENTER STARTING VALUES FOR PARAMETERS
B( 1 )= 167.85
B( 2 )= .895
B( 3 )= 167.85
B( 4 )= .895
B( 5 )=-99.69
B( 6 )= 167.85
B( 7 )= .895
B( 8 )=-74.33
B( 9 )= 167.85
B( 10 )= .895
B( 11 )=-4.8
B( 12 )=-1.03
B( 13 )=-1
B( 14 )= 3.42
B( 15 )=-68.155
B( 16 )=-.73
B( 17 )=-.12
B( 18 )=-20.85
B( 19 )= 1.2
B( 20 )=-2.85
B( 21 )= 31.6
B( 22 )=-20.66
B( 23 )=-8.55
STEPSIZE= 1
 0  1  772798
 1  21  5.76721E-04
```

```
  2   31    5.76718E-04
  3   31    5.76718E-04
  4   42    5.76716E-04
  5   42    5.76716E-04
  6   45    5.76713E-04
  7   45    5.76713E-04
  8   48    5.76711E-04
  9   48    5.76711E-04
CONVERGED TO 5.76711E-04    # ITNS= 10    # EVALS= 50
# EFES= 290
B( 1 )= 167.85      G( 1 )= .148611
B( 2 )= .900395      G( 2 )= 194.637
B( 3 )= 167.85      G( 3 )= .148611
B( 4 )= .900721      G( 4 )= 206.407
B( 5 )=-99.69      G( 5 )= .148626
B( 6 )= 167.85      G( 6 )= .148611
B( 7 )= .900675      G( 7 )= 204.746
B( 8 )=-74.33      G( 8 )= .148626
B( 9 )= 167.85      G( 9 )= .148611
B( 10 )= .900153      G( 10 )= 185.92
B( 11 )=-4.79994      G( 11 )= 2.22923
B( 12 )=-1.02951      G( 12 )= 17.7145
B( 13 )=-.999114      G( 13 )= 31.9523
B( 14 )= 3.42012      G( 14 )= 4.42516
B( 15 )=-68.1549      G( 15 )= 2.22924
B( 16 )=-.726679      G( 16 )= 119.818
B( 17 )=-.114809      G( 17 )= 187.255
B( 18 )=-20.8499      G( 18 )= 3.5103
B( 19 )= 1.21137      G( 19 )= 410.326
B( 20 )=-2.84145      G( 20 )= 308.376
B( 21 )= 31.6001      G( 21 )= 4.42516
B( 22 )=-20.6598      G( 22 )= 4.96376
B( 23 )=-8.54979      G( 23 )= 7.66536

STOP AT 0911
*SIZE
USED: 3626  BYTES
LEFT: 5760  BYTES
*
```

An earlier solution to this problem, obtained using a Data General NOVA operating in 23 binary digit arithmetic, had identical values for the parameters B but quite different values for the gradient components G. The convergence pattern of the program was also slightly different and had the following form:

```
0   1    772741
1   21    5.59179E-4
2   22    5.59179E-4
CONVERGED TO  5.59179E-4    # ITNS= 3    # EVALS= 29
```

In the above output, the quantities printed are the number of iterations (gradient evaluations), the number of function evaluations and the lowest function value found so far. The sensitivity of the gradient and the convergence pattern to relatively small changes in arithmetic is, in my experience, quite common for algorithms of this type.

Chapter 17

MINIMISING A NONLINEAR SUM OF SQUARES

17.1. INTRODUCTION

The mathematical problem to be considered here is that of minimising

$$S(\boldsymbol{x}) = \sum_{i=1}^{m} [f_i(\boldsymbol{x})]^2 \tag{17.1}$$

with respect to the parameters x_j, $j = 1, 2, \ldots, n$ (collected for convenience as the vector $\boldsymbol{x}$), where at least one of the functions $f_i(\boldsymbol{x})$ is nonlinear in $\boldsymbol{x}$. Note that by collecting the m functions $f_i(\boldsymbol{x})$, $i = 1, 2, \ldots, m$, as a vector $\boldsymbol{f}$, we get

$$S(\boldsymbol{x}) = \boldsymbol{f}^{\mathrm{T}}\boldsymbol{f}. \tag{17.2}$$

The minimisation of a nonlinear sum-of-squares function is a sufficiently widespread activity to have developed special methods for its solution. The principal reason for this is that it arises whenever a least-squares criterion is used to fit a nonlinear model to data. For instance, let y_i represent the weight of some laboratory animal at week i after birth and suppose that it is desired to model this by some function of the week number i, which will be denoted $y(i, \boldsymbol{x})$, where $\boldsymbol{x}$ is the set of parameters which will be varied to fit the model to the data. If the criterion of fit is that the sum of squared deviations from the data is to be minimised (least squares) then the *objective* is to minimise (17.1) where

$$f_i(\boldsymbol{x}) = y(i, \boldsymbol{x}) - y_i \tag{17.3}$$

or, in the case that confidence weightings are available for each data point,

$$f_i(\boldsymbol{x}) = [y(i, \boldsymbol{x}) - y_i]w_i \tag{17.4}$$

where w_i, $i = 1, 2, \ldots, m$, are the weightings. As a particular example of a growth function, consider the three-parameter logistic function (Oliver 1964)

$$y(i, \boldsymbol{x}) = y(i, x_1, x_2, x_3) = x_i/[1 + \exp(x_2 + ix_3)]. \tag{17.5}$$

Note that the form of the residuals chosen in (17.3) and (17.4) is the negative of the usual 'actual minus fitted' used in most of the statistical literature. The reason for this is to make the derivatives of $f_i(\boldsymbol{x})$ coincide with those of $y(i, \boldsymbol{x})$.

The minimisation of $S(\boldsymbol{x})$ could, of course, be approached by an algorithm for the minimisation of a general function of n variables. Bard (1970) suggests that this is not as efficient as methods which recognise the sum-of-squares form of $S(\boldsymbol{x})$, though more recently Biggs (1975) and McKeown (1974) have found contrary results. In the paragraphs below, algorithms will be described which take explicit note of the sum-of-squares form of $S(\boldsymbol{x})$, since these are relatively simple and, as building blocks, use algorithms for linear least-squares computations which have already been discussed in earlier chapters.

17.2. TWO METHODS

Almost immediately two possible routes to minimising $S(\boldsymbol{x})$ suggest themselves.

The Cauchy steepest descents method

Find the gradient $2\boldsymbol{v}(\boldsymbol{x})$ of $S(\boldsymbol{x})$ and step downhill along it. (The reason for the factor 2 will become apparent shortly.) Suppose that t represents the step length along the gradient, then for some t we have

$$S(\boldsymbol{x}-t\boldsymbol{v})<S(\boldsymbol{x}) \tag{17.6}$$

except at a local minimum or a saddle point. The *steepest descents* method replaces $\boldsymbol{x}$ by $(\boldsymbol{x}-t\boldsymbol{v})$ and repeats the process from the new point. The iteration is continued until a t cannot be found for which (17.6) is satisfied. The method, which was suggested by Cauchy (1848), is then taken to have converged. It can be shown always to converge if $S(\boldsymbol{x})$ is *convex*, that is, if

$$S(c\boldsymbol{x}_1+(1-c)\boldsymbol{x}_2)\leqslant cS(\boldsymbol{x}_1)+(1-c)S(\boldsymbol{x}_2) \tag{17.7}$$

for $0<c<1$. Even for non-convex functions which are bounded from below, the steepest descents method will find a local minimum or saddle point. All the preceding results are, of course, subject to the provision that the function and gradient are computed exactly (an almost impossible requirement). In practice, however, convergence is so slow as to disqualify the method of steepest descents on its own as a candidate for minimising functions.

Often the cause of this slowness of convergence is the tendency of the method to take pairs of steps which are virtually opposites, and which are both essentially perpendicular to the direction in which the minimum is to be found. In a two-parameter example we may think of a narrow valley with the minimum somewhere along its length. Suppose our starting point is somewhere on the side of the valley but not near this minimum. The gradient will be such that the direction of steepest descent is towards the floor of the valley. However, the step taken can easily traverse the valley. The situation is then similar to the original one, and it is possible to step back across the valley almost to our starting point with only a very slight motion *along* the valley toward the solution point. One can picture the process as following a path similar to that which would be followed by a marble or ball-bearing rolling over the valley-shaped surface.

To illustrate the slow convergence, a modified version of steepest descents was programmed in BASIC on a Data General NOVA minicomputer having machine precision 2^{-22}. The modification consisted of step doubling if a step is successful. The step length is divided by 4 if a step is unsuccessful. This reduction in step size is repeated until either a smaller sum of squares is found or the step is so small that none of the parameters change. As a test problem, consider the Rosenbrock banana-shaped valley:

$$S(\boldsymbol{x})=S(x_1, x_2)=100(x_2-x_1^2)^2+(1-x_1)^2$$

starting with

$$S(-1{\cdot}2, 1)=24{\cdot}1999$$

(as evaluated). The steepest descents program above required 232 computations of the derivative and 2248 evaluations of $S(\boldsymbol{x})$ to find

$$S(1{\cdot}00144, 1{\cdot}0029) = 2{\cdot}1 \times 10^{-6}.$$

The program was restarted with this point and stopped manually after 468 derivative and 4027 sum-of-squares computations, where

$$S(1{\cdot}00084, 1{\cdot}00168) = 7{\cdot}1 \times 10^{-7}.$$

By comparison, the Marquardt method to be described below requires 24 derivative and 32 sum-of-squares evaluations to reach

$$S(1, 1) = 1{\cdot}4 \times 10^{-14}.$$

(There are some rounding errors in the display of x_1, x_2 or in the computation of $S(\boldsymbol{x})$, since $S(1, 1) = 0$ is the solution to the Rosenbrock problem.)

The Gauss–Newton method

At the minimum the gradient $\boldsymbol{v}(\boldsymbol{x})$ must be null. The functions $v_j(\boldsymbol{x})$, $j = 1, 2, \ldots, n$, provide a set of n nonlinear functions in n unknowns $\boldsymbol{x}$ such that

$$\boldsymbol{v}(\boldsymbol{x}) = \boldsymbol{0} \tag{17.8}$$

the solution of which is a stationary point of the function $S(\boldsymbol{x})$, that is, a local maximum or minimum or a saddle point. The particular form (17.1) or (17.2) of $S(\boldsymbol{x})$ gives gradient components

$$2v_j(\boldsymbol{x}) = 2\sum_{i=1}^{m} f_i(\boldsymbol{x})\partial f_i(\boldsymbol{x})/\partial x_j \tag{17.9}$$

which reduces to

$$v_j(\boldsymbol{x}) = \sum_{i=1}^{m} f_i(\boldsymbol{x})J_{ij}(\boldsymbol{x}) \tag{17.10}$$

or

$$\boldsymbol{v} = \mathbf{J}^{\mathrm{T}}\boldsymbol{f} \tag{17.11}$$

by defining the Jacobian matrix $\mathbf{J}$ by

$$J_{ij} = \partial f_i(\boldsymbol{x})/\partial x_j. \tag{17.12}$$

Some approximation must now be made to simplify the equations (17.8). Consider the Taylor expansion of $v_j(\boldsymbol{x})$ about $\boldsymbol{x}$

$$v_j(\boldsymbol{x} + \boldsymbol{q}) = v_j(\boldsymbol{x}) + \sum_{k=1}^{n} q_k \partial v_j(\boldsymbol{x})/\partial x_k + (\text{terms in } q^2). \tag{17.13}$$

If the terms in q^2 (that is, those involving $q_k q_j$ for $k, j = 1, 2, \ldots, n$) are assumed to be negligible and $v_j(\boldsymbol{x} + \boldsymbol{q})$ is taken as zero because it is assumed to be the solution, then

$$\sum_{k=1}^{n} q_k \partial v_j(\boldsymbol{x})/\partial x_k = -v_j(\boldsymbol{x}) \tag{17.14}$$

for each $j=1, 2, \ldots, n$. From (17.10) and (17.12), therefore, we have

$$\partial v_j(\boldsymbol{x})/\partial x_k = \sum_{i=1}^{m} [J_{ik}(\boldsymbol{x})J_{ij}(\boldsymbol{x}) + f_i(\boldsymbol{x})\partial^2 f_i(x)/\partial x_j \partial x_k]. \tag{17.15}$$

To apply the Newton–Raphson iteration (defined by equation (15.9)) to the solution of (17.8) thus requires the second derivatives of f with respect to the parameters, a total of mn^2 computations. On the grounds that near the minimum the functions should be 'small' (a very cavalier assumption), the second term in the summation (17.15) is neglected, so that the partial derivatives of $\boldsymbol{v}$ are approximated by the matrix $\mathbf{J}^{\mathrm{T}}\mathbf{J}$, reducing (17.14) to

$$\mathbf{J}^{\mathrm{T}}\mathbf{J}\boldsymbol{q} = -\boldsymbol{v} = -\mathbf{J}^{\mathrm{T}}\boldsymbol{f}. \tag{17.16}$$

These are simply normal equations for a linearised least-squares problem evaluated locally, and the *Gauss–Newton* iteration proceeds by replacing $\boldsymbol{x}$ by $(\boldsymbol{x}+\boldsymbol{q})$ and repeating until either $\boldsymbol{q}$ is smaller than some predetermined tolerance or

$$S(\boldsymbol{x}+\boldsymbol{q}) \geqslant S(\boldsymbol{x}). \tag{17.17}$$

17.3. HARTLEY'S MODIFICATION

As it has been stated, the Gauss–Newton iteration can quite easily fail since the second term in the summation (17.15) is often not negligible, especially in the case when the residuals $\boldsymbol{f}$ are large. However, if instead of replacing $\boldsymbol{x}$ by $(\boldsymbol{x}+\boldsymbol{q})$ we use $(\boldsymbol{x}+t\boldsymbol{q})$, where t is a step-length parameter, it is possible to show that if $\mathbf{J}$ is of full rank ($\mathbf{J}^{\mathrm{T}}\mathbf{J}$ is positive definite and hence non-singular), then this modified Gauss–Newton algorithm always proceeds towards a minimum. For in this case, we have

$$\boldsymbol{q} = -(\mathbf{J}^{\mathrm{T}}\mathbf{J})^{-1}\mathbf{J}^{\mathrm{T}}\boldsymbol{f} \tag{17.18}$$

so that the inner product between $\boldsymbol{q}$ and the direction of the steepest descent $-\mathbf{J}^{\mathrm{T}}\boldsymbol{f}$ is

$$-\boldsymbol{q}^{\mathrm{T}}\boldsymbol{v} = \boldsymbol{f}^{\mathrm{T}}\mathbf{J}(\mathbf{J}^{\mathrm{T}}\mathbf{J})^{-1}\mathbf{J}^{\mathrm{T}}\boldsymbol{f} > 0 \tag{17.19}$$

since $(\mathbf{J}^{\mathrm{T}}\mathbf{J})^{-1}$ is positive definite by virtue of the positive definiteness of $\mathbf{J}^{\mathrm{T}}\mathbf{J}$. Thus the cosine of the angle between $\boldsymbol{q}$ and $-\boldsymbol{v}$ is always positive, guaranteeing that the search is downhill. In practice, however, angles very close to 90° are observed, which may imply slow convergence.

The modified Gauss–Newton procedure above has been widely used to minimise sum-of-squares functions (Hartley 1961). Usually the one-dimensional search at each iteration is accomplished by fitting a quadratic polynomial to the sum-of-squares surface along the search direction and evaluating the function at the estimated minimum of this quadratic. This requires at least three sum-of-squares function evaluations per iteration and, furthermore, the apparent simplicity of this parabolic inverse interpolation to accomplish a linear search when stated in words hides a variety of strategems which must be incorporated to prevent moves either in the wrong direction or to points where $S(\boldsymbol{x}+\boldsymbol{q}) > S(\boldsymbol{x})$. When the function is sufficiently well behaved for the unmodified Gauss–Newton algorithm to work successfully, the linear search introduces some waste of effort. Furthermore, when there are large differences in scale between the elements of the Jacobian **J** and/or it is difficult to evaluate these accurately, then the condition of $\mathbf{J}^T\mathbf{J}$ may deteriorate so that positive definiteness can no longer be guaranteed computationally and the result (17.19) is inapplicable. Nevertheless, I have found that a carefully coded implementation of Hartley's (1961) ideas using the Choleski decomposition and back-solution for consistent (but possibly singular) systems of linear equations is a very compact and efficient procedure for solving the nonlinear least-squares problem. However, it has proved neither so simple to implement nor so generally reliable as the method which follows and, by virtue of its modifications to handle singular $\mathbf{J}^T\mathbf{J}$, no longer satisfies the conditions of the convergence result (17.19).

17.4. MARQUARDT'S METHOD

The problems of both scale and singularity of $\mathbf{J}^T\mathbf{J}$ are attacked simultaneously by Marquardt (1963). Consider solutions $\boldsymbol{q}$ to the equations

$$(\mathbf{J}^T\mathbf{J}+e\mathbf{1})\boldsymbol{q}=-\mathbf{J}^T\boldsymbol{f} \tag{17.20}$$

where e is some parameter. Then as e becomes very large relative to the norm of $\mathbf{J}^T\mathbf{J}$, $\boldsymbol{q}$ tends towards the steepest descents direction, while when e is very small compared to this norm, the Gauss–Newton solution is obtained. Furthermore, the scaling of the parameters

$$\boldsymbol{x}'=\mathbf{D}\boldsymbol{x} \tag{17.21}$$

where **D** is a diagonal matrix having positive diagonal elements, implies a transformed Jacobian such that

$$\mathbf{J}'=\mathbf{J}\mathbf{D}^{-1} \tag{17.22}$$

and equations (17.20) become

$$[(\mathbf{J}')^T\mathbf{J}'+e\mathbf{1}]\boldsymbol{q}'=-(\mathbf{J}')^T\boldsymbol{f} \tag{17.23a}$$

$$=(\mathbf{D}^{-1}\mathbf{J}^T\mathbf{J}\mathbf{D}^{-1}+e\mathbf{1})\mathbf{D}\boldsymbol{q}=-\mathbf{D}^{-1}\mathbf{J}^T\boldsymbol{f} \tag{17.23b}$$

or

$$(\mathbf{J}^T\mathbf{J}+e\mathbf{D}^2)\boldsymbol{q}=-\mathbf{J}^T\boldsymbol{f} \tag{17.24}$$

so that the scaling may be accomplished implicitly by solving (17.24) instead of (17.23a).

Marquardt (1963) and Levenberg (1944) have suggested the scaling choice

$$D_{ii}^2=(\mathbf{J}^T\mathbf{J})_{ii}. \tag{17.25}$$

However, to circumvent failures when one of the diagonal elements of $\mathbf{J}^T\mathbf{J}$ is zero, I prefer to use

$$D_{ii}^2=(\mathbf{J}^T\mathbf{J})_{ii}+\phi \tag{17.26}$$

where ϕ is some number chosen to ensure the scale is not too small (see Nash 1977). A value of $\phi=1$ seems satisfactory for even the most pathological problems. The matrix $\mathbf{J}^T\mathbf{J}+e\mathbf{D}^2$ is always positive definite, and by choosing e

large enough can, moreover, be made *computationally* positive definite so that the simplest forms of the Choleski decomposition and back-solution can be employed. That is to say, the Choleski decomposition is not completed for non-positive definite matrices. Marquardt (1963) suggests starting the iteration with $e=0{\cdot}1$, reducing it by a factor of 10 before each step if the preceding solution $\boldsymbol{q}$ has given

$$S(\boldsymbol{x}+\boldsymbol{q})<S(\boldsymbol{x})$$

and $\boldsymbol{x}$ has been replaced by $(\boldsymbol{x}+\boldsymbol{q})$. If

$$S(\boldsymbol{x}+\boldsymbol{q})\geqslant S(\boldsymbol{x})$$

then e is increased by a factor of 10 and the solution of equations (17.24) repeated. (In the algorithm below, e is called lambda.)

17.5. CRITIQUE AND EVALUATION

By and large, this procedure is reliable and efficient. Because the bias e is reduced after each successful step, however, there is a tendency to see the following scenario enacted by a computer at each iteration, that is, at each evaluation of $\mathbf{J}$:

(i) reduce e, find $S(\boldsymbol{x}+\boldsymbol{q})\geqslant S(\boldsymbol{x})$;
(ii) increase e, find $S(\boldsymbol{x}+\boldsymbol{q})<S(\boldsymbol{x})$, so replace $\boldsymbol{x}$ by $(\boldsymbol{x}+\boldsymbol{q})$ and proceed to (i).

This procedure would be more efficient if e were not altered. In other examples one hopes to take advantage of the rapid convergence of the Gauss–Newton part of the Marquardt equations by reducing e, so a compromise is called for. I retain 10 as the factor for increasing e, but use $0{\cdot}4$ to effect the reduction. A further safeguard which should be included is a check to ensure that e does not approach zero computationally. Before this modification was introduced into my program, it proceeded to solve a difficult problem by a long series of approximate Gauss–Newton iterations and then encountered a region on the sum-of-squares surface where steepest descents steps were needed. During the early iterations e underflowed to zero, and since $\mathbf{J}^{\mathrm{T}}\mathbf{J}$ was singular, the program made many futile attempts to increase e before it was stopped manually.

The practitioner interested in implementing the Marquardt algorithm will find these modifications included in the description which follows.

Algorithm 23. Modified Marquardt method for minimising a nonlinear sum-of-squares function

```
procedure modmrt( n : integer; {number of residuals and number of parameters}
                  var Bvec : rvector; {parameter vector}
                  var X : rvector; {derivatives and best parameters}
                  var Fmin : real; {minimum value of function}
                      Workdata : probdata);
{alg23.pas == modified Nash Marquardt nonlinear least squares minimisation
      method.
                  Copyright 1988 J.C.Nash
}
var
```

Algorithm 23. Modified Marquardt method for minimising a nonlinear sum-of-squares function (cont.)

```
    a, c: smatvec;
    delta, v : rvector;
    dec, eps, inc, lambda, p, phi, res : real;
    count, i, ifn, igrad, j, k, nn2, q : integer;
    notcomp, singmat, calcmat: boolean;
begin
    writeln('alg23.pas -- Nash Marquardt nonlinear least squares');
    with Workdata do
    begin {STEP 0 partly in procedure call}
        if nlls = false then halt; {cannot proceed if we do not have a nonlinear
                        least squares problem available}
        Fmin:=big; {safety setting of the minimal function value}
        inc:=10.0; {increase factor for damping coefficient lambda}
        dec:=0.4; {decrease factor for damping coefficient lambda}
        eps:=calceps; {machine precision}
        lambda:=0.0001; {initialize damping factor}
        phi:=1.0; {set the Nash damping factor}
        ifn:=0; igrad:=0; {set the function and gradient evaluation counts}
        calcmat:=true; {to force calculation of the J-transpose * J matrix and
                        J-transpose * residual on the first iteration}
        nn2:=(n*(n+1)) div 2; {elements in the triangular form of the inner
                        product matrix -- ensure this is an integer}
        p:=0.0; {STEP1}
        for i:=1 to m do
        begin
            res:=nlres(i, n, Bvec, notcomp); {the residual}
            {writeln('res[',i,']=',res);
            if notcomp then halt; {safety check on initial evaluation}
            p:=p+res*res; {sum of squares accumulation}
        end;
        ifn:=ifn+1; {count the function evaluation}
        Fmin:=p; {to save best sum of squares so far}
        count:=0; {to avoid convergence immediately}
        {STEP 2}
        while count<n do
        begin {main Marquardt iteration}
            {NOTE: in this version we do not reduce the damping parameter here.
            The structure of Pascal lends itself better to adjusting the damping
            parameter below.}
            if calcmat then
            begin {Compute sum of squares and cross-products matrix}
                writeln(igrad,' ',ifn,' sum of squares=',Fmin);
                for i:=1 to n do
                begin
                    write(Bvec[i]:10:5,' ');
                    if (7 * (i div 7) = i) and (i<n) then writeln;
                end; {loop on i}
                writeln;
                igrad:=igrad+1; {STEP 3}
                for j:=1 to nn2 do a[j]:=0.0;
                for j:=1 to n do v[j]:=0.0;
                for i:=1 to m do {STEP 4}
```

Algorithm 23. Modified Marquardt method for minimising a nonlinear sum-of-squares function (cont.)

```
            begin
                nljac(i, n, Bvec, X); {puts i'th row of Jacobian in X}
                res:=nlres(i, n, Bvec, notcomp); {This calculation is not really
                    necessary. The results of the sum of squares calculation
                    can be saved in a residual vector to avoid the
                    recalculation. However, this way saves storing a possibly
                    large vector. NOTE: we ignore notcomp here, since the
                    computation of the residuals has already been proven
                    possible at STEP 1.}
                for j:=1 to n do
                begin
                    v[j]:=v[j]+X[j]*res; {to accumulate the gradient}
                    q:=(j*(j-1)) div 2; {to store the correct position in the
                    row-order vector form of the lower triangle of a symmetric
                    matrix}
                    for k:=1 to j do a[q+k]:=a[q+k]+X[j]*X[k];
                end; {loop on j}
            end; {loop on i}
            for j:=1 to nn2 do c[j]:=a[j]; {STEP 5 -- copy a and b}
            for j:=1 to n do X[j]:=Bvec[j]; {to save the best parameters}
        end; {if calcmat}
        writeln('LAMDA =',lambda:8); {STEP 6}
        for j:=1 to n do
        begin
            q:=(j*(j+1)) div 2;
            a[q]:=c[q]*(1.0+lambda)+phi*lambda; {damping added}
            delta[j]:=-v[j]; {to set RHS in equations 17.24}
            if j>1 then
                for i:=1 to (j-1) do a[q-i]:=c[q-i];
        end; {loop on j}
        notcomp:=false; {to avoid undefined value}
        Choldcmp(n, a, singmat); {STEP 7 -- Choleski factorization}
        if (not singmat) then {matrix successfully decomposed}
        begin {STEP 8 -- Choleski back-substitution}
            Cholback(n, a, delta); {delta is the change in the parameters}
            count:=0; {to count parameters unchanged in update}
            for i:=1 to n do {STEP 9}
            begin
                Bvec[i]:=X[i]+delta[i]; {new = old + update}
                if (reltest + Bvec[i])=(reltest+X[i]) then count:=count+1;
                {Here the global constant reltest is used to test for the
                equality of the new and old parameters.}
            end; {loop on i over parameters}
            if count<n then {STEP 10: parameters have been changed}
            begin {compute the sum of squares, checking computability}
                p:=0.0; i:=0; {initialization}
                repeat
                    i:=i+1; res:=nlres(i,n,Bvec,notcomp);
                    if (not notcomp) then p:=p+res*res;
                until notcomp or (i>=n);
                ifn:=ifn+1; {count the function evaluation}
            end;{if count<n}
```

Algorithm 23. Modified Marquardt method for minimising a nonlinear sum-of-squares function (cont.)

```
          end; {if not singmat}
          if count<n then
             if (not singmat) and (not notcomp) and (p<Fmin) then
             begin {successful in reducing sum of squares}
                lambda:=lambda*dec; {to decrease the damping parameter}
                Fmin:=p; {to save best sum of squares so far}
                calcmat:=true; {to perform new iteration}
             end
          else {somehow have not succeeded in reducing sum of
                      squares: matrix is singular, new point is not
                      a feasible set of parameters, or sum of squares
                      at new point not lower}
          begin
             lambda:=lambda*inc;{to increase the damping parameter}
             if lambda<eps*eps then lambda:=eps; {safety check}
             calcmat:=false; {since we simply renew search from the same
                point in the parameter space with new search step}
          end; {adjustment of damping factor}
          {This also ends 'if count<n'}
       end; {while count<n}
    end; {with Workdata}
end; {alg23.pas == modmrt}
```

17.6. RELATED METHODS

Fletcher (1971) has proposed a Marquardt method in which the bias e (lambda) is computed on the basis of some heuristic rules rather than adjusted arbitrarily. His program, incorporated in the Harwell and NAG libraries, has had widespread usage and seems highly reliable. On straightforward problems I have found it very efficient, and in terms of evaluations of $\mathbf{J}$ and $S(\boldsymbol{b})$ generally better than a FORTRAN version of algorithm 23 even on difficult problems. However, in a small sample of runs, the cost of running the Fletcher program was higher than that of running the FORTRAN algorithm 23 (Nash 1977).

Jones (1970) uses an explicit search along spirals between the Gauss–Newton and steepest descents directions. He reports very favourable performance for this procedure and some of my colleagues who have employed it speak well of it. Unfortunately, it does not seem possible to make it sufficiently compact for a 'small' computer.

Since the first edition was published, algorithm 23 has proved to be highly reliable. Nevertheless, I have at times been interested in using other approaches to minimising a nonlinear sum of squares. In LEQB05 (Nash 1984b), the singular-value decomposition is used to solve the Gauss–Newton linear least-squares problem of equation (17.16) without the formation of the Jacobian inner-products matrix. In Nash and Walker-Smith (1987), we added the facility to impose bounds constraints on the parameters. We also considered changing the method of solving equation (17.16) from the current Choleski method to a scaled Choleski method, but found that this change slowed the iterations.

Example 17.1. Marquardt's minimisation of a nonlinear sum of squares

The following output from a Data General NOVA (machine precision $=2^{-22}$) shows the solution of the problem stated in example 12.2 from the starting point $b^{(0)} = (200, 30, -0\cdot4)^T$. The program which ran the example below used a test within the function (residual) subroutine which set a flag to indicate the function was not computable if the argument of the exponential function within the expression for the logistic model exceeded 50. Without such a test, algorithm 23 (and indeed the other minimisers) may fail to converge to reasonable points in the parameter space.

```
NEW
LOAD ENHMRT
LOAD ENHHBS
RUN
ENHMRT FEB 16 76
REVISED MARQUARDT
# OF VARIABLES ? 1
# OF DATA POINTS ? 12
# OF PARAMETERS ? 3
ENTER VARIABLES
VARIABLE 1 :
? 5.308 ? 7.24 ? 9.638 ? 12.866 ? 17.069
? 23.192 ? 31.443 ? 38.558 ? 50.156 ? 62.948
? 75.995 ? 91.972
ENTER STARTING VALUES FOR PARAMETERS
B( 1 )=? 200
B( 3 )= ? 30
B( 3 )= -.4

ITN 1  SS= 23586.3
LAMBDA= .00004
ITN 2  SS= 327.692
LAMBDA= .000016
ITN 3  SS= 51.1076
LAMBDA= 6.4E-6
ITN 4  SS=2.65555
LAMBDA= 2.56E-6
ITN 5  SS= 2.58732
LAMBDA= 1.024E-6
ITN 6  SS= 2.58727
LAMBDA= 4.096E-7
LAMBDA= 4.096E-6
LAMBDA= 4.096E-5
LAMBDA= 4.096E-4
LAMBDA= .004096
ITN 7  SS= 2.58726
LAMBDA= 1.6384E-3
LAMBDA= .016384
CONVERGED TO SS= 2.58726   # ITNS=  7   # EVALNS=  12
SIGMA^2=   .287473
B( 1 )= 196.186   STD ERR= 11.3068   GRAD( 1 )= -7.18236E-6
B( 2 )= 49.0916   STD ERR= 1.68843   GRAD( 2 )= 1.84178E-5
B( 3 )= -.31357   STD ERR= 6.8632E-3   GRAD( 3 )= 8.48389E-3

RESIDUALS
1.18942E-2   -3.27625E-2   9.20258E-2   .208776   .392632
-5.75943E-2   -1.10573   .71579   -.107643   -.348396
.652573   -.287567
```

The derivatives in the above example were computed analytically. Using numerical approximation of the Jacobian as in §18.2 gives, from the same starting point,

$$\boldsymbol{b}^* = (196{\cdot}251, 49{\cdot}1012, -0{\cdot}313692)^{\mathrm{T}}$$

with $S(\boldsymbol{b}^*) = 2{\cdot}6113$ in 7 iterations and 16 function evaluations. By comparison, if the starting point is

$$\boldsymbol{b}^{(0)} = (1, 1, 1)^{\mathrm{T}} \qquad S(\boldsymbol{b}^{(0)}) = 24349{\cdot}5$$

then using analytic derivatives yields

$$\boldsymbol{b}^* = (196{\cdot}151, 49{\cdot}0876, -0{\cdot}313589)^{\mathrm{T}}$$

with $S(\boldsymbol{b}^*) = 2{\cdot}58726$ in 19 iterations and 33 evaluations, while numerically approximated derivatives give

$$\boldsymbol{b}^* = (194{\cdot}503, 48{\cdot}8935, -0{\cdot}314545)^{\mathrm{T}}$$

with $S(\boldsymbol{b}^*) = 2{\cdot}59579$ in 20 iterations and 36 evaluations.

Chapter 18

LEFT-OVERS

18.1. INTRODUCTION

This chapter is entitled 'left-overs' because each of the topics—approximation of derivatives, constrained optimisation and comparison of minimisation algorithms—has not so far been covered, though none is quite large enough in the current treatment to stand as a chapter on its own. Certainly a lot more could be said on each, and I am acutely aware that my knowledge (and particularly my experience) is insufficient to allow me to say it. As far as I am aware, very little work has been done on the development of compact methods for the mathematical programming problem, that is, constrained minimisation with many constraints. This is a line of research which surely has benefits for large machines, but it is also one of the most difficult to pursue due to the nature of the problem. The results of my own work comparing minimisation algorithms are to my knowledge the only study of such methods which has been made on a small computer. With the cautions I have given about results derived from experiments with a single system, the conclusions made in §18.4 are undeniably frail, though they are for the most part very similar to those of other workers who have used larger computers.

18.2. NUMERICAL APPROXIMATION OF DERIVATIVES

In many minimisation problems, the analytic computation of partial derivatives is impossible or extremely tedious. Furthermore, the program code to compute

$$\partial S(\boldsymbol{b})/\partial b_j \tag{18.1}$$

in a general unconstrained minimisation problem or

$$\partial f_i(\boldsymbol{b})/\partial b_j \tag{18.2}$$

in a nonlinear least-squares problem may be so long as to use up a significant proportion of the working space of a small computer. Moreover, in my experience 9 cases out of 10 of 'failure' of a minimisation program are due to errors in the code used to compute derivatives. The availability of numerical derivatives facilitates a check of such possibilities as well as allowing programs which require derivatives to be applied to problems for which analytic expressions for derivatives are not practical to employ.

In the literature, a great many expressions exist for numerical differentiation of functions by means of interpolation formulae (see, for instance, Ralston 1965). However, in view of the large number of derivative calculations which must be made during the minimisation of a function, these are not useful in the present instance. Recall that

$$\partial S(\boldsymbol{b})/\partial b_j = \lim_{h \to 0}[S(\boldsymbol{b} + h\boldsymbol{e}_j) - S(\boldsymbol{b})]/h \tag{18.3}$$

where $\boldsymbol{e}_j$ is the jth column of the unit matrix of order n ($\boldsymbol{b}$ is presumed to have n elements). For explanatory purposes, the case $n = 1$ will be used. In place of the limit (18.3), it is possible to use the *forward difference*

$$D = [S(b+h) - S(b)]/h \tag{18.4}$$

for some value of h.

Consider the possible sources of error in using D.

(i) For h small, the discrete nature of the representation of numbers in the computer causes severe inaccuracies in the calculation of D. The function S is continuous; its representation is not. In fact it will be a series of steps. Therefore, h cannot be allowed to be small. Another way to think of this is that since most of the digits of b are the same as those of $(b+h)$, any function S which is not varying rapidly will have similar values at b and $(b+h)$, so that the expression (18.4) implies a degree of digit cancellation causing D to be determined inaccurately.
(ii) For h large, the line joining the points $(b, S(b))$ and $(b+h, S(b+h))$ is no longer tangential to the curve at the former. Thus expression (18.4) is in error due to the nonlinearity of the function. Even worse, for some functions there may be a discontinuity between b and $(b+h)$. Checks for such situations are expensive of both human and computer time. The penalty for ignoring them is unfortunately more serious.

As a compromise between these extremes, I suggest letting

$$h = (|b_j| + \varepsilon^{1/2})\varepsilon^{1/2} \tag{18.5}$$

where ε is the machine precision. The parameter has once more been given a subscript to show that the step taken along each parameter axis will in general be different. The value for h given by (18.5) has the pleasing property that it cannot become smaller than the machine precision even if b_j is zero. Neither can it fail to change at least the right-most half of the digits in b_j since it is scaled by the magnitude of the parameter. Table 18.1 shows some typical results for this step-length choice compared with some other values.

Some points to note in table 18.1 are:

(i) The effect of the discontinuity in the tangent function in the computations for $b = 1$ and $b = 1{\cdot}57$ (near $\pi/2$). The less severely affected calculations for $b = -1{\cdot}57$ suggest that in some cases the *backward difference*

$$D = [S(b) - S(b-h)]/h \tag{18.6}$$

may be preferred.
(ii) In approximating the derivative of $\exp(0{\cdot}001)$ using $h = 1{\cdot}93024\text{E}-6$ as in equation (18.5), the system used for the calculations *printed* identical values for $\exp(b)$ and $\exp(b+h)$ even though the internal representations were different

TABLE 18.1. Derivative approximations computed by formula (18.4) on a Data General NOVA. (Extended BASIC system. Machine precision = 16^{-5}.)

	log(b)			exp(b)				tan(b)			
h	$b=0·001$	$b=1$	$b=100$	$b=-10$	$b=0·001$	$b=1$	$b=100$	$b=0$	$b=1$	$b=1·57$	$b=-1·57$
1	6·90876	0·693147	0·95064E−3	7·80101E−5	1·72	4·67077	4·6188E43	1·55741	−3·74245	−1256·61	1255·32
0·0625	66·4166	0·969993	9·99451E−3	4·68483E−5	1·03294	2·80502	2·77371E43	1·0013	3·80045	−20354·4	19843
3·90625E−3	407·171	0·998032	1·00098E−2	4·54858E−5	1·00293	2·72363	2·69068E43	1	3·44702	−403844	267094
2·44141E−4	894·754	0·999756	1·17188E−2	4·52995E−5	1	2·71875	2·67435E43	0·999999	3·42578	2·2655E6	1·2074E6
1·52588E−5	992·437	1	0·0625	4·3869E−5	1	2·75	1·81193E43	0·999999	3·5	1·57835E6	1·59459E6
9·53674E−7	1000	1	0	3·05176E−5	1	2	0	0·999999†	5	1·24006E6	2·47322E6
1·93024E−6	999·012†				0·988142†						
9·77516E−4		0·999512†				2·72†			3·43317†		
9·76572E−2			9·9999E−3†				2·82359E43†				
9·76658E−3				4·56229E−5†							
1·53416E−3										−1·70216E6†	539026†
Analytic derivative‡	1000	1	0·01	4·54E−5	1·001	2·71828	2·68805E43	1	3·42552	1·57744E6	1·57744E6

† h computed by formula (18.5).

‡ The analytic derivative reported has been evaluated on the NOVA system and may be in error to the extent that the special function routines of this system are faulty.

enough to generate a reasonable derivative approximation. Note that both smaller and larger values of h generated better approximations for the derivative of this function. The exponential function is changing only very slowly near this point.

18.3. CONSTRAINED OPTIMISATION

The general constrained function minimisation problem is stated: minimise $S(\boldsymbol{b})$ with respect to the parameters b_i, $i = 1, 2, \ldots, n$, subject to the constraints

$$c_j(\boldsymbol{b}) = 0 \qquad j = 1, 2, \ldots, m \tag{18.7}$$

and

$$h_k(\boldsymbol{b}) \leqslant 0 \qquad k = 1, 2, \ldots, q. \tag{18.8}$$

In general, if the constraints $\boldsymbol{c}$ are independent, m must be less than n, since via solution of each constraint for one of the parameters $\boldsymbol{b}$ in terms of the others, the dimensionality of the problem may be reduced. The inequality restrictions $\boldsymbol{h}$, on the other hand, reduce the size of the domain in which the solution can be found without necessarily reducing the dimensionality. Thus there is no formal bound to the number, q, of such constraints. Note, however, that the two inequalities

$$\begin{aligned} h(\boldsymbol{b}) &\leqslant 0 \\ h(\boldsymbol{b}) &\geqslant 0 \end{aligned} \tag{18.9}$$

are obviously equivalent to a single equality constraint. Moreover, a very simple change to the constraints (18.9) to give

$$\begin{aligned} h(\boldsymbol{b}) + \varepsilon &\leqslant 0 \\ h(\boldsymbol{b}) - \varepsilon &\geqslant 0 \end{aligned} \tag{18.10}$$

for $\varepsilon > 0$ shows that the inequality constraints may be such that they can never be satisfied simultaneously. Problems which have such sets of constraints are termed *infeasible*. While mutually contradicting constraints such as (18.10) are quite obvious, in general it is not trivial to detect infeasible problems so that their detection can be regarded as one of the tasks which any solution method should be able to accomplish.

There are a number of effective techniques for dealing with constraints in minimisation problems (see, for instance, Gill and Murray 1974). The problem is by and large a difficult one, and the programs to solve it generally long and complicated. For this reason, several of the more mathematically sophisticated methods, such as Lagrange multiplier or gradient projection techniques, will not be considered here. In fact, all the procedures proposed are quite simple and all involve modifying the objective function $S(\boldsymbol{b})$ so that the resulting function has its unconstrained minimum at or near the constrained minimum of the original function. Thus no new algorithms will be introduced in this section.

Elimination or substitution

The equality constraints (18.7) provide m relationships between the n parameters $\boldsymbol{b}$. Therefore, it may be possible to solve for as many as m of the b's in terms of the other $(n-m)$. This yields a new problem involving fewer parameters which automatically satisfy the constraints.

Simple inequality constraints may frequently be removed by *substitutions* which satisfy them. In particular, if the constraint is simply

$$b_k \geqslant 0 \tag{18.11}$$

then a substitution of $\tilde{b}_k^2$ for b_k suffices. If, as a further case, we have

$$v \geqslant b_k \geqslant u \tag{18.12}$$

then we can replace b_k by

$$\left(\frac{u+v}{2}+\frac{v-u}{2}\sin \tilde{b}_k\right). \tag{18.13}$$

It should be noted that while the constraint has obviously disappeared, these substitutions tend to complicate the resulting unconstrained minimisation by introducing local minima and multiple solutions. Furthermore, in many cases suitable substitutions are very difficult to construct.

Penalty functions

The basis of the penalty function approach is simply to add to the function $S(\boldsymbol{b})$ some positive quantity reflecting the degree to which a constraint is violated. Such penalties may be global, that is, added in everywhere, or partial, that is, only added in where the constraint is violated. While there are many possibilities, here we will consider only two very simple choices. These are, for equality constraints, the penalty functions

$$w_j c_j^2(\boldsymbol{b}) \tag{18.14}$$

and for inequalities, the partial penalty

$$W_k h_k^2(\boldsymbol{b}) H(h_k(\boldsymbol{b})) \tag{18.15}$$

where H is the Heaviside function

$$H(x)=\begin{cases}1 & \text{for } x>0\\ 0 & \text{for } x\leqslant 0.\end{cases} \tag{18.16}$$

The quantities w_j, $j=1, 2, \ldots, m$, and W_k, $k=1, 2, \ldots, q$, are weights which have to be assigned to each of the constraints. The function

$$S'(\boldsymbol{b})=S(\boldsymbol{b})+\sum_{j=1}^{m} w_j c_j^2(\boldsymbol{b})+\sum_{k=1}^{q} W_k h_k^2(\boldsymbol{b}) H(h_k(\boldsymbol{b})) \tag{18.17}$$

then has an unconstrained minimum which approaches the constrained minimum of $S(\boldsymbol{b})$ as the weights $\boldsymbol{w}$, $\boldsymbol{W}$ become large.

The two methods outlined above, while obviously acceptable as theoretical approaches to constrained minimisation, may nonetheless suffer serious difficulties in the finite-length arithmetic of the computer. Consider first the example: minimise

$$(b_1+b_2-2)^2 \tag{18.18a}$$

subject to

$$b_1^2 + b_2^2 = 1. \tag{18.18b}$$

This can be solved by elimination. However, in order to perform the elimination it is necessary to decide whether

$$b_1 = (1 - b_2^2)^{1/2} \tag{18.19}$$

or

$$b_1 = -(1 - b_2^2)^{1/2}. \tag{18.20}$$

The first choice leads to the constrained minimum at $b_1 = b_2 = 2^{-1/2}$. The second leads to the constrained maximum at $b_1 = b_2 = -2^{-1/2}$. This problem is quite easily solved approximately by means of the penalty (18.14) and any of the unconstrained minimisation methods.

A somewhat different question concerning elimination arises in the following problem due to Dr Z Hassan who wished to estimate demand equations for commodities of which stocks are kept: minimise

$$S(\boldsymbol{b}) = \sum_{i=1}^{m} \left(\sum_{j=2}^{6} (b_j y_{ij}) + b_1 - y_{i1} \right)^2 \tag{18.21}$$

subject to

$$b_3 b_6 = b_4 b_5. \tag{18.22}$$

The data for this problem are given in table 18.2. The decision that must now be made is which variable is to be eliminated via (18.22); for instance, b_6 can be found as

$$b_6 = b_4 b_5 / b_3. \tag{18.23}$$

The behaviour of the Marquardt algorithm 23 on each of the four unconstrained minimisation problems which can result from elimination in this fashion is shown in table 18.3. Numerical approximation of the Jacobian elements was used to save some effort in running these comparisons. Note that in two cases the algorithm has failed to reach the minimum. The starting point for the iterations was $b_j = 1$, $j = 1, 2, \ldots, 6$, in every case, and these failures are most likely due to the large differences in scale between the variables. Certainly, this poor scaling is responsible for the failure of the variable metric and conjugate gradients algorithms when the problem is solved by eliminating b_6. (Analytic derivatives were used in these cases.)

The penalty function approach avoids the necessity of choosing which parameter is eliminated. The lower half of table 18.3 presents the results of computations with the Marquardt-like algorithm 23. Similar results were obtained using the Nelder–Mead and variable metric algorithms, but the conjugate gradients method failed to converge to the true minimum. Note that as the penalty weighting w is increased the minimum function value increases. This will always be the case if a constraint is active, since enforcement of the constraint pushes the solution 'up the hill'.

Usually the penalty method will involve more computational effort than the elimination approach (a) because there are more parameters in the resulting

TABLE 18.2. Data for the problem of Z Hassan specified by (18.21) and (18.22). Column j below gives the observations y_{ij}, for rows $i = 1, 2, \ldots, m$, for $m = 26$.

1	2	3	4	5	6
286·75	309·935	−40·4026	1132·66	0·1417	0·6429
274·857	286·75	1·3707	1092·26	0·01626	0·7846
286·756	274·857	43·1876	1093·63	0·01755	0·8009
283·461	286·756	−20·0324	1136·81	0·11485	0·8184
286·05	283·461	31·2226	1116·78	0·001937	0·9333
295·925	286·05	47·2799	1148	−0·0354	0·9352
299·863	295·925	4·8855	1195·28	0·00221	0·8998
305·198	299·863	62·22	1200·17	0·00131	0·902
317·953	305·198	57·3661	1262·39	0·01156	0·9034
317·941	317·953	3·4828	1319·76	0·03982	0·9149
312·646	317·941	7·0303	1323·24	0·03795	0·9547
319·625	312·646	38·7177	1330·27	−0·00737	0·9927
324·063	319·625	15·1204	1368·99	0·004141	0·9853
318·566	324·063	21·3098	1384·11	0·01053	0·9895
320·239	318·566	42·7881	1405·42	0·021	1
319·582	320·239	45·7464	1448·21	0·03255	1·021
326·646	319·582	57·9923	1493·95	0·016911	1·0536
330·788	326·646	65·0383	1551·94	0·0308	1·0705
326·205	330·788	51·8661	1616·98	0·069821	1·1013
336·785	326·205	67·0433	1668·85	0·01746	1·1711
333·414	336·785	39·6747	1735·89	0·045153	1·1885
341·555	333·414	49·061	1775·57	0·03982	1·2337
352·068	341·555	18·4491	1824·63	0·02095	1·2735
367·147	352·068	74·5368	1843·08	0·01427	1·2945
378·424	367·147	106·711	1917·61	0·10113	1·3088
385·381	378·424	134·671	2024·32	0·21467	1·4099

unconstrained problem, in our example six instead of five, and (b) because the unconstrained problem must be solved for each increase of the weighting $\boldsymbol{w}$. Furthermore, the ultimate de-scaling of the problem as $\boldsymbol{w}$ is made large may cause slow convergence of the iterative algorithms.

In order to see how the penalty function approach works for inequality constraints where there is no corresponding elimination, consider the following problem (Dixon 1972, p 92): minimise

$$2(b_1^2+b_2^2-b_1b_2-3b_1) \tag{18.24}$$

subject to

$$3b_1+4b_2\leqslant 6 \tag{18.25}$$

and

$$-b_1+4b_2\leqslant 2. \tag{18.26}$$

The constraints were weighted equally in (18.15) and added to (18.24). The resulting function was minimised using the Nelder–Mead algorithm starting from $b_1=b_2=0$ with a step of 0·01 to generate the initial simplex. The results are presented in table 18.4 together with the alternative method of assigning the

TABLE 18.3. Solutions found for Z Hassan problem via Marquardt-type algorithm using numerical approximation of the Jacobian and elimination of one parameter via equation (18.14). The values in italics are for the eliminated parameter. All calculations run in BASIC on a Data General NOVA in 23-bit binary arithmetic.

b_1	b_2	b_3	b_4	b_5	b_6	Eliminated	Sum of squares†
−6·58771	0·910573	*1·58847E−2*	−2·54972E−2	−46·1524	74·0813	b_3	706·745 (104)
80·6448	0·587883	0·155615	*3·27508E−2*	0·541558	0·113976	b_4	749·862 (170)
43·9342	0·76762	0·167034	0·026591	*−59·629*	−9·49263	b_5	606·163 (67)
46·0481	0·757074	0·167033	2·76222E−2	−58·976	*−9·75284*	b_6	606·127 (67)
With analytic derivatives							
45·3623	0·760703	0·167029	2·72248E−2	−59·1436	*−9·64008*	b_6	606·106 (65)
Penalty method; analytic derivatives; initial $w = 100$							
44·9836	0·761652	0·165848	2·67492E−2	−58·9751	−8·85664	—	604·923 (73)
Increase to $w = 1E4$							
45·353	0·760732	0·167005	2·72233E−2	−59·1574	−9·63664	—	606·097 (+48)
Increase to $w = 1E6$							
45·3508	0·760759	0·167023	2·72204E−2	−59·1504	−9·63989	—	606·108 (+22)

† Figures in brackets below each sum of squares denote total number of equivalent function evaluations ($=(n+1)*$(number of Jacobian calculations)+(number of function calculations)) to convergence.

function a very large value whenever one or more of the constraints is violated. In this last approach it has been reported that the simplex may tend to 'collapse' or 'flatten' against the constraint. Swann discusses some of the devices used to counteract this tendency in the book edited by Gill and Murray (1974). Dixon (1972, chap 6) gives a discussion of various methods for constrained optimisation with a particular mention of some of the convergence properties of the penalty function techniques with respect to the weighting factors $\boldsymbol{w}$ and $\mathbf{W}$.

TABLE 18.4. Results of solution of problem (18.24)–(18.26) by the Nelder–Mead algorithm.

Weighting	Function value	Number of evaluations to converge	b_1	b_2
10	−5·35397	113	1·46161	0·40779
1E4	−5·35136	167	1·45933	0·40558
Set function very large	−5·35135	121	1·45924	0·405569

Calculations performed on a Data General NOVA in 23-bit binary arithmetic.

18.4. A COMPARISON OF FUNCTION MINIMISATION AND NONLINEAR LEAST-SQUARES METHODS

It is quite difficult to compare the four basic algorithms presented in the preceding chapters in order to say that one of them is 'best.' The reason for this is that one algorithm may solve some problems very easily but run out of space on another. An algorithm which requires gradients may be harder to use than another which needs only function values, hence requiring more human time and effort even though it saves computer time. Alternatively, it may only be necessary to improve an approximation to the minimum, so that a method which will continue until it has exhausted all the means by which it may reduce the function value may very well be unnecessarily persistent.

Despite all these types of question, I have run an extensive comparison of the methods which have been presented. In this, each method was applied to 79 test problems, all of which were specified in a sum-of-squares form

$$S = \boldsymbol{f}^{\mathrm{T}}\boldsymbol{f}. \tag{18.27}$$

Gradients, where required, were computed via

$$\mathbf{g} = -\mathbf{J}^{\mathrm{T}}\boldsymbol{f}. \tag{18.28}$$

Nash (1976) presents 77 of these problems, some of which have been used as examples in this text†. The convergence criteria in each algorithm were as stringent as possible to allow execution to continue as long as the function was being reduced. In some cases, this meant that the programs had to be stopped manually when convergence was extremely slow. Some judgement was then used to decide if a satisfactory solution had been found. In programs using numerical approximation to the derivative, the same kind of rather subjective judgement was used to decide whether a solution was acceptable based on the *computed* gradient. This means that the results given below for minimisation methods employing the approximation detailed in §18.2 have used a more liberal definition of success than that which was applied to programs using analytic or no derivatives. The question this raises is: do we judge a program by what it is intended to do? Or do we test to see if it finds some globally correct answer? I have chosen the former position, but the latter is equally valid. Such differences of viewpoint unfortunately give rise to many disputes and controversies in this field.

Having now described what has been done, a measure of success is needed. For reliability, we can compare the number of problems for which acceptable (rather than correct) solutions have been found to the total number of problems run. Note that this total is not necessarily 79 because some problems have an initial set of parameters corresponding to a local maximum and the methods which use gradient information will not generally be able to proceed from such points. Other problems may be too large to fit in the machine used for the tests—a partition of a Data General NOVA which uses 23-bit binary arithmetic. Also some of the algorithms have intermediate steps which cause overflow or underflow to occur because of the particular scale of the problems at hand.

† The two remaining functions are given by Osborne (1972).

To measure efficiency of an algorithm, we could time the execution. This is generally quite difficult with a time-shared machine and is in any case highly dependent on the compiler or interpreter in use as well as on a number of programming details, and may be affected quite markedly by special hardware features of some machines. Therefore, a measure of work done is needed which does not require reference to the machine and I have chosen to use *equivalent function evaluations* (efe's). This is the number of function evaluations required to minimise a function plus, in the case of algorithms requiring derivatives, the number of function evaluations which would have been required to approximate the derivatives numerically. For a problem of order n, a program requiring ig gradient evaluations and ifn function evaluations is then assigned

$$\mathrm{efe} = (n+1) * \mathrm{ig} + \mathrm{ifn} \tag{18.29}$$

equivalent function evaluations. I use the factor $(n+1)$ rather than n because of the particular structure of my programs. The use of equivalent function evaluations, and in particular the choice of multiplier for ig, biases my appraisal against methods using derivatives, since by and large the derivatives are *not n* times more work to compute than the function. Hillstrom (1976) presents a more comprehensive approach to comparing algorithms for function minimisation, though in the end he is still forced to make some subjective judgements.

Having decided to use equivalent function evaluations as the measure of efficiency, we still have a problem in determining how to use them, since the omission of some problems for each of the methods means that a method which has successfully solved a problem involving many parameters (the maximum in any of the tests was 20, but most were of order five or less) may appear less efficient than another method which was unable to tackle this problem. To take account of this, we could determine the average number of equivalent function evaluations to convergence per parameter, either by averaging this quantity over the successful applications of a method or by dividing the total number of efe's for all successful cases by the total number of parameters involved. In practice, it was decided to keep both measures, since their ratio gives a crude measure of the performance of algorithms as problem order increases.

To understand this, consider that the work required to minimise the function in any algorithm is proportional to some power, a, of the order n, thus

$$w = pn^a. \tag{18.30}$$

The expected value of the total work over all problems divided by the total number of parameters is approximated by

$$\bar{w}_1 \simeq \left(\int pn^a \, \mathrm{d}n\right)\left(\int n \, \mathrm{d}n\right)^{-1} = 2pn^{(a-1)}/(a+1). \tag{18.31}$$

The average of w/n over all problems, on the other hand, is approximately

$$\bar{w}_0 \simeq \left(\int pn^{(a-1)} \, \mathrm{d}n\right)\left(\int \mathrm{d}n\right)^{-1} = pn^{(a-1)}/a. \tag{18.32}$$

Hence the ratio

$$r = \bar{w}_0/\bar{w}_1 = (a+1)/(2a) \tag{18.33}$$

gives

$$a = 1/(2r-1) \tag{18.34}$$

as an estimate of the degree of the relationship between work and problem order, n, of a given method. The limited extent of the tests and the approximation of sums by integrals, however, mean that the results of such an analysis are no more than a guide to the behaviour of algorithms. The results of the tests are presented in table 18.5.

The conclusions which may be drawn from the table are loosely as follows.

(i) The Marquardt algorithm 23 is generally the most reliable and efficient. Particularly if problems having 'large' residuals, which cause the Gauss–Newton approximation (17.16) to be invalid, are solved by other methods or by increasing the parameter phi in algorithm 23, it is extremely efficient, as might be expected since it takes advantage of the sum-of-squares form.
(ii) The Marquardt algorithm using a numerical approximation (18.4) for the Jacobian is even more efficient than its analytic-derivative counterpart on those problems it can solve. It is less reliable, of course, than algorithms using analytic derivatives. Note, however, that in terms of the number of parameters determined successfully, only the variable metric algorithm and the Marquardt algorithm are more effective.
(iii) The Nelder–Mead algorithm is the most reliable of the derivative-free methods in terms of number of problems solved successfully. However, it is also one of the least efficient, depending on the choice of measure w_1 or w_0, though in some ways this is due to the very strict convergence criterion and the use of the axial search procedure. Unfortunately, without the axial search, the number of problems giving rise to 'failures' of the algorithm is 11 instead of four, so I cannot recommend a loosening of the convergence criteria except when the properties of the function to be minimised are extremely well known.
(iv) The conjugate gradients algorithm, because of its short code length and low working-space requirements, should be considered whenever the number of parameters to be minimised is large, especially if the derivatives are inexpensive to compute. The reliability and efficiency of conjugate gradients are lower than those measured for variable metric and Marquardt methods. However, this study, by using equivalent function evaluations and by ignoring the overhead imposed by each of the methods, is biased quite heavily against conjugate gradients, and I would echo Fletcher's comment (in Murray 1972, p 82) that 'the algorithm is extremely reliable and well worth trying'.

As a further aid to the comparison of the algorithms, this chapter is concluded with three examples to illustrate their behaviour.

Example 18.1. Optimal operation of a public lottery

In example 12.1 a function minimisation problem has been described which arises in an attempt to operate a lottery in a way which maximises revenue per unit

TABLE 18.5. Comparison of algorithm performance as measured by equivalent function evaluations (efe's).

Algorithm	19+20	21	21	22	22	23	23	23
Type	Nelder–Mead	Variable metric	With numerically approximated gradient	Conjugate gradients	With numerically approximated gradient	Marquardt	With numerically approximated Jacobian	Omitting problem 34†
Code length‡	1242	1068		1059		1231		
Array elements	n^2+4n+2	n^2+5n		$5n$		n^2+5n		
(*a*) number of successful runs	68	68	51	66	52	76	61	75
(*b*) total efe's	33394	18292	10797	29158	16065	26419	8021	14399
(*c*) total parameters	230	261	184	236	196	322	255	318
(*d*) $w_1 = (b)/(c)$	145·19	70·08	58·68	123·55	81·96	82·05	31·45	45·28
(*e*) w_0 = average efe's per parameter	141·97	76·72	88·25	154·78	78·96	106·01	36·57	67·36
(*f*) $r = w_0/w_1$	0·98	1·09	1·50	1·25	0·96	1·29	1·16	1·49
(*g*) $a = 1/(2r-1)$	1·05	0·84	0·50	0·66	1·08	0·63	0·75	0·51
(*h*) number of failures	6	3	20	5	16	0	8	0
(*i*) number of problems not run	5	8	8	8	11	3	10	4
(*j*) successes as percentage of problems run	92	96	72	93	76	100	88	100

† Problem 34 of the set of 79 has been designed to have residuals f so that the second derivatives of these residuals cannot be dropped from equation (17.15) to make the Gauss–Newton approximation. The failure of the approximation in this case is reflected in the very slow (12000 efe's) convergence of algorithm 23.

‡ On Data General NOVA (23-bit mantissa).

time. Perry and Soland (1975) derive analytic solutions to this problem, but both to check their results and determine how difficult the problem might be if such a solution were not possible, the following data were run through the Nelder–Mead simplex algorithm 19 on a Data General NOVA operating in 23-bit binary arithmetic.

$$K_1 = 3{\cdot}82821 \qquad K_2 = 0{\cdot}416 \qquad K_3 = 5{\cdot}24263 \qquad F = 8{\cdot}78602$$

$$\alpha = 0{\cdot}23047 \qquad \beta = 0{\cdot}12 \qquad \gamma = 0{\cdot}648 \qquad \delta = 1{\cdot}116.$$

Starting at the suggested values $\boldsymbol{b}^{\mathrm{T}} = (7, 4, 300, 1621)$, $S(\boldsymbol{b}) = -77{\cdot}1569$, the algorithm took 187 function evaluations to find $S(\boldsymbol{b}^*) = -77{\cdot}1602$ at $(\boldsymbol{b}^*)^{\mathrm{T}} = (6{\cdot}99741, 3{\cdot}99607, 300{\cdot}004, 1621{\cdot}11)$. The very slow convergence here is cause for some concern, since the start and finish points are very close together.

From $\boldsymbol{b}^{\mathrm{T}} = (1, 1, 300, 1621)$, $S(\boldsymbol{b}) = 707{\cdot}155$, the Nelder–Mead procedure took 888 function evaluations to $\boldsymbol{b}^* = (6{\cdot}97865, 3{\cdot}99625, 296{\cdot}117, 1619{\cdot}92)^{\mathrm{T}}$ and $S(\boldsymbol{b}^*) = -77{\cdot}1578$ where it was stopped manually. However, S was less than -77 after only 54 evaluations, so once again convergence appears very slow near the minimum. Finally, from $\boldsymbol{b}^{\mathrm{T}} = (1, 1, 1, 1)$, $S(\boldsymbol{b}) = 5{\cdot}93981$, the algorithm converged to $S(\boldsymbol{b}^*) = -77{\cdot}1078$ in 736 evaluations with $\boldsymbol{b}^* = (11{\cdot}1905, 3{\cdot}99003, 481{\cdot}054, 2593{\cdot}67)^{\mathrm{T}}$. In other words, if the model of the lottery operation, in particular the production function for the number of tickets sold, is valid, there is an alternative solution which 'maximises' revenue per unit time. There may, in fact, be several alternatives.

If we attempt the minimisation of $S(\boldsymbol{b})$ using the variable metric algorithm 21 and analytic derivatives, we obtain the following results.

Initial point $\boldsymbol{b}$	$S(\boldsymbol{b})$	Final point $\boldsymbol{b}^*$	$S(\boldsymbol{b}^*)$	efe's
(*a*) 7, 4, 300, 1621	−77·1569	6·99509, 4·00185, 300, 1621	−77·1574	46
(*b*) 1, 1, 300, 1621	707·155	591·676, 563·079, 501·378, 1821·55	0·703075	157
(*c*) 1, 1, 1, 1	5·93981	6·99695, 3·99902, 318·797, 1605·5	−77·0697	252

(The efe's are equivalent function evaluations; see §18.4 for an explanation.) In case (*b*), the price per ticket (second parameter) is clearly exorbitant and the duration of the draw (first parameter) over a year and a half. The first prize (third parameter, measured in units 1000 times as large as the price per ticket) is relatively small. Worse, the revenue ($-S$) per unit time is negative! Yet the derivatives with respect to each parameter at this solution are small. An additional fact to be noted is that the algorithm was not able to function normally, that is, at each step algorithm 21 attempts to update an iteration matrix. However, under certain conditions described at the end of §15.3, it is inadvisable to do this and the method reverts to steepest descent. In case (*b*) above, this occurred in 23 of the 25 attempts to perform the update, indicating that the problem is very far from being well approximated by a quadratic form. This is hardly surprising. The matrix of second partial derivatives of S is certain to depend very much on the parameters due to the fractional powers (α, β, γ, δ) which appear. Thus it is unlikely to be 'approximately constant' in most regions of the parameter space as required of the Hessian in §15.2. This behaviour is repeated in the early iterations of case (c) above.

In conclusion, then, this problem presents several of the main difficulties which may arise in function minimisation:

(i) it is highly nonlinear;
(ii) there are alternative optima; and
(iii) there is a possible scaling instability in that parameters 3 and 4 (v and w) take values in the range 200–2000, whereas parameters 1 and 2 (t and p) are in the range 1–10.

These are problems which affect the choice and outcome of minimisation procedures. The discussion leaves unanswered all questions concerning the reliability of the model or the difficulty of incorporating other parameters, for instance to take account of advertising or competition, which will undoubtedly cause the function to be more difficult to minimise.

Example 18.2. Market equilibrium and the nonlinear equations that result

In example 12.3 the reconciliation of the market equations for supply

$$q = Kp^{\alpha}$$

and demand

$$q = Zp^{-\beta}$$

has given rise to a pair of nonlinear equations. It has been my experience that such systems are less common than minimisation problems, unless the latter are solved by zeroing the partial derivatives simultaneously, a practice which generally makes work and sometimes trouble. One's clients have to be trained to present a problem in its crude form. Therefore, I have not given any special method in this monograph for simultaneous nonlinear equations, which can be written

$$\boldsymbol{f}(\boldsymbol{b}) = \boldsymbol{0} \tag{12.5}$$

preferring to solve them via the minimisation of

$$\boldsymbol{f}^{\mathrm{T}}\boldsymbol{f} = S(\boldsymbol{b}) \tag{12.4}$$

which is a nonlinear least-squares problem. This does have a drawback, however, in that the problem has in some sense been 'squared', and criticisms of the same kind as have been made in chapter 5 against the formation of the sum-of-squares and cross-products matrix for linear least-squares problems can be made against solving nonlinear equations as nonlinear least-squares problems. Nonetheless, a compact nonlinear-equation code will have to await the time and energy for its development. For the present problem we can create the residuals

$$f_1 = q - Kp^{\alpha}$$
$$f_2 = \ln(q) - \ln(Z) + \beta \ln(p).$$

The second residual is the likely form in which the demand function would be estimated. To obtain a concrete and familiar form, substitute

$$q = b_1 \qquad p = b_2 \qquad K = 1$$
$$\alpha = 1{\cdot}5 \qquad \beta = 1{\cdot}2 \qquad Z = \exp(2)$$

so that

$$f_1 = b_1 - b_2^{1{\cdot}5}$$
$$f_2 = \ln(b_1) - 2 + 1{\cdot}2 \ln(b_2).$$

Now minimising the sum of squares

$$S(\boldsymbol{b}) = f_1^2 + f_2^2$$

should give the desired solution.

The Marquardt algorithm 23 with numerical approximation of the Jacobian as in §18.2 gives

$$p = b_2 = 2{\cdot}09647 \qquad q = b_1 = 3{\cdot}03773$$

with $S = 5{\cdot}28328\text{E}-6$ after five evaluations of the Jacobian and 11 evaluations of S. This is effectively 26 function evaluations. The conjugate gradients algorithm 22 with numerical approximation of the gradient gave

$$p = 2{\cdot}09739 \qquad q = 3{\cdot}03747 \qquad S = 2{\cdot}33526\text{E}-8$$

after 67 sum-of-squares evaluations. For both these runs, the starting point chosen was $b_1 = b_2 = 1$. All the calculations were run with a Data General NOVA in 23-bit binary arithmetic.

Example 18.3. Magnetic roots

Brown and Gearhart (1971) raise the possibility that certain nonlinear-equation systems have 'magnetic roots' to which algorithms will converge even if starting points are given close to other roots. One such system they call the cubic-parabola:

$$b_2 = b_1(4b_1^2 - 3)$$
$$b_2 = b_1^2.$$

To solve this by means of algorithms 19, 21, 22 and 23, the residuals

$$f_1 = b_1(4b_1^2 - 3) - b_2$$
$$f_2 = b_1^2 - b_2$$

were formed and the following function minimised:

$$S(\boldsymbol{b}) = f_1^2 + f_2^2.$$

The roots of the system are

$$\text{R}_1\text{: } b_1 = 0 = b_2$$
$$\text{R}_2\text{: } b_1 = 0 = b_2$$
$$\text{R}_3\text{: } b_1 = -0{\cdot}75 \qquad b_2 = 0{\cdot}5625.$$

To test the claimed magnetic-root properties of this system, 24 starting points were generated, namely the eight points about each root formed by an axial step of $\pm 0{\cdot}5$ and $\pm 0{\cdot}1$. In every case the starting point was still nearest to the root used to generate it.

All the algorithms converged to the root expected when the starting point was only 0·1 from the root. With the following exceptions they all converged to the expected root when the distance was 0·5.

(i) The Marquardt algorithm 23 converged to R_3 from $(-0{\cdot}5, 0) = (b_1, b_2)$ instead of to R_1.
(ii) The Nelder–Mead algorithm 19 found R_2 from $(0{\cdot}5, 0)$ instead of R_1.
(iii) The conjugate gradients algorithm 22 found R_3 and the variable metric algorithm 21 found R_1 when started from $(1{\cdot}5, 1)$, to which R_2 is closest.
(iv) All algorithms found R_1 instead of R_3 when started from $(-0{\cdot}25, 0{\cdot}5625)$.
(v) The conjugate gradients algorithm also found R_1 instead of R_3 from $(-1{\cdot}25, 0{\cdot}5625)$.

Note that all the material in this chapter is from the first edition of the book. However, I believe it is still relevant today. We have, as mentioned in chapter 17, added bounds constraints capability to our minimisation codes included in Nash and Walker-Smith (1987). Also the performance figures in this chapter relate to BASIC implementations of the original algorithms. Thus some of the results will alter. In particular, I believe the present conjugate gradients method would appear to perform better than that used in the generation of table 18.5. Interested readers should refer to Nash and Nash (1988) for a more modern investigation of the performance of compact function minimisation algorithms.

Chapter 19

THE CONJUGATE GRADIENTS METHOD APPLIED TO PROBLEMS IN LINEAR ALGEBRA

19.1. INTRODUCTION

This monograph concludes by applying the conjugate gradients method, developed in chapter 16 for the minimisation of nonlinear functions, to linear equations, linear least-squares and algebraic eigenvalue problems. The methods suggested may not be the most efficient or effective of their type, since this subject area has not attracted a great deal of careful research. In fact much of the work which has been performed on the sparse algebraic eigenvalue problem has been carried out by those scientists and engineers in search of solutions. Stewart (1976) has prepared an extensive bibliography on the large, sparse, generalised symmetric matrix eigenvalue problem in which it is unfortunately difficult to find many reports that do more than describe a method. Thorough or even perfunctory testing is often omitted and convergence is rarely demonstrated, let alone proved. The work of Professor Axel Ruhe and his co-workers at Umea is a notable exception to this generality. Partly, the lack of testing is due to the sheer size of the matrices that may be involved in real problems and the cost of finding eigensolutions.

The linear equations and least-squares problems have enjoyed a more diligent study. A number of studies have been made of the conjugate gradients method for linear-equation systems with positive definite coefficient matrices, of which one is that of Reid (1971). Related methods have been developed in particular by Paige and Saunders (1975) who relate the conjugate gradients methods to the Lanczos algorithm for the algebraic eigenproblem. The Lanczos algorithm has been left out of this work because I feel it to be a tool best used by someone prepared to tolerate its quirks. This sentiment accords with the statement of Kahan and Parlett (1976):'The urge to write a universal Lanczos program should be resisted, at least until the process is better understood.' However, in the hands of an expert, it is a very powerful method for finding the eigenvalues of a large symmetric matrix. For indefinite coefficient matrices, however, I would expect the Paige–Saunders method to be preferred, by virtue of its design. In preparing the first edition of this book, I experimented briefly with some FORTRAN codes for several methods for iterative solution of linear equations and least-squares problems, finding no clear advantage for any one approach, though I did not focus on indefinite matrices. Therefore, the treatment which follows will stay with conjugate gradients, which has the advantage of introducing no fundamentally new ideas.

It must be pointed out that the general-purpose minimisation algorithm 22 does not perform very well on linear least-squares or Rayleigh quotient minimisations.

In some tests run by S G Nash and myself, the inexact line searches led to very slow convergence in a number of cases, even though early progress may have been rapid (Nash and Nash 1977).

19.2. SOLUTION OF LINEAR EQUATIONS AND LEAST-SQUARES PROBLEMS BY CONJUGATE GRADIENTS

The conjugate gradients algorithm 22 can be modified to solve systems of linear equations and linear least-squares problems. Indeed, the conjugate gradients methods have been derived by considering their application to the quadratic form

$$S(\boldsymbol{b}) = \tfrac{1}{2}\boldsymbol{b}^{\mathrm{T}}\mathbf{H}\boldsymbol{b} - \boldsymbol{c}^{\mathrm{T}}\boldsymbol{b} + (\text{any scalar}) \tag{15.11}$$

where $\mathbf{H}$ is symmetric and positive definite. The minimum at $\boldsymbol{b}'$ has the zero gradient

$$\boldsymbol{g} = \mathbf{H}\boldsymbol{b}' - \boldsymbol{c} = \mathbf{0} \tag{15.17}$$

so that $\boldsymbol{b}'$ solves the linear equations

$$\mathbf{H}\boldsymbol{b}' = \boldsymbol{c}. \tag{19.1}$$

The linear least-squares problem

$$\mathbf{A}\boldsymbol{b} \simeq \boldsymbol{f} \tag{19.2}$$

can be solved by noting that $\mathbf{A}^{\mathrm{T}}\mathbf{A}$ is non-negative definite. For the present purpose we shall assume that it is positive definite. Hestenes (1975) shows that it is only necessary for $\mathbf{H}$ to be non-negative definite to produce a least-squares solution of (19.1). Thus identification of

$$\mathbf{H} = \mathbf{A}^{\mathrm{T}}\mathbf{A} \tag{19.3a}$$

$$\boldsymbol{c} = \mathbf{A}^{\mathrm{T}}\boldsymbol{f} \tag{19.3b}$$

permits least-squares problems to be solved via the normal equations.

The particular changes which can be made to the general-purpose minimising algorithm of §§16.1 and 16.2 to suit these linear systems of equations better are that (*a*) the linear search and (*b*) the gradient calculation can be performed explicitly and simply. The latter is accomplished as follows. First, if the initial point is taken to be

$$\boldsymbol{b}^{(0)} = \mathbf{0} \tag{19.4}$$

then the initial gradient, from (15.17), is

$$\boldsymbol{g}_1 = -\boldsymbol{c}. \tag{19.5}$$

If the linear search along the search direction $\boldsymbol{t}_j$ yields a step-length factor k_j, then from (15.18)

$$\boldsymbol{g}_{j+1} = \boldsymbol{g}_j + k_j\mathbf{H}\boldsymbol{t}_j. \tag{19.6}$$

This with (19.5) defines a recurrence scheme which avoids the necessity of multiplication of $\boldsymbol{b}$ by $\mathbf{H}$ to compute the gradient, though $\mathbf{H}\boldsymbol{t}_j$ must be formed. However, this is needed in any case in the linear search to which attention will now be directed.

Substitution of

$$\boldsymbol{b}_{j+1}=\boldsymbol{b}_j+k_j\boldsymbol{t}_j \tag{19.7}$$

into (15.11) gives an equation for k_j, permitting the optimal step length to be computed. For convenience in explaining this, the subscript j will be omitted. Thus from substituting, we obtain

$$\phi(k)=\tfrac{1}{2}(\boldsymbol{b}+k\boldsymbol{t})^{\mathrm{T}}\mathbf{H}(\boldsymbol{b}+k\boldsymbol{t})-\boldsymbol{c}^{\mathrm{T}}(\boldsymbol{b}+k\boldsymbol{t})+(\text{any scalar}). \tag{19.8}$$

The derivative of this with respect to k is

$$\begin{aligned}\phi'(k)&=\boldsymbol{t}^{\mathrm{T}}\mathbf{H}(\boldsymbol{b}+k\boldsymbol{t})-\boldsymbol{c}^{\mathrm{T}}\boldsymbol{t}\\&=\boldsymbol{t}^{\mathrm{T}}(\mathbf{H}\boldsymbol{b}-\boldsymbol{c})+k\boldsymbol{t}^{\mathrm{T}}\mathbf{H}\boldsymbol{t}\\&=\boldsymbol{t}^{\mathrm{T}}\mathbf{g}+k\boldsymbol{t}^{\mathrm{T}}\mathbf{H}\boldsymbol{t}\end{aligned} \tag{19.9}$$

so that $\phi'(k)=0$ implies

$$k=-\boldsymbol{t}^{\mathrm{T}}\mathbf{g}/\boldsymbol{t}^{\mathrm{T}}\mathbf{H}\boldsymbol{t}. \tag{19.10}$$

But from the same line of reasoning as that which follows equation (16.6), the accurate line searches imply that the function has been minimised on a hyperplane spanned by the gradient directions which go to make up $\boldsymbol{t}$ except for $\mathbf{g}$ itself; thus

$$k=+\mathbf{g}^{\mathrm{T}}\mathbf{g}/\boldsymbol{t}^{\mathrm{T}}\mathbf{H}\boldsymbol{t}. \tag{19.11}$$

Note the sign has changed since $\boldsymbol{t}$ will have component in direction $\mathbf{g}$ with a coefficient of -1.

The recurrence formula for this problem can be generated by one of the formulae (16.8), (16.10) or (16.11). The last of these is the simplest and is the one that has been chosen. Because the problem has a genuine quadratic form it may be taken to have converged after n steps if the gradient does not become small earlier. However, the algorithm which follows restarts the iteration in case rounding errors during the conjugate gradients steps have caused the solution to be missed.

Algorithm 24. Solution of a consistent set of linear equations by conjugate gradients

```
procedure lecg( n : integer; {order of the problem}
                    H : rmatrix; {coefficient matrix in linear equations}
                    C : rvector; {right hand side vector}
                var Bvec : rvector; {the solution vector -- on input, we
                    must supply some guess to this vector}
                var itcount : integer; {on input, a limit to the number of
                    conjugate gradients iterations allowed; on output,
                    the number of iterations used (negative if the
                    limit is exceeded)}
                var ssmin : real); {the approximate minimum sum of squared
                    residuals for the solution}
{alg24.pas == linear equations by conjugate gradients
    This implementation uses an explicit matrix H. However,
    we only need the result of multiplying H times a vector,
    that is, where the procedure call
            matmul( n, H, t, v);
```

Algorithm 24. Solution of a consistent set of linear equations by conjugate gradients (cont.)

```
        computes the matrix - vector product H t and places it in
        v, we really only need have
                    matmulH(n, t, v),
        with the matrix being implicit. In many situations, the
        elements of H are simple to provide or calculate so that
        its explicit storage is not required. Such modifications
        allow array H to be removed from the calling sequence
        of this procedure.
                    Copyright 1988 J.C.Nash
}
var
    count, i, itn, itlimit : integer;
    eps, g2, oldg2, s2, step, steplim, t2, tol : real;
    g, t, v : rvector;
begin
{ writeln('alg24.pas -- linear equations solution by conjugate gradients');}
    {Because alg24.pas is called repeatedly in dr24ii.pas, we do not always
        want to print the algorithm banner. }
    itlimit := itcount; {to save the iteration limit -- STEP 0}
    itcount := 0; {to initialize the iteration count}
    eps := calceps; {machine precision}
    steplim := 1.0/sqrt(eps); {a limit to the size of the step which may
                        be taken in changing the solution Bvec}
    g2 := 0.0; {Compute norm of right hand side of equations.}
    for i := 1 to n do g2 := g2+abs(C[i]); tol := g2*eps*eps*n;
    {STEP 1: compute the initial residual vector}
    {Note: we do not count this initial calculation of the residual vector.}
    matmul(n, H, Bvec, g);
    for i := 1 to n do g[i] := g[i]-C[i];{this loop need not be separate if
        explicit matrix multiplication is used} {This loop computes the
        residual vector. Note that it simplifies if the initial guess to the
        solution Bvec is a null vector. However, this would restrict the
        algorithm to one major cycle (steps 4 to 11)}
    g2 := 0.0; {STEP 2}
    for i := 1 to n do
    begin
        g2 := g2+g[i]*g[i]; t[i] := -g[i];
    end; {loop on i}
        {writeln('initial gradient norm squared = ',g2);}
    ssmin := big; {to ensure no accidental halt -- STEP 3}
    while (g2>tol) and (itcount<itlimit) and (ssmin>0.0) do
    {Convergence test -- beginning of main work loop in method.}
    begin {STEP 4 -- removed}
        {STEP 5 -- here in the form of explicit statements}
        itcount := itcount+1; {to count matrix multiplications}
        matmul( n, H, t, v);
        t2 := 0.0; {STEP 6 -- t2 will contain the product t-transpose * H * t}
        for i := 1 to n do t2 := t2+t[i]*v[i];
        step := g2/t2; oldg2 := g2; {STEP 7}
        if abs(step)>steplim then
        begin
            writeln('Step too large -- coefficient matrix indefinite?');
```

Algorithm 24. Solution of a consistent set of linear equations by conjugate gradients (cont.)

```
          ssmin := -big; {to indicate this failure}
        end
        else
        begin
          {The step length has been computed and the gradient norm saved.}
          g2 := 0.0; count := 0; {STEP 8}
          for i := 1 to n do
          begin
            g[i] := g[i]+step*v[i]; {to update the gradient/residual}
            t2 := Bvec[i]; Bvec[i] := t2+step*t[i]; {to update the solution}
            if Bvec[i]=t2 then count := count+1;
            g2 := g2+g[i]*g[i]; {accumulate gradient norm}
          end; {loop on i -- updating}
          if count<n then {STEP 9}
          begin
            if g2>tol then
            begin {STEP 10 -- recurrence relation to give next search direction.}
              t2 := g2/oldg2;
              for i := 1 to n do t[i] := t2*t[i]-g[i];
            end; {if g2>tol}
          end; {if count<n}
          {writeln(' Iteration ',itcount,' count=',count,' residual ss=',g2);}
          ssmin := g2; {to return the value of the estimated sum of squares}
        end; {else on stepsize failure}
      end; {while g2>tol }
      if itcount>=itlimit then itcount := -itcount; {to notify calling program
          of the convergence failure.}
      {After n cycles, the conjugate gradients algorithm will, in exact
        arithmetic, have converged. However, in floating-point arithmetic
        the solution may still need refinement. For best results, the
        residual vector g should be recomputed at this point, in the same
        manner as it was initially computed. There are a number of ways
        this can be programmed. Here we are satisfied to merely point out
        the issue.}
    end; {alg24.pas == lecg}
```

The above algorithm requires five working vectors to store the problem and intermediate results as well as the solution. This is exclusive of any space needed to store or generate the coefficient matrix.

Example 19.1. Solution of Fröberg's differential equation (example 2.2)

Recall that the finite difference approximation of a second-order boundary value problem gives rise to a set of linear equations whose coefficient matrix is easily generated. That is to say, there is no need to build the set of equations explicitly if the coefficient matrix satisfies the other conditions of the conjugate gradients method. It turns out that it is negative definite, so that we can solve the equations

$$(-\mathbf{A})\boldsymbol{x} = -\boldsymbol{b} \tag{19.12}$$

by means of the conjugate gradients algorithm 24. Alternatively, if the coefficient matrix is not definite, the normal equations

$$\mathbf{A}^{\mathrm{T}}\mathbf{A}x = \mathbf{A}^{\mathrm{T}}b \tag{19.13}$$

provide a non-negative definite matrix $\mathbf{A}^{\mathrm{T}}\mathbf{A}$. Finally, the problem can be approached in a completely different way. Equation (2.13) can be rewritten

$$y_{j+1} = 7jh^3 + [2 - h^2/(1 + j^2h^2)]y_j - y_{j-1}. \tag{19.14}$$

TABLE 19.1. Comparison of three compact methods for solving Froberg's differential equation by finite difference approximation.

Order of problem	Algorithm 24 and equation (19.12)	Algorithm 24 and equation (19.13)	Shooting method
	Largest deviation from true solution		
4	2·62E−6	2·88E−5	1·97E−6
10	8·34E−6	1·02E−2	1·14E−5
50	2·03E−4	0·930	2·03E−4
	Approximate time for the three problems (min)		
	0·005	0·028	0·003

All computations performed in six hexadecimal digit arithmetic on an IBM 370/168 computer.

Thus, since y_0 is fixed as zero by (2.9), if the value y_1 is known, all the points of the solution can be computed. But (2.10) requires

$$y_{n+1} = 2 \tag{19.15}$$

thus we can consider the difference

$$f(y_1) = y_{n+1} - 2 \tag{19.16}$$

to generate a root-finding problem. This process is called a *shooting method* since we aim at the value of y_{n+1} desired by choosing y_1. Table 19.1 compares the three methods suggested for $n = 4$, 10 and 50. The main comparison is between the values found for the deviation from the true solution

$$y(x) = x + x^3 \tag{19.17}$$

or

$$y(x_j) = jh(1 + j^2h^2). \tag{19.18}$$

It is to be noted that the use of the conjugate gradients method with the normal equations (19.13) is unsuccessful, since we have unfortunately increased the ill conditioning of the equations in the manner discussed in chapter 5. The other two methods offer comparable accuracy, but the shooting method, applying algorithm 18 to find the root of equation (19.16) starting with the interval [0, 0·5], is somewhat simpler and faster. In fact, it could probably be solved using a trial-and-error method to find the root on a pocket calculator.

Example 19.2. Surveying-data fitting

The output below illustrates the solution of a linear least-squares problem of the type described in example 2.4. No weighting of the observations is employed here, though in practice one would probably weight each height-difference observation by some factor inversely related to the distance of the observation position from the points whose height difference is measured. The problem given here was generated by artificially perturbing the differences between the heights $\boldsymbol{b} = (0, 100, 121, 96)^T$. The quantities G printed are the residuals of the normal equations.

```
RUN
SURVEYING LEAST SQUARES
# OF POINTS? 4
# OF OBSERVATIONS? 5
HEIGHT DIFF BETWEEN? 1 AND? 2=? -99.99
HEIGHT DIFF BETWEEN? 2 AND? 3=? -21.03
HEIGHT DIFF BETWEEN? 3 AND? 4=? 24.98
HEIGHT DIFF BETWEEN? 1 AND? 3=? -121.02
HEIGHT DIFF BETWEEN? 2 AND? 4=? 3.99
B( 1 )=-79.2575   G= 2.61738E-6
B( 2 )= 20.7375   G=-2.26933E-6
B( 3 )= 41.7575   G=-6.05617E-6
B( 4 )= 16.7625   G=-5.73596E-6
DIFF( 1 )=-99.995
DIFF( 2 )=-21.02
DIFF( 3 )= 24.995
DIFF( 4 )=-121.015
DIFF( 5 )= 3.97501
# MATRIX PRODUCTS= 4
HEIGHTS FORM B(1)=0
2                  99.995
3                  121.015
4                  96.02
```

The software diskette contains the data file EX24LS1.CNM *which, used with the driver* DR24LS.PAS, *will execute this example.*

As a test of the method of conjugate gradients (algorithm 24) in solving least-squares problems of the above type, a number of examples were generated using all possible combinations of n heights. These heights were produced using a pseudo-random-number generator which produces numbers in the interval (0, 1). All $m = n*(n-1)/2$ height differences were then computed and perturbed by pseudo-random values formed from the output of the above-mentioned generator minus 0·5 scaled to some range externally specified. Therefore if S1 is the scale factor for the heights and S2 the scale factor for the perturbation and the function RND(X) gives a number in (0, 1), heights are computed using

$$S1*RND(X)$$

and perturbations on the height differences using

$$S2*[RND(X)-0{\cdot}5].$$

Table 19.2 gives a summary of these calculations. It is important that the convergence tolerance be scaled to the problem. Thus I have used

$$tol = n * S1 * S1 * eps * eps$$

where eps is the machine precision. The points to be noted in the solutions are as follows:

(i) the rapid convergence in terms of matrix–vector products needed to arrive at a solution (recall that the problem is singular since no 'zero' of height has been specified), and
(ii) the reduction in the variance of the computed height differences from their known values.

All the computations were performed on a Data General NOVA operating in 23-bit binary arithmetic.

TABLE 19.2. Surveying-data fitting by algorithm 24.

n	$m = n(n-1)/2$	Matrix products	Height scale S1	Perturbation scale S2	Perturbation variance†	Variance of computed height differences†	Variance reduction factor
4	6	3	100	0·01	9·15E−6	1·11E−6	0·12
4	6	2	100	1	9·15E−2	1·11E−2	0·12
4	6	2	10000	1	9·15E−2	1·11E−2	0·12
10	45	3	1000	0·1	9·25E−4	2·02E−4	0·22
20	190	3	1000	0·1	8·37E−4	6·96E−5	0·08
20	190	3	1000	100	836·6	69·43	0·08
20	190	5	1000	2000	334631	26668·7	0·08

† From 'known' values.

19.3. INVERSE ITERATION BY ALGORITHM 24

The algorithm just presented, since it solves systems of linear equations, can be employed to perform inverse iteration on symmetric matrix eigenproblems via either of the schemes (9.10) or (9.12), that is, the ordinary and generalised symmetric matrix eigenproblems. The only difficulties arise because the matrix

$$\mathbf{A}' = \mathbf{A} - s\mathbf{B} \tag{19.19}$$

where s is some shift, is not positive definite. (Only the generalised problem will be treated.) In fact, it is bound to be indefinite if an intermediate eigenvalue is sought. Superficially, this can be rectified immediately by solving the least-squares problem

$$(\mathbf{A}')^{\mathrm{T}}(\mathbf{A}')\mathbf{y}_i = (\mathbf{A}')^{\mathrm{T}}\mathbf{B}\mathbf{x}_i \tag{19.20}$$

in place of (9.12a). However, as with the Froberg problem of example 19.1, this is done at the peril of worsening the condition of the problem. Since Hestenes (1975) has pointed out that the conjugate gradients method may work for indefinite systems—it is simply no longer supported by a convergence theorem—

we may be tempted to proceed with inverse iteration via conjugate gradients for any real symmetric problem.

Ruhe and Wiberg (1972) warn against allowing too large an increase in the norm of **y** in a single step of algorithm 24, and present techniques for coping with the situation. Of these, the one recommended amounts only to a modification of the shift. However, since Ruhe and Wiberg were interested in refining eigenvectors already quite close to exact, I feel that an *ad hoc* shift may do just as well if a sudden increase in the size of the vector **y**, that is, a large step length k, is observed.

Thus my suggestion for solution of the generalised symmetric matrix eigenvalue problem by inverse iteration using the conjugate gradients algorithm 24 is as follows.

(i) Before each iteration, the norm (any norm will do) of the residual vector

$$\boldsymbol{r} = (\mathbf{A} - e\mathbf{B})\boldsymbol{x} \tag{19.21}$$

should be computed and this norm compared to some user-defined tolerance as a convergence criterion. While this is less stringent than the test made at STEPs 14 and 15 of algorithm 10, it provides a constant running check of the closeness of the current trial solution $(e, \boldsymbol{x})$ to an eigensolution. Note that a similar calculation could be performed in algorithm 10 but would involve keeping copies of the matrices **A** and **B** in the computer memory. It is relatively easy to incorporate a restart procedure into inverse iteration so that tighter tolerances can be entered without discarding the current approximate solution. Furthermore, by using $\boldsymbol{b} = \mathbf{0}$ as the starting vector in algorithm 24 at each iteration and only permitting n conjugate gradient steps or less (by deleting STEP 12 of the algorithm), the matrix–vector multiplication of STEP 1 of algorithm 24 can be made implicit in the computation of residuals (19.21) since

$$\mathbf{c} = \mathbf{B}\boldsymbol{x}. \tag{19.22}$$

Note that the matrix **H** to be employed at STEP 5 of the algorithm is

$$\mathbf{H} = (\mathbf{A} - s\mathbf{B}) = \mathbf{A}'. \tag{19.23}$$

(ii) To avoid too large an increase in the size of the elements of $\boldsymbol{b}$, STEP 7 of algorithm 24 should include a test of the size of the step-length parameter k. I use the test

If ABS(k) ≥ 1/SQR(eps), then . . .

where eps is the machine precision, to permit the shift s to be altered by the user. I remain unconvinced that satisfactory simple automatic methods yet exist to calculate the adjustment to the shift without risking convergence to an eigensolution other than that desired. The same argument applies against using the Rayleigh quotient to provide a value for the shift s. However, since the Rayleigh quotient is a good estimate of the eigenvalue (see § 10.2, p 100), it is a good idea to compute it.

(iii) In order to permit the solution $\boldsymbol{b}$ of

$$\mathbf{H}\boldsymbol{b} = (\mathbf{A} - s\mathbf{B})\boldsymbol{b} = \mathbf{B}\boldsymbol{x} = \boldsymbol{c} \tag{19.24}$$

to be used to compute the eigenvalue approximation

$$e \simeq s + x_m/b_m \tag{19.25}$$

where b_m is the largest element in magnitude in $\boldsymbol{b}$ (but keeps its sign!!), I use the infinity norm (9.14) in computing the next iterate $\boldsymbol{x}$ from $\boldsymbol{b}$. To obtain an eigenvector normalised so that

$$\boldsymbol{x}^T\mathbf{B}\boldsymbol{x} = 1 \tag{19.26}$$

one has only to form

$$\boldsymbol{x}^T\mathbf{c} = \boldsymbol{x}^T\mathbf{B}\boldsymbol{x}. \tag{19.27}$$

At the point where the norm of $\boldsymbol{r}$ is compared to the tolerance to determine if the algorithm has converged, $\mathbf{c}$ is once again available from the computation (19.21). The residual norm $\|\boldsymbol{r}\|$ should be divided by $(\boldsymbol{x}^T\mathbf{B}\boldsymbol{x})^{1/2}$ if $\boldsymbol{x}$ is normalised by this quantity.

Since only the solution procedure for the linear equations arising in inverse iteration has been changed from algorithm 10 (apart from the convergence criterion), the method outlined in the above suggestions will not converge any faster than a more conventional approach to inverse iteration. Indeed for problems which are ill conditioned with respect to solution via the conjugate gradients algorithm 24, which nominally requires that the coefficient matrix $\mathbf{H}$ be non-negative definite, the present method may take many more iterations. Nevertheless, despite the handicap that the shift s cannot be too close to the eigenvalue or the solution vector $\boldsymbol{b}$ will 'blow up', thereby upsetting the conjugate gradients process, inverse iteration performed in the manner described does provide a tool for finding or improving eigensolutions of large matrices on a small computer. Examples are given at the end of §19.4.

19.4. EIGENSOLUTIONS BY MINIMISING THE RAYLEIGH QUOTIENT

Consider two symmetric matrices $\mathbf{A}$ and $\mathbf{B}$ where $\mathbf{B}$ is also positive definite. The Rayleigh quotient defined by

$$R = \boldsymbol{x}^T\mathbf{A}\boldsymbol{x}/\boldsymbol{x}^T\mathbf{B}\boldsymbol{x} \tag{19.28}$$

then takes on its stationary values (that is, the values at which the partial derivatives with respect to the components of $\boldsymbol{x}$ are zero) at the eigensolutions of

$$\mathbf{A}\boldsymbol{x} = e\mathbf{B}\boldsymbol{x}. \tag{2.63}$$

In particular, the maximum and minimum values of R are the extreme eigenvalues of the problem (2.63). This is easily seen by expanding

$$\boldsymbol{x} = \sum_{j=1}^{n} c_j \boldsymbol{\phi}_j \tag{19.29}$$

where $\boldsymbol{\phi}_j$ is the jth eigenvector corresponding to the eigenvalue e_j. Then we have

$$R = \left(\sum_i \sum_j c_i c_j e_j \boldsymbol{\phi}_i \mathbf{B} \boldsymbol{\phi}_j\right)\left(\sum_i \sum_j c_i c_j \boldsymbol{\phi}_i \mathbf{B} \boldsymbol{\phi}_j\right)^{-1} = \left(\sum_i c_i^2 e_i\right)\left(\sum c_i^2\right)^{-1}. \tag{19.30}$$

If

$$e_1 \geqslant e_2 \geqslant \ldots \geqslant e_n \tag{19.31}$$

then the minimum value of R is e_n and occurs when $\boldsymbol{x}$ is proportional to $\boldsymbol{\phi}_n$. The maximum value is e_1. Alternatively, this value can be obtained via minimisation of $-R$. Furthermore, if $\mathbf{B}$ is the identity, then minimising the Rayleigh quotient

$$R' = \boldsymbol{x}^{\mathrm{T}}(\mathbf{A} - k\mathbf{1}_n)^2\boldsymbol{x}/\boldsymbol{x}^{\mathrm{T}}\boldsymbol{x} \tag{19.32}$$

will give the eigensolution having its eigenvalue closest to k.

While any of the general methods for minimising a function may be applied to this problem, concern for storage requirements suggests the use of the conjugate gradients procedure. Unfortunately, the general-purpose algorithm 22 may converge only very slowly. This is due (a) to the inaccuracy of the linear search, and (b) to loss of conjugacy between the search directions $\boldsymbol{t}_j$, $j = 1, 2, \ldots, n$. Both these problems are exacerbated by the fact that the Rayleigh quotient is homogeneous of degree zero, which means that the Rayleigh quotient takes the same value for any vector $C\boldsymbol{x}$, where C is some non-zero constant. This causes the Hessian of the Rayleigh quotient to be singular, thus violating the conditions normally required for the conjugate gradients algorithm. Bradbury and Fletcher (1966) address this difficulty by setting to unity the element of largest magnitude in the current eigenvector approximation and adjusting the other elements accordingly. This adjustment is made at each iteration. However, Geradin (1971) has tackled the problem more directly, that is, by examining the Hessian itself and attempting to construct search directions which are mutually conjugate with respect to it. This treatment, though surely not the last word on the subject, is essentially repeated here. The implementation details are my own.

Firstly, consider the linear search subproblem, which in the current case can be solved analytically. It is desired to minimise

$$R = (\boldsymbol{x} + k\boldsymbol{t})^{\mathrm{T}}\mathbf{A}(\boldsymbol{x} + k\boldsymbol{t})/(\boldsymbol{x} + k\boldsymbol{t})^{\mathrm{T}}\mathbf{B}(\boldsymbol{x} + k\boldsymbol{t}) \tag{19.33}$$

with respect to k. For convenience this will be rewritten

$$R = N(k)/D(k) \tag{19.34}$$

with N and D used to denote the numerator and denominator, respectively. Differentiating with respect to k gives

$$\mathrm{d}R/\mathrm{d}k = 0 = (D\mathrm{d}N/\mathrm{d}k - N\mathrm{d}D/\mathrm{d}k)/D^2. \tag{19.35}$$

Because of the positive definiteness of $\mathbf{B}$, D can never be zero unless

$$\boldsymbol{x} + k\boldsymbol{t} = \mathbf{0}. \tag{19.36}$$

Therefore, ignoring this last possibility, we set the numerator of expression (19.35) to zero to obtain the quadratic equation

$$uk^2 + vk + w = 0 \tag{19.37}$$

where

$$u = (\boldsymbol{t}^{\mathrm{T}}\mathbf{A}\boldsymbol{t})(\boldsymbol{x}^{\mathrm{T}}\mathbf{B}\boldsymbol{t}) - (\boldsymbol{x}^{\mathrm{T}}\mathbf{A}\boldsymbol{t})(\boldsymbol{t}^{\mathrm{T}}\mathbf{B}\boldsymbol{t}) \tag{19.38}$$

$$v = (\boldsymbol{t}^T\mathbf{A}\boldsymbol{t})(\boldsymbol{x}^T\mathbf{B}\boldsymbol{x}) - (\boldsymbol{x}^T\mathbf{A}\boldsymbol{x})(\boldsymbol{t}^T\mathbf{B}\boldsymbol{t}) \tag{19.39}$$

$$w = (\boldsymbol{x}^T\mathbf{A}\boldsymbol{t})(\boldsymbol{x}^T\mathbf{B}\boldsymbol{x}) - (\boldsymbol{x}^T\mathbf{A}\boldsymbol{x})(\boldsymbol{x}^T\mathbf{B}\boldsymbol{t}). \tag{19.40}$$

Note that by symmetry

$$\boldsymbol{x}^T\mathbf{A}\boldsymbol{t} = \boldsymbol{t}^T\mathbf{A}\boldsymbol{x} \tag{19.41}$$

and

$$\boldsymbol{x}^T\mathbf{B}\boldsymbol{t} = \boldsymbol{t}^T\mathbf{B}\boldsymbol{x}. \tag{19.42}$$

Therefore, only six inner products are needed in the evaluation of u, v and w. These are

$$(\boldsymbol{x}^T\mathbf{A}\boldsymbol{x}) \qquad (\boldsymbol{x}^T\mathbf{A}\boldsymbol{t}) \qquad \text{and} \qquad (\boldsymbol{t}^T\mathbf{A}\boldsymbol{t})$$

and

$$(\boldsymbol{x}^T\mathbf{B}\boldsymbol{x}) \qquad (\boldsymbol{x}^T\mathbf{B}\boldsymbol{t}) \qquad \text{and} \qquad (\boldsymbol{t}^T\mathbf{B}\boldsymbol{t}).$$

The quadratic equation (19.37) has two roots, of which only one will correspond to a minimum. Since

$$\psi(k) = 0{\cdot}5D^2(\mathrm{d}R/\mathrm{d}k) = uk^2 + vk + w \tag{19.43}$$

we get

$$\frac{\mathrm{d}\psi}{\mathrm{d}k} = 0{\cdot}5D^2\left(\frac{\mathrm{d}^2R}{\mathrm{d}k^2}\right) + D\left(\frac{\mathrm{d}D}{\mathrm{d}k}\right)\left(\frac{\mathrm{d}R}{\mathrm{d}k}\right) = 2uk + v. \tag{19.44}$$

At either of the roots of (19.37), this reduces to

$$0{\cdot}5D^2(\mathrm{d}^2R/\mathrm{d}k^2) = 2uk + v \tag{19.45}$$

so that a minimum has been found if $\mathrm{d}^2R/\mathrm{d}k^2$ is positive, or

$$2uk + v > 0. \tag{19.46}$$

Substitution of both roots from the quadratic equation formula shows that the desired root is

$$k = [-v + (v^2 - 4uw)^{1/2}]/(2u). \tag{19.47}$$

If v is negative, (19.47) can be used to evaluate the root. However, to avoid digit cancellation when v is positive,

$$k = -2w/[v + (v^2 - 4uw)^{1/2}] \tag{19.48}$$

should be used. The linear search subproblem has therefore been resolved in a straightforward manner in this particular case.

The second aspect of particularising the conjugate gradients algorithm to the minimisation of the Rayleigh quotient is the generation of the next conjugate direction. Note that the gradient of the Rayleigh quotient at $\boldsymbol{x}$ is given by

$$\mathbf{g} = 2(\mathbf{A}\boldsymbol{x} - R\mathbf{B}\boldsymbol{x})/(\boldsymbol{x}^T\mathbf{B}\boldsymbol{x}) \tag{19.49}$$

and the local Hessian by

$$\mathbf{H} = 2(\mathbf{A} - R\mathbf{B} - \mathbf{B}\boldsymbol{x}\mathbf{g}^T - \mathbf{g}\boldsymbol{x}^T\mathbf{B})/(\boldsymbol{x}^T\mathbf{B}\boldsymbol{x}). \tag{19.50}$$

Substituting

$$q = -g \tag{16.5}$$

and the Hessian (19.50) into the expression

$$z = g^T Ht / t^T Ht \tag{16.4}$$

for the parameter in the two-term conjugate gradient recurrence, and noting that

$$g^T t = 0 \tag{19.51}$$

by virtue of the 'exact' linear searches, gives

$$z = [g^T(A - RB)t - (g^T g)(x^T Bt)]/[t^T(A - RB)t]. \tag{19.52}$$

The work of Geradin (1971) implies that z should be evaluated by computing the inner products within the square brackets and subtracting. I prefer to perform the subtraction within each element of the inner product to reduce the effect of digit cancellation.

Finally, condition (19.51) permits the calculation of the new value of the Rayleigh quotient to be simplified. Instead of the expression which results from expanding (19.33), we have from (19.49) evaluted at $(x + kt)$, with (19.51), the expression

$$R = (t^T Ax + kt^T At)/(t^T Bx + kt^T Bt). \tag{19.53}$$

This expression is not used in the algorithm. Fried (1972) has suggested several other formulae for the recurrence parameter z of equation (19.52). At the time of writing, too little comparative testing has been carried out to suggest that one such formula is superior to any other.

Algorithm 25. Rayleigh quotient minimisation by conjugate gradients

```
procedure rqmcg( n : integer; {order of matrices}
                  A, B : rmatrix; {matrices defining eigenproblem}
              var X : rvector; {eigenvector approximation, on both
                  input and output to this procedure}
              var ipr : integer; {on input, a limit to the number of
                  matrix products allowed, on output, the number of
                  matrix products used}
              var rq : real); {Rayleigh quotient = eigenvalue approx.}
{alg25.pas == Rayleigh quotient minimization by conjugate gradients
   Minimize Rayleigh quotient
              X-transpose A X / X-transpose B X
      thereby solving generalized symmetric matrix eigenproblem
              A X = rq B X
      for minimal eigenvalue rq and its associated eigenvector.
      A and B are assumed symmetric, with B positive definite.
      While we supply explicit matrices here, only matrix products
      are needed of the form v = A u, w = B u.
                 Copyright 1988 J.C.Nash
}
var
   count, i, itn, itlimit : integer;
```

Algorithm 25. Rayleigh quotient minimisation by conjugate gradients (cont.)

```
    avec, bvec, yvec, zvec, g, t : rvector;
    beta, d, eps, g2, gg, oldg2, pa, pn, s2, step : real;
    t2, ta, tabt, tat, tb, tbt, tol, u, v, w, xat, xax, xbt, xbx : real;
    conv, fail : boolean;
begin
    writeln('alg25.pas -- Rayleigh quotient minimisation');
    itlimit := ipr; {to save the iteration limit} {STEP 0}
    fail := false; {Algorithm has yet to fail.}
    conv := false; {Algorithm has yet to converge.}
    ipr := 0; {to initialize the iteration count}
    eps := calceps;
    tol := n*n*eps*eps; {a convergence tolerance}
    {The convergence tolerance, tol, should ideally be chosen relative
    to the norm of the B matrix when used for deciding if matrix B is
    singular. It needs a different scaling when deciding if the gradient
    is "small". Here we have chosen a compromise value. Performance of
    this method could be improved in production codes by judicious choice
    of the tolerances.}
    pa := big; {a very large initial value for the 'minimum' eigenvalue}
    while (ipr<=itlimit) and (not conv) do
    begin {Main body of algorithm}
        matmul(n, A, X, avec); {STEP 1}
        matmul(n, B, X, bvec); {initial matrix multiplication}
        ipr := ipr+1; {to count the number of products used}
        {STEP 2: Now form the starting Rayleigh quotient}
        xax := 0.0; xbx := 0.0; {accumulators for numerator and denominator
                        of the Rayleigh quotient}
        for i := 1 to n do
        begin
            xax := xax+X[i]*avec[i]; xbx := xbx+X[i]*bvec[i];
        end; {loop on i for Rayleigh quotient}
        if xbx<=tol then halt; {STEP 3: safety check to avoid zero divide.
                This may imply a singular matrix B, or an inappropriate
                starting vector X i.e. one which is in the null space of B
                (implying B is singular!) or which is itself null.}
        rq := xax/xbx; {the Rayleigh quotient -- STEP 4}
        write(ipr,' products -- ev approx. =',rq:18);
        if rq<pa then {Principal convergence check, since if the Rayleigh
            quotient has not been decreased in a major cycle, and we must
            presume that the minimum has been found. Note that this test
            requires that we initialize pa to a large number.}
        begin { body of algorithm -- STEP 5}
            pa := rq; {to save the lowest value so far}
            gg := 0.0; {to initialize gradient norm. Now calculate gradient.}
            for i := 1 to n do {STEP 6}
            begin
                g[i] := 2.0*(avec[i]-rq*bvec[i])/xbx; gg := gg+g[i]*g[i];
            end; {gradient calculation}
            writeln(' squared gradient norm =',gg:8);
            if gg>tol then {STEP 7}
            {Test to see if algorithm has converged. This test is unscaled
                and in some problems it is necessary to scale the tolerance tol
```

Algorithm 25. Rayleigh quotient minimisation by conjugate gradients (cont.)

```
        to the size of the numbers in the problem.}
    begin {conjugate gradients search for improved eigenvector}
      {Now generate the first search direction.}
      for i := 1 to n do t[i] := -g[i]; {STEP 8}
      itn := 0; {STEP 9}
      repeat {Major cg loop}
        itn := itn+1; {to count the conjugate gradient iterations}
        matmul(n, A, t, yvec); {STEP 10}
        matmul(n, B, t, zvec); ipr := ipr+1;
        tat := 0.0; tbt := 0.0; xat := 0.0; xbt := 0.0; {STEP 11}
        for i := 1 to n do
        begin
          xat := xat+X[i]*yvec[i]; tat := tat+t[i]*yvec[i];
          xbt := xbt+X[i]*zvec[i]; tbt := tbt+t[i]*zvec[i];
        end;
        {STEP 12 -- formation and solution of quadratic equation}
        u := tat*xbt-xat*tbt; v := tat*xbx-xax*tbt;
        w := xat*xbx-xax*xbt; d := v*v-4.0*u*w; {the discriminant}
        if d<0.0 then halt; {STEP 13 -- safety check}
        {Note: we may want a more imaginative response to this result
          of the computations. Here we will assume imaginary roots of
          the quadradic cannot arise, but perform the check.}
        d := sqrt(d); {Now compute the roots in a stable manner -- STEP 14}
        if v>0.0 then step := -2.0*w/(v+d) else step := 0.5*(d-v)/u;
        {STEP 15 -- update vectors}
        count := 0; {to count the number of unchanged vector components}
        xax := 0.0; xbx := 0.0;
        for i := 1 to n do
        begin
          avec[i] := avec[i]+step*yvec[i];
          bvec[i] := bvec[i]+step*zvec[i];
          w := X[i]; X[i] := w+step*t[i];
          if (reltest+w)=(reltest+X[i]) then count := count+1;
          xax := xax+X[i]*avec[i]; xbx := xbx+X[i]*bvec[i];
        end; {loop on i}
        if xbx<=tol then halt {to avoid zero divide if B singular}
          else pn := xax/xbx; {STEP 16}
        if (count<n) and (pn<rq) then {STEPS 17 & 18}
        begin
          rq := pn; gg := 0.0; {STEP 19}
          for i := 1 to n do
          begin
            g[i] := 2.0*(avec[i]-pn*bvec[i])/xbx; gg := gg+g[i]*g[i];
          end; {loop on i}
          if gg>tol then {STEP 20}
          begin {STEP 21}
            xbt := 0.0; for i := 1 to n do xbt := xbt+X[i]*zvec[i];
            {STEP 22 -- compute formula (19.52)}
            tabt := 0.0; beta := 0.0;
            for i := 1 to n do
            begin
              w := yvec[i]-pn*zvec[i]; tabt := tabt+t[i]*w;
```

Algorithm 25. Rayleigh quotient minimisation by conjugate gradients (cont.)

```
                              beta := beta+g[i]*(w-g[i]*xbt); {to form the numerator}
                           end; {loop on i}
                           beta := beta/tabt; {STEP 23}
                           {Now perform the recurrence for the next search vector.}
                           for i := 1 to n do t[i] := beta*t[i]-g[i];
                        end; {if gg>tol}
                     end {if (count<n) and (pn<rq)}
                     {Note: pn is computed from update information and may not be
                        precise. We may wish to recalculate from raw data.}
                     else {count=n or pn>=rq so cannot proceed}
                     begin {rest of STEPS 17 & 18}
                        if itn=1 then conv := true; {We cannot proceed in either
                        reducing the eigenvalue approximation or changing the
                        eigenvector, so must assume convergence if we are using
                        the gradient (steepest descent) direction of search.}
                        itn := n+1;{to force end of cg cycle}
                     end; {STEP 24}
                  until (itn>=n) or (count=n) or (gg<=tol) or conv; {end of cg loop}
               end {if gg>tol}
               else conv := true; {The gradient norm is small, so we presume an
                        eigensolution has been found.}
            end {if rq<pa}
            else {we have not reduced Rayleigh quotient in a major cg cycle}
            begin
               conv := true; {if we cannot reduce the Rayleigh quotient}
            end;
            ta := 0.0; {Normalize eigenvector at each major cycle}
            for i := 1 to n do ta := ta+sqr(X[i]); ta := 1.0/sqrt(ta);
            for i := 1 to n do X[i] := ta*X[i];
         end;{ while (ipr<=itlimit) and (not conv) }
         if ipr>itlimit then ipr := -ipr;{to inform calling program limit exceeded}
         writeln;
      end; {alg25.pas == rqmcg}
```

Example 19.3. *Conjugate gradients for inverse iteration and Rayleigh quotient minimisation*

Table 19.3 presents approximations to the minimal and maximal eigensolutions of the order-10 matrix eigenproblem (2.63) having as **A** the Moler matrix and as **B** the Frank matrix (appendix 1). The following notes apply to the table.

(i) The maximal (largest eigenvalue) eigensolution is computed using (−**A**) instead of **A** in algorithm 25.
(ii) Algorithm 15 computes all eigensolutions for the problem. The maximum absolute residual quoted is computed in my program over all these solutions, not simply for the eigenvalue and eigenvector given.
(iii) It was necessary to halt algorithm 10 manually for the case involving a shift of 8·8. This is discussed briefly in §9.3 (p 109).
(iv) The three iterative algorithms were started with an initial vector of ones.

TABLE 19.3. (*a*) Minimal and (*b*) maximal eigensolutions of $\mathbf{A}x = e\mathbf{B}x$ for $\mathbf{A}$ = Moler matrix, $\mathbf{B}$ = Frank matrix (order 10).

	Algorithm 10	Algorithm 15	Section 19.3	Algorithm 25
(*a*) *Minimal eigensolution*				
Shift	0	—	0	—
Eigenvalue	2·1458E−6	2·53754E−6	2·14552E−6	—
Iterations or sweeps	4	7	3	—
Matrix products	—	—	23	26
Rayleigh quotient	—	—	2·1455E−6	2·14512E−6
Eigenvector:	0·433017	−0·433015	0·433017	0·433017
	0·21651	−0·216509	0·21651	0·21651
	0·108258	−0·108257	0·108257	0·108257
	5·41338E−2	−5·41331E−2	5·41337E−2	5·41337E−2
	2·70768E−2	−2·70767E−2	2·70768E−2	2·70768E−2
	1·35582E−2	−1·35583E−2	1·35583E−2	1·35582E−2
	6·81877E−3	−6·81877E−3	6·81877E−3	6·81879E−3
	3·48868E−3	−3·48891E−3	3·48866E−3	3·48869E−3
	1·90292E−3	−1·90299E−3	1·9029E−3	1·90291E−3
	1·26861E−3	−1·26864E−3	1·26858E−3	1·26859E−3
Maximum residual	2·17929E−7	<8·7738E−5	—	—
Error sum of squares r^Tr	—	—	2·0558E−13	4·62709E−11
Gradient norm² g^Tg	—	—	—	9·62214E−15
(*b*) *Maximal eigensolution*				
Shift	8·8	—	8·8	—
Eigenvalue	8·81652	8·81644	8·8165	—
Iterations or sweeps	(see notes)	7	16	—
Matrix products	—	—	166	96
Rayleigh quotient	—	—	—	8·81651
Eigenvector:	0·217765	−0·217764	0·219309	0·219343
	−0·459921	0·459918	−0·462607	−0·462759
	0·659884	−0·659877	0·662815	0·663062
	−0·799308	0·799302	−0·802111	−0·801759
	0·865401	−0·865396	0·867203	0·866363
	−0·852101	0·8521	−0·85142	−0·851188
	0·760628	−0·760632	0·757186	0·757946
	−0·599375	0·599376	−0·594834	−0·595627
	0·383132	−0·383132	0·379815	0·379727
	−0·131739	0·131739	−0·130648	−0·130327
Maximum residual	7·62939E−6	<8·7738E−5	—	—
Error sum of squares r^Tr	—	—	4·9575E−6	5·73166E−3
Gradient norm² g^Tg	—	—	—	5·82802E−9

(v) Different measures of convergence and different tolerances have been used in the computations, which were all performed on a Data General NOVA in 23-bit binary arithmetic. That these measures are different is due to the various operating characteristics of the programs involved.

Example 19.4. Negative definite property of Fröberg's matrix

In example 19.1 the coefficient matrix arising in the linear equations 'turns out to be negative definite'. In practice, to determine this property the eigenvalues of the matrix could be computed. Algorithm 25 is quite convenient in this respect, since a matrix **A** having a positive minimal eigenvalue is positive definite. Conversely, if the smallest eigenvalue of (−**A**) is positive, **A** is negative definite. The minimum eigenvalues of Fröberg coefficient matrices of various orders were therefore computed. (The matrices were multiplied by −1.)

Order	Rayleigh quotient of (−**A**)	Matrix products needed	Gradient norm^2 $\mathbf{g}^T\mathbf{g}$
4	0·350144	5	1·80074E−13
10	7·44406E−2	11	2·08522E−10
50	3·48733E−3	26	1·9187E−10
100	8·89398E−4	49	7·23679E−9

These calculations were performed on a Data General NOVA in 23-bit binary arithmetic.

Note that because the Fröberg matrices are tridiagonal, other techniques may be preferable in this specific instance (Wilkinson and Reinsch 1971).

Appendix 1

NINE TEST MATRICES

In order to test programs for the algebraic eigenproblem and linear equations, it is useful to have a set of easily generated matrices whose properties are known. The following nine real symmetric matrices can be used for this purpose.

Hilbert segment of order n

$$A_{ij} = 1/(i+j-1).$$

This matrix is notorious for its logarithmically distributed eigenvalues. While it can be shown in theory to be positive definite, in practice it is so ill conditioned that most eigenvalue or linear-equation algorithms fail for some value of $n < 20$.

Ding Dong matrix

$$A_{ij} = 0{\cdot}5/(n-i-j+1{\cdot}5).$$

The name and matrix were invented by Dr F N Ris of IBM, Thomas J Watson Research Centre, while he and the author were both students at Oxford. This Cauchy matrix has few trailing zeros in any elements, so is always represented inexactly in the machine. However, it is very stable under inversion by elimination methods. Its eigenvalues have the property of clustering near $\pm\pi/2$.

Moler matrix

$$A_{ii} = i$$
$$A_{ij} = \min(i, j) - 2 \qquad \text{for } i \neq j.$$

Professor Cleve Moler devised this simple matrix. It has the very simple Choleski decomposition given in example 7.1, so is positive definite. Nevertheless, it has one small eigenvalue and often upsets elimination methods for solving linear-equation systems.

Frank matrix

$$A_{ij} = \min(i, j).$$

A reasonably well behaved matrix.

Bordered matrix

$$A_{ii} = 1$$
$$A_{in} = A_{ni} = 2^{1-i} \qquad \text{for } i \neq n$$
$$A_{ij} = 0 \qquad \text{otherwise.}$$

The matrix has $(n-2)$ eigenvalues at 1. Wilkinson (1965, pp 94–7) gives some discussion of this property. The high degree of degeneracy and the form of the 'border' were designed to give difficulties to a specialised algorithm for matrices of this form in which I have been interested from time to time.

Diagonal matrix

$$A_{ii} = i$$
$$A_{ij} = 0 \qquad \text{for } i \neq j.$$

This matrix permits solutions to eigenvalue and linear-equation problems to be computed trivially. It is included in this set because I have known several programs to fail to run correctly when confronted with it. Sometimes programs are unable to solve trivial problems because their designers feel they are 'too easy.' Note that the ordering is 'wrong' for algorithms 13 and 14.

Wilkinson W+ matrix

$$A_{ii} = [n/2] + 1 - \min(i, n - i + 1) \qquad \text{for } i = 1, 2, \ldots, n$$
$$A_{i,i+1} = A_{i+1,i} = 1 \qquad \text{for } i = 1, 2, \ldots, (n-1)$$
$$A_{ij} = 0 \qquad \text{for } |j - i| > 1$$

where $[b]$ is the largest integer less than or equal to b. The W+ matrix (Wilkinson 1965, p 308) is normally given odd order. This tridiagonal matrix then has several pairs of close eigenvalues despite the fact that no superdiagonal element is small. Wilkinson points out that the separation between the two largest eigenvalues is of the order of $(n!)^{-2}$ so that the power method will be unable to separate them unless n is very small.

Wilkinson W− matrix

$$A_{ii} = [n/2] + 1 - i \qquad \text{for } i = 1, 2, \ldots, n$$
$$A_{i,i+1} = A_{i+1,1} \qquad \text{for } i = 1, 2, \ldots, (n-1)$$
$$A_{ij} = 0 \qquad \text{for } |j - i| > 1$$

where $[b]$ is the largest integer less than or equal to b. For odd order, this matrix has eigenvalues which are pairs of equal magnitude but opposite sign. The magnitudes of these are very close to some of those of the corresponding W+ matrix.

Ones

$$A_{ij} = 1 \qquad \text{for all } i,j.$$

This matrix is singular. It has only rank one, that is, $(n-1)$ zero eigenvalues.

The matrices described here may all be generated by the Pascal procedure MATRIXIN.PAS, which is on the software diskette. This procedure also allows for keyboard entry of matrices.

Appendix 2

LIST OF ALGORITHMS

Appendix 3

LIST OF EXAMPLES

Appendix 4

FILES ON THE SOFTWARE DISKETTE

The files on the diskette fall into several categories. For the new user of the diskette, we strongly recommend looking at the file

README.CNM

which contains notes of any errors or additions to material in either the book or the diskette. This can be displayed by issuing a command

TYPE (drive:)README.CNM

where drive: is the disk drive specification for the location of the README.CNM file. The file may also be printed, or viewed with a text editor.

The algorithms (without comments) are in the files which follow. Only ALG03A.PAS has not appeared on the pages of the book.

ALG01.PAS
ALG02.PAS
ALG03.PAS
ALG03A.PAS
ALG04.PAS
ALG05.PAS
ALG06.PAS
ALG07.PAS
ALG08.PAS
ALG09.PAS
ALG10.PAS
ALG11.PAS
ALG12.PAS
ALG13.PAS
ALG14.PAS
ALG15.PAS
ALG16.PAS
ALG17.PAS
ALG18.PAS
ALG19.PAS
ALG20.PAS
ALG21.PAS
ALG22.PAS

ALG23.PAS
ALG24.PAS
ALG25.PAS
ALG26.PAS
ALG27.PAS

The following files are driver programs to run examples of use of the algorithms.

DR0102.PAS —algorithms 1 and 2, svd and least-squares solution
DR03.PAS —algorithm 3, columnwise Givens' reduction
DR03A.PAS —algorithm 3a, row-wise Givens' reduction
DR04.PAS —algorithm 4, Givens' reduction, svd and least-squares solution
DR0506.PAS —algorithms 4 and 5, Gauss elimination and back-substitution to solve linear equations
DR0708.PAS —algorithms 7 and 8, Choleski decomposition and back-substitution to solve linear equations of a special form
DR09.PAS —algorithm 9, to invert a symmetric, positive-definite matrix
DR10.PAS —algorithm 10, to find eigensolutions of matrices via inverse iteration using Gauss elimination
DR13.PAS —algorithm 13, eigensolutions of a symmetric matrix via the svd
DR14.PAS —algorithm 14, eigensolutions of a symmetric matrix via a cyclic Jacobi method
DR15.PAS —algorithm 15, solution of a generalised matrix eigenproblem via two applications of the Jacobi method
DR1617.PAS —algorithms 16 and 17, grid search and one-dimensional minimisation
DR1618.PAS —algorithms 16 and 18, grid search and one-dimensional root-finding
DR1920.PAS —algorithms 19 and 20, Nelder–Mead function minimiser and axial search for lower points in the multivariate space
DR21.PAS —algorithm 21, variable metric function minimiser
DR22.PAS —algorithm 22, conjugate gradients function minimiser
DR23.PAS —algorithm 23, modified Marquardt nonlinear least-squares method
DR24II.PAS —algorithm 24, applied to finding eigensolutions of a symmetric matrix by inverse iteration
DR24LE.PAS —algorithm 24, applied to finding solutions of linear equations
DR24LS.PAS —algorithm 24, applied to solving least-squares problems
DR25.PAS —algorithm 25, solutions of a generalised symmetric eigenproblem by conjugate gradients minimisation of the Rayleigh quotient
DR26.PAS —algorithms 26, 11, and 12, to find eigensolutions of a general complex matrix, standardise the eigenvectors and compute residuals
DR27.PAS —algorithm 27, Hooke and Jeeves function minimiser

The following support codes are needed to execute the driver programs:

CALCEPS.PAS	— to compute the machine precision for the Turbo Pascal computing environment in which the program is compiled
CONSTYPE.DEF	— a set of **constant** and **type** specifications common to the codes
CUBEFN.PAS	— a cubic test function of one variable with minimum at 0.81650
FNMIN.PAS	— a main program to run function minimisation procedures
GENEVRES.PAS	— residuals of a generalised eigenvalue problem
GETOBSN.PAS	— a procedure to read a single observation for several variables (one row of a data matrix)
HTANFN.PAS	— the hyperbolic tangent, example 13.2
JJACF.PAS	— Jaffrelot's autocorrelation problem, example 14.1
MATCOPY.PAS	— to copy a matrix
MATMUL.PAS	— to multiply two matrices
MATRIXIN.PAS	— to create or read in matrices
PSVDRES.PAS	— to print singular-value decomposition results
QUADFN.PAS	— real valued test function of x for [1D] minimisation and root-finding
RAYQUO.PAS	— to compute the Rayleigh quotient for a generalised eigenvalue problem
RESIDS.PAS	— to compute residuals for linear equations and least-squares problems
ROSEN.PAS	— to set up and compute function and derivative information for the Rosenbrock banana-shaped valley test problem
SPENDFN.PAS	— the expenditure example, illustrated in example 12.5 and example 13.1
STARTUP.PAS	— code to read the names of and open console image and/or console control files for driver programs. This common code segment is *not* a complete procedure, so cannot be *included* in Turbo Pascal 5.0 programs.
SVDTST.PAS	— to compute various tests of a singular-value decomposition
TDSTAMP.PAS	— to provide a time and date stamp for output (files). This code makes calls to the operating system and is useful only for MS-DOS computing environments. In Turbo Pascal 5.0, there are utility functions which avoid the DOS call.
VECTORIN.PAS	— to create or read in a vector

The following files provide control information and data to the driver programs. Their names can be provided in response to the question

File for input of control data ([cr] for keyboard)?

Be sure to include the filename extension (.CNM). The nomenclature follows that for the DR*.PAS files. In some cases additional examples have been provided. For these files a brief description is provided in the following list of control files.

EX0102.CNM
EX03.CNM
EX03A.CNM
EX04.CNM
EX0506.CNM
EX0506S.CNM — a set of equations with a singular coefficient matrix
EX0708.CNM
EX09.CNM
EX10.CNM
EX13.CNM
EX14.CNM
EX15.CNM
EX1617.CNM
EX1618.CNM
EX19.CNM
EX1920.CNM
EX1920J.CNM — data for the Jaffrelot problem (JJACF.PAS), example 14.1
EX21.CNM
EX22.CNM
EX23.CNM
EX24II.CNM
EX24LE.CNM
EX24LS.CNM
EX24LS1.CNM — data for example 19.2
EX25.CNM
EX26.CNM
EX26A.CNM
EX27J.CNM — data for the Jaffrelot problem (JJACF.PAS), example 14.1.
EX27R.CNM — console control file for the regular test problem, the Rosenbrock test function (ROSEN.PAS)

If the driver programs have been loaded and compiled to saved executable (.COM) files, then we can execute these programs by typing their names, e.g. DR0102. The user must then enter command information from the keyboard. This is not difficult, but it is sometimes useful to be able to issue such commands from a file. Such a BATch command file (.BAT extension) is commonly used in MS-DOS systems. In the driver programs we have included compiler directives to make this even easier to use by allowing command input to come from a file. A batch file EXAMPLE.BAT which could run drivers for algorithms 1 through 6 would have the form

```
rem EXAMPLE.BAT
rem runs Nash Algorithms 1 through 6 automatically
DR0102 <DR0102X.
DR03A <DR03AX.
DR03 <DR03X.
DR04 <DR04X.
DR0506 <DR0506X.
```

The files which end in an 'X.' contain information to control the drivers, in fact, they contain the names of the EX*.CNM control files. This facility is provided to allow for very rapid testing of all the codes at once (the technical term for this is 'regression testing'). Note that console image files having names of the form OUT0102 are created, which correspond in form to the driver names, i.e. DR0102.PAS. The command line files present on the disk are:

DR0102X.	DR03AX.	DR03X.	DR04X.	DR0506X.	DR0708X.
DR09X.	DR10X.	DR13X.	DR14X.	DR15X.	DR1617X.
DR1618X.	DR19X.	DR21X.	DR22X.	DR23X.	DR24IIX.
DR24LEX.	DR24LSX.	DR25X.	DR26X.	DR27X.	

Users may wish to note that there are a number of deficiencies with version 3.01a of Turbo Pascal. I have experienced some difficulty in halting programs with the Control-C or Control-Break keystrokes, in particular when the program is waiting for input. In some instances, attempts to halt the program seem to interfere with the files on disk, and the 'working' algorithm file has been over-written! On some occasions, the leftmost characters entered from the keyboard are erased by READ instructions. From the point of view of a software developer, the absence of a facility to compile under command of a BATch command file is a nuisance. Despite these faults, the system is relatively easy to use. Many of the faults of Turbo Pascal 3.01a have been addressed in later versions of the product. We anticipate that a diskette of the present codes adapted for version 5.0 of Turbo Pascal will be available about the time the book is published. Turbo Pascal 5.0 is, however, a much 'larger' system in terms of memory requirements.

BIBLIOGRAPHY

ABRAMOWITZ M and STEGUN I A 1965 *Handbook of Mathematical Functions with Formulas, Graphs and Mathematical Tables* (New York: Dover)

ACTON F S 1970 *Numerical Methods that Work* (New York: Harper and Row)

BARD Y 1967 *Nonlinear Parameter Estimation and Programming* (New York: IBM New York Scientific Center)

—— 1970 Comparison of gradient methods for the solution of nonlinear parameter estimation problems *SIAM J. Numer. Anal.* **7** 157–86

—— 1974 *Nonlinear Parameter Estimation* (New York/London: Academic)

BATES D M and WATTS D G 1980 Relative curvature measures of nonlinearity *J. R. Stat. Soc.* B **42** 1–25

—— 1981a A relative offset orthogonality convergence criterion for nonlinear least squares *Technometrics* **23** 179–83

—— 1988 *Nonlinear Least Squares* (New York: Wiley)

BAUER F L and REINSCH C 1971 Inversion of positive definite matrices by the Gauss-Jordan method in linear algebra *Handbook for Automatic Computation* vol 2, eds J H Wilkinson and C Reinsch (Berlin: Springer) contribution 1/3 (1971)

BEALE E M L 1972 A derivation of conjugate gradients *Numerical Methods for Nonlinear Optimization* ed. F A Lootsma (London: Academic)

BELSLEY D A, KUH E and WELSCH R E 1980 *Regression Diagnostics: Identifying Influential Data and Sources of Collinearity* (New York/Toronto: Wiley)

BIGGS M C 1975 Some recent matrix updating methods for minimising sums of squared terms *Hatfield Polytechnic, Numerical Optimization Centre, Technical Report* 67

BOOKER T H 1985 Singular value decomposition using a Jacobi algorithm with an unbounded angle of rotation *PhD Thesis* (Washington, DC: The American University)

BOWDLER H J, MARTIN R S, PETERS G and WILKINSON J H 1966 Solution of real and complex systems of linear equations *Numer. Math.* **8** 217–34; also in *Linear Algebra, Handbook for Automatic Computation* vol 2, eds J H Wilkinson and C Reinsch (Berlin: Springer) contribution 1/7 (1971)

BOX G E P 1957 Evolutionary operation: a method for increasing industrial productivity *Appl. Stat.* **6** 81–101

BOX M J 1965 A new method of constrained optimization and a comparison with other methods *Comput. J.* **8** 42–52

BOX M J, DAVIES D and SWANN W H 1971 *Techniques d'optimisation non linéaire, Monographie No* 5 (Paris: Entreprise Moderne D' Edition) Original English edition (London: Oliver and Boyd)

BRADBURY W W and FLETCHER R 1966 New iterative methods for solution of the eigenproblem *Numer. Math.* **9** 259–67

BREMMERMANN H 1970 A method of unconstrained global optimization *Math. Biosci.* **9** 1–15

BRENT R P 1973 *Algorithms for Minimization Without Derivatives* (Englewood Cliffs, NJ: Prentice–Hall)

BROWN K M and GEARHART W B 1971 Deflation techniques for the calculation of further solutions of nonlinear systems *Numer. Math.* **16** 334–42

BROYDEN C G 1970a The convergence of a class of double-rank minimization algorithms, pt 1 *J. Inst. Maths Applics* **6** 76–90

—— 1970b The convergence of a class of double-rank minimization algorithms, pt 2 *J. Inst. Maths Applics* **6** 222–31

—— 1972 Quasi-Newton methods *Numerical methods for Unconstrained Optimization* ed. W Murray (London: Academic) pp 87–106

Bunch J R and Neilsen C P 1978 Updating the singular value decomposition *Numerische Mathematik* **31** 111–28

Bunch J R and Rose D J (eds) 1976 *Sparse Matrix Computation* (New York: Academic)

Businger P A 1970 Updating a singular value decomposition (ALGOL programming contribution, No 26) *BIT* **10** 376–85

Caceci M S and Cacheris W P 1984 Fitting curves to data (the Simplex algorithm is the answer) *Byte* **9** 340–62

Cauchy A 1848 Méthode générale pour la résolution des systèmes d'équations simultanées *C. R. Acad. Sci., Paris* **27** 536–8

Chambers J M 1969 A computer system for fitting models to data *Appl. Stat.* **18** 249–63

—— 1971 Regression updating *J. Am. Stat. Assoc.* **66** 744–8

—— 1973 Fitting nonlinear models: numerical techniques *Biometrika* **60** 1–13

Chartres B A 1962 Adaptation of the Jacobi methods for a computer with magnetic tape backing store *Comput. J.* **5** 51–60

Cody W J and Waite W 1980 *Software Manual for the Elementary Functions* (Englewood Cliffs, NJ: Prentice–Hall)

Conn A R 1985 Nonlinear programming, exact penalty functions and projection techniques for non-smooth functions *Boggs, Byrd and Schnabel* pp 3–25

Coonen J T 1984 Contributions to a proposed standard for binary floating-point arithmetic *PhD Dissertation* University of California, Berkeley

Craig R J and Evans J W *c.*1980 A comparison of Nelder-Mead type simplex search procedures *Technical Report No* 146 (Lexington, KY: Dept of Statistics, Univ. of Kentucky)

Craig R J, Evans J W and Allen D M 1980 The simplex-search in non-linear estimation *Technical Report No* 155 (Lexington, KY: Dept of Statistics, Univ. of Kentucky)

Curry H B 1944 The method of steepest descent for non-linear minimization problems *Q. Appl. Math.* **2** 258–61

Dahlquist G and Björck A 1974 *Numerical Methods* (translated by N Anderson) (Englewood Cliffs, NJ: Prentice–Hall)

Dantzig G B 1979 Comments on Khachian's algorithm for linear programming *Technical Report No* SOL 79-22 (Standford, CA: Systems Optimization Laboratory, Stanford Univ.)

Davidon W C 1959 Variable metric method for minimization *Physics and Mathematics, AEC Research and Development Report No* ANL-5990 (Lemont, IL: Argonne National Laboratory)

—— 1976 New least-square algorithms *J. Optim. Theory Applic.* **18** 187–97

—— 1977 Fast least squares algorithms *Am. J. Phys.* **45** 260–2

Dembo R S, Eisenstat S C and Steihaug T 1982 Inexact Newton methods *SIAM J. Numer. Anal.* **19** 400–8

Dembo R S and Steihaug T 1983 Truncated-Newton algorithms for large-scale unconstrained optimization *Math. Prog.* **26** 190–212

Dennis J E Jr, Gay D M and Welsch R E 1981 An adaptive nonlinear least-squares algorithm *ACM Trans. Math. Softw.* **7** 348–68

Dennis J E Jr and Schnabel R 1983 *Numerical Methods for Unconstrained Optimization and Nonlinear Equations* (Englewood Cliffs, NJ: Prentice–Hall)

Dixon L C W 1972 *Nonlinear Optimisation* (London: The English Universities Press)

Dixon L C W and Szegö G P (eds) 1975 *Toward Global Optimization* (Amsterdam/Oxford: North-Holland and New York: American Elsevier)

—— (eds) 1978 *Toward Global Optimization 2* (Amsterdam/Oxford: North-Holland and New York: American Elsevier)

Donaldson J R and Schnabel R B 1987 Computational experience with confidence regions and confidence intervals for nonlinear least squares *Technometrics* **29** 67–82

Dongarra and Grosse 1987 Distribution of software by electronic mail *Commun. ACM* **30** 403–7

Draper N R and Smith H 1981 *Applied Regression Analysis* 2nd edn (New York/Toronto: Wiley)

Eason E D and Fenton R G 1972 Testing and evaluation of numerical methods for design optimization *Report No* UTME-TP7204 (Toronto, Ont.: Dept of Mechanical Engineering, Univ. of Toronto)

—— 1973 A comparison of numerical optimization methods for engineering design *Trans. ASME J. Eng. Ind.* paper 73-DET-17, pp 1–5

Evans D J (ed.) 1974 *Software for Numerical Mathematics* (London: Academic)

Evans J W and Craig R J 1979 Function minimization using a modified Nelder-Mead simplex search procedure *Technical Report No* 144 (Lexington, KY: Dept of Statistics, Univ. of Kentucky)

Fiacco A V and McCormick G P 1964 Computational algorithm for the sequential unconstrained minimization technique for nonlinear programming *Mgmt Sci.* **10** 601–17

—— 1966 Extensions of SUMT for nonlinear programming: equality constraints and extrapolation *Mgmt Sci.* **12** 816–28

Finkbeiner D T 1966 *Introduction to Matrices and Linear Transformations* (San Francisco: Freeman)

Fletcher R 1969 *Optimization Proceedings of a Symposium of the Institute of Mathematics and its Applications, Univ. of Keele, 1968* (London: Academic)

—— 1970 A new approach to variable metric algorithms *Comput. J.* **13** 317–22

—— 1971 A modified Marquardt subroutine for nonlinear least squares *Report No* AERE-R 6799 (Harwell, UK: Mathematics Branch, Theoretical Physics Division, Atomic Energy Research Establishment)

—— 1972 A FORTRAN subroutine for minimization by the method of conjugate gradients *Report No* AERE-R 7073 (Harwell, UK: Theoretical Physics Division, Atomic Energy Research Establishment)

—— 1980a *Practical Methods of Optimization* vol 1: *Unconstrained Optimization* (New York/Toronto: Wiley)

—— 1980b *Practical Methods of Optimization* vol 2: *Constrained Optimization* (New York/Toronto: Wiley)

Fletcher R and Powell M J D 1963 A rapidly convergent descent method for minimization *Comput. J.* **6** 163–8

Fletcher R and Reeves C M 1964 Function minimization by conjugate gradients *Comput. J.* **7** 149–54

Ford B and Hall G 1974 The generalized eigenvalue problem in quantum chemistry *Comput. Phys. Commun.* **8** 337–48

Forsythe G E and Henrici P 1960 The cyclic Jacobi method for computing the principal values of a complex matrix *Trans. Am. Math. Soc.* **94** 1–23

Forsythe G E, Malcolm M A and Moler C E 1977 *Computer Methods for Mathematical Computations* (Englewood Cliffs, NJ: Prentice–Hall)

Fried I 1972 Optimal gradient minimization scheme for finite element eigenproblems *J. Sound Vib.* **20** 333–42

Fröberg C 1965 *Introduction to Numerical Analysis* (Reading, Mass: Addison-Wesley) 2nd edn, 1969

Gallant A R 1975 Nonlinear regression *Am. Stat.* **29** 74–81

Gass S I 1964 *Linear Programming* 2nd edn (New York/Toronto: McGraw-Hill)

Gauss K F 1809 *Theoria Motus Corporum Coelestiam Werke Bd.* **7** 240–54

Gay D M 1983 Remark on algorithm 573 (NL2SOL: an adaptive nonlinear least squares algorithm) *ACM Trans. Math. Softw.* **9** 139

Gentleman W M 1973 Least squares computations by Givens' transformations without square roots *J. Inst. Maths Applics* **12** 329–36

Gentleman W M and Marovich S B 1974 More on algorithms that reveal properties of floating point arithmetic units *Commun. ACM* **17** 276–7

Geradin M 1971 The computational efficiency of a new minimization algorithm for eigenvalue analysis *J. Sound Vib.* **19** 319–31

Gill P E and Murray W (eds) 1974 *Numerical Methods for Constrained Optimization* (London: Academic)

—— 1978 Algorithms for the solution of the nonlinear least squares problem *SIAM J. Numer. Anal.* **15** 977–92

Gill P E, Murray W and Wright M H 1981 *Practical Optimization* (London: Academic)

Golub G H and Pereyra V 1973 The differentiation of pseudo-inverses and nonlinear least squares problems whose variables separate *SIAM J. Numer. Anal.* **10** 413–32

Golub G H and Styan G P H 1973 Numerical computations for univariate linear models *J. Stat. Comput. Simul.* **2** 253–74

Golub G H and Van Loan C F 1983 *Matrix Computations* (Baltimore, MD: Johns Hopkins University Press)

Gregory R T and Karney D L 1969 *Matrices for Testing Computational Algorithms* (New York: Wiley Interscience)

Hadley G 1962 *Linear Programming* (Reading, MA: Addison-Wesley)

Hammarling S 1974 A note on modifications to the Givens' plane rotation *J. Inst. Maths Applics* **13** 215–18

Hartley H O 1948 The estimation of nonlinear parameters by 'internal least squares' *Biometrika* **35** 32–45

—— 1961 The modified Gauss-Newton method for the fitting of non-linear regression functions by least squares *Technometrics* **3** 269–80

Hartley H O and Booker A 1965 Nonlinear least squares estimation *Ann. Math. Stat.* **36** 638–50

Healy M J R 1968 Triangular decomposition of a symmetric matrix (algorithm AS6) *Appl Stat.* **17** 195–7

Henrici P 1964 *Elements of Numerical Analysis* (New York: Wiley)

Hestenes M R 1958 Inversion of matrices by biorthogonalization and related results *J. Soc. Ind. Appl. Math.* **5** 51–90

—— 1975 Pseudoinverses and conjugate gradients *Commun. ACM* **18** 40–3

Hestenes M R and Stiefel E 1952 Methods of conjugate gradients for solving linear systems *J. Res. Nat. Bur. Stand.* **49** 409–36

Hillstrom K E 1976 A simulation test approach to the evaluation and comparison of unconstrained nonlinear optimization algorithms *Argonne National Laboratory Report* ANL-76-20

Hock W and Schittkowski K 1981 Test examples for nonlinear programming codes *Lecture Notes in Economics and Mathematical Systems 187* (Berlin: Springer)

Holt J N and Fletcher R 1979 An algorithm for constrained nonlinear least squares *J. Inst. Maths Applics* **23** 449–63

Hooke R and Jeeves T A 1961 'Direct Search' solution of numerical and statistical problems *J. ACM* **8** 212–29

Jacobi C G J 1846 Uber ein leichtes Verfahren, die in der Theorie der Sakularstorungen vorkommenden Gleichungen numerisch aufzulosen *Crelle's J.* **30** 51–94

Jacoby S L S, Kowalik J S and Pizzo J T 1972 *Iterative Methods for Nonlinear Optimization Problems* (Englewood Cliffs, NJ: Prentice–Hall)

Jenkins M A and Traub J F 1975 Principles for testing polynomial zero-finding programs *ACM Trans. Math. Softw.* **1** 26–34

Jones A 1970 Spiral – a new algorithm for non-linear parameter estimation using least squares *Comput. J.* **13** 301–8

Kahaner D, Moler C and Nash S G 1989 *Numerical Analysis and Software* (Englewood Cliffs, NJ: Prentice–Hall)

Kahaner D and Parlett B N 1976 How far should you go with the Lanczos process? *Sparse Matrix Computations* eds J R Bunch and D J Rose (New York: Academic) pp 131–44

Kaiser H F 1972 The JK method: a procedure for finding the eigenvectors and eigenvalues of a real symmetric matrix *Comput. J.* **15** 271–3

Karmarkar N 1984 A new polynomial time algorithm for linear programming *Combinatorica* **4** 373–95

Karpinski R 1985 PARANOIA: a floating-point benchmark *Byte* **10**(2) 223–35 (February)

Kaufman L 1975 A variable projection method for solving separable nonlinear least squares problems *BIT* **15** 49–57

Kendall M G 1973 *Time-series* (London: Griffin)

Kendall M G and Stewart A 1958–66 *The Advanced Theory of Statistics* vols 1–3 (London: Griffin)

Kennedy W J Jr and Gentle J E 1980 *Statistical Computing* (New York: Marcel Dekker)

Kernighan B W and Plauger P J 1974 *The Elements of Programming Style* (New York: McGraw-Hill)

Kirkpatrick S, Gelatt C D Jr and Vecchi M P 1983 Optimization by simulated annealing *Science* **220** (4598) 671–80

Kowalik J and Osborne M R 1968 *Methods for Unconstrained Optimization Problems* (New York: American Elsevier)

Kuester J L and Mize H H 1973 *Optimization Techniques with FORTRAN* (New York/London/Toronto: McGraw-Hill)

Kulisch U 1987 *Pascal SC: A Pascal extension for scientific computation* (Stuttgart: B G Teubner and Chichester: Wiley)

Lanczos C 1956 *Applied Analysis* (Englewood Cliffs, NJ: Prentice–Hall)

Lawson C L and Hanson R J 1974 *Solving Least Squares Problems* (Englewood Cliffs, NJ: Prentice–Hall)

Levenberg K 1944 A method for the solution of certain non-linear problems in least squares *Q. Appl. Math.* **2** 164–8

LOOTSMA F A (ed.) 1972 *Numerical Methods for Non-Linear Optimization* (London/New York: Academic)

MAINDONALD J H 1984 *Statistical Computation* (New York: Wiley)

MALCOLM M A 1972 Algorithms to reveal properties of floating-point arithmetic *Commun. ACM* **15** 949–51

MARQUARDT D W 1963 An algorithm for least-squares estimation of nonlinear parameters *J. SIAM* **11** 431–41

—— 1970 Generalized inverses, ridge regression, biased linear estimation, and nonlinear estimation *Technometrics* **12** 591–612

MCKEOWN J J 1973 A comparison of methods for solving nonlinear parameter estimation problems *Identification & System Parameter Estimation, Proc. 3rd IFAC Symp.* ed. P Eykhoff (The Hague: Delft) pp 12–15

—— 1974 Specialised versus general purpose algorithms for minimising functions that are sums of squared terms *Hatfield Polytechnic, Numerical Optimization Centre Technical Report No* 50, Issue 2

MEYER R R and ROTH P M 1972 Modified damped least squares: an algorithm for non-linear estimation *J. Inst. Math. Applic.* **9** 218–33

MOLER C M and VAN LOAN C F 1978 Nineteen dubious ways to compute the exponential of a matrix *SIAM Rev.* **20** 801–36

MORÉ J J, GARBOW B S and HILLSTROM K E 1981 Testing unconstrained optimization software *ACM Trans. Math. Softw.* **7** 17–41

MOSTOW G D and SAMPSON J H 1969 *Linear Algebra* (New York: McGraw-Hill)

MURRAY W (ed.) 1972 *Numerical Methods for Unconstrained Optimization* (London: Academic)

NASH J C 1974 The Hermitian matrix eigenproblem HX = eSx using compact array storage *Comput. Phys. Commun.* **8** 85–94

—— 1975 A one-sided transformation method for the singular value decomposition and algebraic eigenproblem *Comput. J.* **18** 74–6

—— 1976 *An Annotated Bibliography on Methods for Nonlinear Least Squares Problems Including Test Problems* (microfiche) (Ottawa: Nash Information Services)

—— 1977 Minimizing a nonlinear sum of squares function on a small computer *J. Inst. Maths Applics* **19** 231–7

—— 1979a *Compact Numerical Methods for Computers: Linear Algebra and Function Minimisation* (Bristol: Hilger and New York: Halsted)

—— 1979b Accuracy of least squares computer programs: another reminder: comment *Am. J. Ag. Econ.* **61** 703–9

—— 1980 Problèmes mathématiques soulevés par les modèles économiques *Can. J. Ag. Econ.* **28** 51–7

—— 1981 Nonlinear estimation using a microcomputer *Computer Science and Statistics: Proceedings of the 13th Symposium on the Interface* ed. W F Eddy (New York: Springer) pp 363–6

—— 1984a *Effective Scientific Problem Solving with Small Computers* (Reston, VA: Reston Publishing) (all rights now held by J C Nash)

—— 1984b *LEQB05: User Guide – A Very Small Linear Algorithm Package* (Ottawa, Ont.: Nash Information Services Inc.)

—— 1985 Design and implementation of a very small linear algebra program package *Commun. ACM* **28** 89–94

—— 1986a Review: IMSL MATH/PC-LIBRARY *Am. Stat.* **40** 301–3

—— 1986b Review: IMSL STAT/PC-LIBRARY *Am. Stat.* **40** 303–6

—— 1986c Microcomputers, standards, and engineering calculations *Proc. 5th Canadian Conf. Engineering Education, Univ. of Western Ontario, May 12-13, 1986* pp 302–16

NASH J C and LEFKOVITCH L P 1976 Principal components and regression by singular value decomposition on a small computer *Appl. Stat.* **25** 210–16

—— 1977 *Programs for Sequentially Updated Principal Components and Regression by Singular Value Decomposition* (Ottawa: Nash Information Services)

NASH J C and NASH S G 1977 Conjugate gradient methods for solving algebraic eigenproblems *Proc. Symp. Minicomputers and Large Scale Computation, Montreal* ed. P Lykos (New York: American Chemical Society) pp 24–32

—— 1988 Compact algorithms for function minimisation *Asia-Pacific J. Op. Res.* **5** 173–92

NASH J C and SHLIEN S 1987 Simple algorithms for the partial singular value decomposition *Comput. J.* **30** 268–75

NASH J C and TEETER N J 1975 Building models: an example from the Canadian dairy industry *Can. Farm. Econ.* **10** 17–24

NASH J C and WALKER-SMITH M 1986 Using compact and portable function minimization codes in forecasting applications *INFOR* **24** 158–68

—— 1987 *Nonlinear Parameter Estimation, an Integrated System in Basic* (New York: Marcel Dekker)

NASH J C and WANG R L C 1986 Algorithm 645 Subroutines for testing programs that compute the generalized inverse of a matrix *ACM Trans. Math. Softw.* **12** 274–7

NASH S G 1982 Truncated-Newton methods *Report No* STAN-CS-82-906 (Stanford, CA: Dept of Computer Science, Stanford Univ.)

—— 1983 Truncated-Newton methods for large-scale function minimization *Applications of Nonlinear Programming to Optimization and Control* ed. H E Rauch (Oxford: Pergamon) pp 91–100

—— 1984 Newton-type minimization via the Lanczos method *SIAM J. Numer. Anal.* **21** 770–88

—— 1985a Preconditioning of truncated-Newton methods *SIAM J. Sci. Stat. Comp.* **6** 599–616

—— 1985b Solving nonlinear programming problems using truncated-Newton techniques *Boggs, Byrd and Schnabel* pp 119–36

NASH S G and RUST B 1986 Regression problems with bounded residuals *Technical Report No* 478 (Baltimore, MD: Dept of Mathematical Sciences, The Johns Hopkins University)

NELDER J A and MEAD R 1965 A simplex method for function minimization *Comput. J.* **7** 308–13

NEWING R A and CUNNINGHAM J 1967 *Quantum Mechanics* (Edinburgh: Oliver and Boyd)

OLIVER F R 1964 Methods of estimating the logistic growth function *Appl. Stat.* **13** 57–66

—— 1966 Aspects of maximum likelihood estimation of the logistic growth function *JASA* **61** 697–705

OLSSON D M and NELSON L S 1975 The Nelder-Mead simplex procedure for function minimization *Technometrics* **17** 45–51; Letters to the Editor 393–4

O'NEILL R 1971 Algorithm AS 47: function minimization using a simplex procedure *Appl. Stat.* **20** 338–45

OSBORNE M R 1972 Some aspects of nonlinear least squares calculations *Numerical Methods for Nonlinear Optimization* ed. F A Lootsma (London: Academic) pp 171–89

PAIGE C C and SAUNDERS M A 1975 Solution of sparse indefinite systems of linear equations *SIAM J. Numer. Anal.* **12** 617–29

PAULING L and WILSON E B 1935 *Introduction to Quantum Mechanics with Applications to Chemistry* (New York: McGraw-Hill)

PENROSE R 1955 A generalized inverse for matrices *Proc. Camb. Phil. Soc.* **51** 406–13

PERRY A and SOLAND R M 1975 Optimal operation of a public lottery *Mgmt. Sci.* **22** 461–9

PETERS G and WILKINSON J H 1971 The calculation of specified eigenvectors by inverse iteration *Linear Algebra, Handbook for Automatic Computation* vol 2, eds J H Wilkinson and C Reinsch (Berlin: Springer) pp 418–39

—— 1975 On the stability of Gauss-Jordan elimination with pivoting *Commun. ACM* **18** 20–4

PIERCE B O and FOSTER R M 1956 *A Short Table of Integrals* 4th edn (New York: Blaisdell)

POLAK E and RIBIERE G 1969 Note sur la convergence de méthodes de directions conjugées *Rev. Fr. Inf. Rech. Oper.* **3** 35–43

POWELL M J D 1962 An iterative method for stationary values of a function of several variables *Comput. J.* **5** 147–51

—— 1964 An efficient method for finding the minimum of a function of several variables without calculating derivatives *Comput. J.* **7** 155–62

—— 1975a Some convergence properties of the conjugate gradient method *CSS Report No* 23 (Harwell, UK: Computer Science and Systems Division, Atomic Energy Research Establishment)

—— 1975b Restart procedures for the conjugate gradient method *CSS Report No* 24 (Harwell, UK: Computer Science and Systems Division, Atomic Energy Research Establishment)

—— 1981 *Nonlinear Optimization* (London: Academic)

PRESS W H, FLANNERY B P, TEUKOLSKY S A and VETTERLING W T (1986/88) *Numerical Recipes (in Fortran/Pascal/C), the Art of Scientific Computing* (Cambridge, UK: Cambridge University Press)

RALSTON A 1965 *A First Course in Numerical Analysis* (New York: McGraw-Hill)

RATKOWSKY D A 1983 *Nonlinear Regression Modelling* (New York: Marcel-Dekker)

REID J K 1971 *Large Sparse Sets of Linear Equations* (London: Academic)

RHEINBOLDT W C 1974 *Methods for Solving Systems of Nonlinear Equations* (Philadelphia: SIAM)

RICE J 1983 *Numerical Methods, Software and Analysis* (New York: McGraw-Hill)

RILEY D D 1988 Structured programming: sixteen years later *J. Pascal, Ada* and *Modula-2* **7** 42–8
ROSENBROCK H H 1960 An automatic method for finding the greatest or least value of a function *Comput. J.* **3** 175–84
ROSS G J S 1971 The efficient use of function minimization in non-linear maximum-likelihood estimation *Appl. Stat.* **19** 205–21
—— 1975 Simple non-linear modelling for the general user *Warsaw: 40th Session of the International Statistical Institute 1–9 September 1975, ISI/BS* Invited Paper 81 pp 1–8
RUHE A and WEDIN P-A 1980 Algorithms for separable nonlinear least squares problems *SIAM Rev.* **22** 318–36
RUHE A and WIBERG T 1972 The method of conjugate gradients used in inverse iteration *BIT* **12** 543–54
RUTISHAUSER H 1966 The Jacobi method for real symmetric matrices *Numer. Math.* **9** 1–10; also in *Linear Algebra, Handbook for Automatic Computation* vol 2, eds J H Wilkinson and C Reinsch (Berlin: Springer) pp 202–11 (1971)
SARGENT R W H and SEBASTIAN D J 1972 Numerical experience with algorithms for unconstrained minimisation *Numerical Methods for Nonlinear Optimization* ed. F A Lootsma (London: Academic) pp 445–68
SCHNABEL R B, KOONTZ J E and WEISS B E 1985 A modular system of algorithms for unconstrained minimization *ACM Trans. Math. Softw.* **11** 419–40
SCHWARZ H R, RUTISHAUSER H and STIEFEL E 1973 *Numerical Analysis of Symmetric Matrices* (Englewood Cliffs, NJ: Prentice–Hall)
SEARLE S R 1971 *Linear Models* (New York: Wiley)
SHANNO D F 1970 Conditioning of quasi-Newton methods for function minimization *Math. Comput.* **24** 647–56
SHEARER J M and WOLFE M A 1985 Alglib, a simple symbol-manipulation package *Commun. ACM* **28** 820–5
SMITH F R Jr and SHANNO D F 1971 An improved Marquardt procedure for nonlinear regressions *Technometrics* **13** 63–74
SORENSON H W 1969 Comparison of some conjugate direction procedures for function minimization *J. Franklin Inst.* **288** 421–41
SPANG H A 1962 A review of minimization techniques for nonlinear functions *SIAM Rev.* **4** 343–65
SPENDLEY W 1969 Nonlinear least squares fitting using a modified Simplex minimization method *Fletcher* pp 259–70
SPENDLEY W, HEXT G R and HIMSWORTH F R 1962 Sequential application of simplex designs in optimization and evolutionary operation *Technometrics* **4** 441–61
STEWART G W 1973 *Introduction to Matrix Computations* (New York: Academic)
—— 1976 A bibliographical tour of the large, sparse generalized eigenvalue problem *Sparse Matrix Computations* eds J R Bunch and D J Rose (New York: Academic) pp 113–30
—— 1987 Collinearity and least squares regression *Stat. Sci.* **2** 68–100
STRANG G 1976 *Linear Algebra and its Applications* (New York: Academic)
SWANN W H 1974 Direct search methods *Numerical Methods for Unconstrained Optimization* ed. W Murray (London/New York: Academic)
SYNGE J L and GRIFFITH B A 1959 *Principles of Mechanics* 3rd edn (New York: McGraw-Hill)
TOINT PH L 1987 On large scale nonlinear least squares calculations *SIAM J. Sci. Stat. Comput.* **8** 416–35
VARGA R S 1962 *Matrix Iterative Analysis* (Englewood Cliffs, NJ: Prentice–Hall)
WILKINSON J H 1961 Error analysis of direct methods of matrix inversion *J. ACM* **8** 281–330
—— 1963 *Rounding Errors in Algebraic Processes* (London: HMSO)
—— 1965 *The Algebraic Eigenvalue Problem* (Oxford: Clarendon)
WILKINSON J H and REINSCH C (eds) 1971 *Linear Algebra, Handbook for Automatic Computation* vol 2 (Berlin: Springer)
WOLFE M A 1978 *Numerical Methods for Unconstrained Optimization, an Introduction* (Wokingham, MA: Van Nostrand–Reinhold)
YOURDON E 1975 *Techniques of Program Structure and Design* (Englewood Cliffs, NJ: Prentice–Hall)
ZAMBARDINO R A 1974 Solutions of systems of linear equations with partial pivoting and reduced storage requirements *Comput. J.* **17** 377–8

INDEX